.NET MAUI Projects

Build multi-platform desktop and mobile apps from scratch
using C# and Visual Studio 2022

Michael Cummings
Daniel Hindrikes
Johan Karlsson

.NET MAUI Projects

Group Product Manager: Rohit Rajkumar

Publishing Product Manager: Kaustubh Manglurkar

Book Project Manager: Aishwarya Mohan

Senior Editor: Rakhi Patel

Technical Editor: K Bimala Singha

Copy Editor: Safis Editing

Proofreader: Safis Editing

Indexer: Pratik Shirodkar

Production Designer: Joshua Misquitta

DevRel Marketing Coordinators: Namita Velgekar and Nivedita Pandey

First published: December 2018

Second edition: June 2020

Third edition: February 2024

Production reference: 2080224

Published by Packt Publishing Ltd.

Grosvenor House

11 St Paul's Square

Birmingham

B3 1RB, UK

ISBN 978-1-83763-491-0

www.packtpub.com

I dedicate this book to my rock, my confidant, and my partner for life, Rita. Your unwavering support and love have been my guiding light through thick and thin. This book is dedicated to you with all my heart.

– Michael Cummings

I dedicate this book to all the people I met during my years as a Xamarin developer who never believed in Xamarin. If you read this book, you will hopefully understand how great Xamarin is!

– Daniel Hindrikes

This book is dedicated to the spider in my basement that motivated me to write a book thick enough to finally kill it with.

– Johan Karlsson

Foreword

I'd never actually made a mobile app before, but that didn't stop me from standing in line, a little starstruck, to get my hands on a preview edition of a Xamarin.Forms book –signed by the author! Microsoft Build 2015 was an amazing conference, but nothing topped the Xamarin afterparty, and getting to meet one of my favorite technical writers was unbelievable. I walked away from the party with an armful of little stuffed monkeys and my personally inscribed book.

I had no idea at the time that I'd be frantically reading that book a few months later to prepare for my interview with the Xamarin.Forms team. I had no idea at the time that I'd spend the next eight and a half years with them – first, building the framework, taking it from 1.5 to 5.0, and expanding from mobile to desktop, too; then, leading and growing the team; and finally, helping to architect the next evolution of cross-platform development with .NET MAUI, .NET for iOS, and .NET for Android.

The mobile and desktop development world changes quickly, and any framework must be ready to bend, shift, and grow along with its underlying platforms. .NET MAUI is meant to be as flexible as the palm trees that sway with island winds. With that goal comes a formidable challenge – how do we build a framework that will stay constant enough for developers to rely on, yet continue to be relevant with the modern landscape?

We use the word "evolution" a lot when we talk about .NET MAUI because that's the most apt term for what it is. Each new version looks familiar – the fundamentals are all there – but if you dig a little deeper, you'll see the new capabilities, boosted performance, and updated expectations that make each version unique. If we did our jobs right, each version will deliver more delight than the last.

I've had the pleasure of working with Michael for several years, and when I learned that he would be writing this book, I wasn't surprised at all. Michael is a natural teacher, and he understands that .NET MAUI (and, of course, .NET for Android and .NET for iOS) is a sensible choice for the simplest of apps as well as more complex and intricate apps. This book walks you through the journey of your basic first app and onward to the real-world scenarios that you're likely to incorporate in your next project.

And even though you can be sure that .NET MAUI, .NET for Android, and .NET for iOS will continue to evolve and that some of the code you find in this book will look a bit different a few years from now, you can also be sure that the concepts and the fundamentals you learn in this book will be valuable for a long time to come. Have some fun with this book, and keep coding!

Samantha Houts

Principal Group Engineering Manager – .NET MAUI/.NET for Android/.NET for iOS

Contributors

About the authors

Michael Cummings is a senior development engineer at Microsoft. He currently works on the Visual Studio tooling for building WPF, Xamarin.Forms, and .NET MAUI apps. He has experience as a developer and architect with a focus on app development, design, deployment, and the business process as it relates to technology. Michael has dabbled in graphics and game programming since the days of the TI99-4/A. He contributes to open source projects, including AXIOM, a .NET 3D rendering engine, and Planeshift, a 3D **Massive Multiplayer Online Role-Playing Game (MMORPG)**. Michael lives in Lexington, MA, with his wife and their dog. When not working with technology, he enjoys watching movies, trying out new recipes, and the occasional game of full-contact racquetball.

I want to give a special thank you to my faithful companion, Smidgeon, who was always on point to be the needed distraction. You are missed.

Thank you to all the members of the teams I work with at Microsoft, for your patience and knowledge.

Additionally, thank you to the incredible team at Packt and the reviewers for this edition of the book, Pietro Libro and Robby Gunawan Sutanto; your comments and suggestions were invaluable.

Daniel Hindrikes is a developer and architect with a passion for developing mobile apps powered by the cloud. Daniel fell in love with Xamarin in the early days of Xamarin when he realized that he could use C# even for iOS and Android apps and that he could share code with the Windows apps he was also building. But Daniel started to build mobile apps long before that, working on Android apps with Java and even Java ME apps (a long, long time ago).

Daniel enjoys sharing his knowledge and can be found speaking at conferences, blogging, or recording the podcast *The Code Behind*. Daniel works at tretton37 in Sweden and has experience working with both local and global customers.

I want to say a special thanks to my family, my wife, Anna-Karin, and our twins, Ella and Willner, all of whom have supported me during the writing process.

I also would like to say thanks to the fantastic team at Packt and our technical reviewers, who helped us write this book and make us look better than we really are.

Johan Karlsson has been working with Xamarin since the days of MonoTouch and Mono for Android, and it all started with writing a game. He is a full stack developer, currently focusing on mobile applications using Xamarin, but has in the past worked a lot with ASP.NET MVC, Visual Basic .NET (not proud), and C# in general. Also, he's created a whole bunch of databases in SQL Server over the years.

Johan works at tretton37 in Sweden and has about 20 years of experience in the trade of assembling ones and zeros.

I want to send a special thanks to my partner in life, Elin. Thanks for being there during this special period of time including (but not limited to) moving together, living through a pandemic, writing a book, and selling a house. And, of course, to my children, Ville and Lisa, for being an inspiration in life!

Also, thanks to Packt and our tech reviewers, who nitpicked our applications apart, making us spend late nights correcting our code.

About the reviewers

Pietro Libro is a seasoned technology enthusiast with over two decades of experience in software development. He has honed his practical problem-solving skills through engagements in public administration, finance, and the automation industries. Simultaneously pursuing university education and professional work, Pietro attained a second-level degree, enriching his practical knowledge. Demonstrating a commitment to continuous learning, he holds numerous certifications and actively imparts his expertise as a technical speaker. His areas of expertise span a diverse range of technologies, with a particular emphasis on software and cloud architecture. Born in Italy, Pietro moved to Switzerland in 2013 and currently works as a Cloud Solution Architect. He resides in Zürich with his wife, Eleonora, and daughter, Giulia.

Robby Gunawan is a proficient full stack C# developer specializing in Windows Forms applications. He has been actively involved in writing commercial software since 2007, contributing his skills to several companies in Indonesia. His journey began with Basic during high school, followed by VB6 in university, and subsequently transitioning to VB.NET. In his professional career, he has adeptly embraced C#. Leveraging his C# background, Robby successfully published his first mobile application for Windows Phone 7 and developed his initial Android application using Xamarin Forms. Beyond software development, he possesses additional expertise as a Windows Server Administrator, Postfix Administrator, and WordPress Designer and Administrator.

Table of Contents

Preface   xiii

Part 1: Introduction

1

Introduction to .NET MAUI   3

Defining native applications   4   mobile   15

.NET mobile   4   Setting up our development
                                                machine   17
Code sharing   5
Use of existing knowledge   6   Setting up a Mac   17
.NET mobile platforms   6   Setting up a Windows machine   23

Exploring the .NET MAUI   .NET mobile productivity tooling   34
framework   9
                                                Xamarin Hot Restart   34
The architecture of .NET MAUI   10   Hot Reload   35
Defining a UI using XAML   11   Windows Subsystem for Android   35
.NET MAUI versus traditional .NET   Summary   36

2

Building Our First .NET MAUI App   37

Technical requirements   37   Updating the .NET MAUI packages   48
Project overview   38   Creating a repository and a
Setting up the project   38   TodoItem model   51
Creating the new project   39   Defining a to-do list item   51
Examining the files   42   Creating a repository and its interface 52

Connecting SQLite to persist data   54

Using MVVM – creating views and ViewModels   59
Defining a ViewModel base class   60
Introducing the CommunityToolkit.
Mvvm library's ObservableObject and
ObservableProperty   61
Creating MainViewModel   63
Creating TodoItemViewModel   64
Creating the ItemViewModel class   65
Creating the MainView view   66
Creating the ItemView view   69
Wiring up dependency injection   70
Making the app run   73

Adding data bindings   74
Navigating from MainView to ItemView
to add a new item   75

Adding new items to the list   77
Binding ListView in MainView   79
Creating a ValueConverter object for the
item's status   82
Navigating to an item using a command   85
Marking an item as complete using a
command   87
Creating the filter toggle function using a
command   88

Laying out the contents   91
Setting an application-wide background
color   91
Laying out the MainView and ListView
items   92

Summary   94

3

Converting a Xamarin.Forms App into .NET MAUI   95

Technical requirements   95
Project overview   96
**Migrating into a blank .NET MAUI
template   96**
Creating a new Xamarin.Forms app   97
Creating a new .NET MAUI app   101
Migrating the MauiMigration app to
MyMauiApp   103

Manual migration overview   110
Converting the Xamarin.Forms projects from
.NET Framework into .NET SDK style   110
Updating code from Xamarin.Forms
to .NET MAUI   113

Updating any incompatible NuGet
packages   116
Addressing any breaking API changes   118
Custom renderers and effects   118
Running the converted app and verifying its
functionality   119

**Installing and running .NET
Upgrade Assistant   119**
Installing .NET Upgrade Assistant   119
Preparing to run .NET Upgrade
Assistant   121
Running .NET Upgrade Assistant   125

Summary   136

Part 2: Basic Projects

4

Building a News App Using .NET MAUI Shell   141

Technical requirements	142	Creating the structure of the app	145
Project overview	142	Creating the news service	158
Building the news app	142	Wiring up the NewsService class	165
Setting up the project	142	Handling navigation	173
		Summary	**179**

5

A Matchmaking App with a Rich UX Using Animations   181

Technical requirements	181	Creating the Swiper control	187
Project overview	182	Testing the control	198
Creating the matchmaking app	182	Wiring up the Swiper control	207
Setting up the project	182	**Summary**	**210**

6

Building a Photo Gallery App Using CollectionView and CarouselView   211

Technical requirements	212	Creating the new project	213
Project overview	212	Importing photos	216
Building the photo gallery app	212	Creating the gallery view	238
		Summary	**252**

7

Building a Location Tracking App Using GPS and Maps   253

Technical requirements	253	**Building the MeTracker app**	**254**
Project overview	254	Creating a repository to save the locations of the users	257

Creating a service for location tracking   261
Setting up the app logic   263
Background location tracking on iOS and
Mac Catalyst   278
Background location tracking with Android   283

Creating a heat map   289
Creating a custom control for the map   294
Refreshing the map when the app resumes   297

Summary   **298**

8

Building a Weather App for Multiple Form Factors   301

Technical requirements   **301**
Project overview   **302**
Building the weather app   **302**
Creating models for the weather data   305
Creating a service to fetch the weather
data   310
Configuring the application platforms so
that they use location services   313
Creating the ViewModel class   315

Creating the view for tablets and desktop
computers   319
Creating the view for phones   327
Adding services and ViewModels to
dependency injection   332
Navigating to different views based on the
form factor   332
Handling states with VisualStateManager   334

Summary   **341**

Part 3: Advanced Projects

9

Setting Up a Backend for a Game Using Azure Services   345

Technical requirements   **345**
Project overview   **346**
An overview of the game   347

**Understanding the different Azure
serverless services**   **348**
Azure SignalR Service   348
Azure Functions   349

Building the serverless backend   **349**
Creating a SignalR service   349
Using Azure Functions as an API   351

**Deploying the functions to
Azure**   **403**
Summary   **405**

10

Building a Real-Time Game   407

Technical requirements	407	Creating the Connect page	439
Project overview	407	Creating the Lobby page	464
Getting started	408	Creating the Match page	479
An overview of the game	408	Testing the completed project	508
Building the game app	409	**Summary**	**513**
Creating the game services	415		

11

Building a Calculator Using .NET MAUI Blazor   515

Technical requirements	515	Creating the Keypad view	526
Project overview	515	Creating the Compute service	536
What is Blazor?	516	Adding memory functions	538
Creating the calculator app	516	Resizing the main window	543
Setting up the project	516	**Summary**	**545**

12

Hot Dog or Not Hot Dog Using Machine Learning   547

Technical requirements	547	Getting started	550
Machine learning	548	Building the Hot Dog or Not Hot Dog application using machine learning	550
Azure Cognitive Services – Custom Vision	548		
Core ML	549		
TensorFlow	549	Training a model	551
ML.Net	549	Building the app	556
The project overview	550	**Summary**	**590**

Index   593

Other Books You May Enjoy   604

Preface

.NET MAUI Projects is a hands-on book in which you get to create nine applications from the ground up. You will gain the fundamental skills you need to set up your environment, and we will explain what .NET Mobile is before we transition into .NET MAUI to really take advantage of truly native cross-platform code.

After reading this book, you will have a real-life understanding of what it takes to create an app that you can build on and that stands the test of time.

We will cover, among other things, upgrading from Xamarin.Forms, animations, consuming REST interfaces, real-time communication using SignalR, and location tracking using a device's GPS. There is also room for machine learning, a touch of .NET Blazor, and the must-have to-do list.

Happy coding!

Who this book is for

This book is for developers who know their way around C# and Visual Studio. You don't have to be a professional programmer, but you should have basic knowledge of object-oriented programming using .NET and C#. The typical reader would be someone who wants to explore how you can use .NET Mobile, and specifically .NET MAUI, to create applications using .NET and C#.

No knowledge of .NET Mobile is required in advance, but it would be a great help if you've worked with traditional .NET Mobile and want to take the step toward .NET MAUI.

What this book covers

Chapter 1, Introduction to .NET MAUI, explains the basic concepts of .NET Mobile and .NET MAUI. It helps you understand the building blocks of how to create a true cross-platform app. It's the only theoretical chapter of the book and will help you get started and set up your development environment.

Chapter 2, Building Our First .NET MAUI App, guides you through the concepts of **Model-View-ViewModel** (**MVVM**) and explains how to use the Inversion of Control pattern to simplify the creation of views and view models. We will create a to-do app that supports navigation, filtering, and the addition of to-do items to a list, and will also render a user interface that takes advantage of the powerful data-binding mechanisms in .NET MAUI.

Chapter 3, Converting a Xamarin.Forms App to .NET MAUI, walks through the steps to convert an existing Xamarin.Forms app running on Mono to a .NET MAUI app running on .NET 7. We will discuss two different methods for converting your Xamarin.Forms application to .NET MAUI. The first method will use a new .NET MAUI project and move our old Xamarin.Forms code into the new project, and the second method will use the **.NET Upgrade Assistant** tool to do some of the upgrades for us.

Chapter 4, Building a News App Using .NET MAUI Shell, explores the default navigation template in .NET MAUI, `Shell`, a standard way to define the structure of .NET MAUI apps. In this chapter, you will learn all you need to know to use `Shell` in a .NET MAUI app.

Chapter 5, A Matchmaking App with a Rich UX Using Animations, lets you dive deeper into how to define a richer user interface with animations and content placement. It also covers the concept of custom controls to encapsulate the user interface into self-contained components.

Chapter 6, Building a Photo Gallery App Using CollectionView and CarouselView, details the .NET MAUI `CollectionView` and `CarouselView` controls. In this chapter, we will use them to build a photo gallery app to learn how to master the controls.

Chapter 7, Building a Location Tracking App Using GPS and Maps, taps into the use of geolocation data from the device's GPS and how to plot this data on a layer on a map. It also explains how to use background services to keep tracking the location over a long period of time to create a heat map of where you spend your time.

Chapter 8, Building a Weather App for Multiple Form Factors, is all about consuming a third-party REST interface and displaying the data in a user-friendly way. We will hook up to a weather service to get the forecast for the current location you are in and display the results in a list.

Chapter 9, Setting Up a Backend for a Game Using Azure Services, is the first of two parts in which we'll set up a game app. This chapter explains how to use Azure services to create a backend that exposes functionality through SignalR to set up a real-time communication channel between apps.

Chapter 10, Building a Real-Time Game, follows on from the previous chapter and covers the frontend of the app – in this case, a .NET MAUI app that connects to the backend and relays messages between users. The chapter focuses on setting up SignalR on the client side and explains how to create a service model that abstracts this communication through messages and events.

Chapter 11, Building a Calculator Using .NET MAUI Blazor, explores a .NET Blazor app embedded within a .NET MAUI app. We will write part of the calculator app using Blazor and host that in .NET MAUI using `BlazorWebView`. We will also communicate between Blazor and .NET MAUI.

Chapter 12, Hot Dog or Not Hot Dog Using Machine Learning, covers the creation of an app that uses machine learning to identify whether an image contains a hot dog or not.

To get the most out of this book

We recommend that you read the first chapter to make sure that you are up to speed with the basic concepts of Xamarin in general. After that, you could pretty much pick any chapter you would like to learn more about. Each chapter is standalone but the chapters are ordered by complexity; the further you are into the book, the more complex the app will be.

The apps are adapted for real-world use but some parts are left out, such as proper error handling and analytics, since they are out of the scope of the book. However, you should get a good grasp of the building blocks of how to create an app.

Having said that, it does help if you have been a C# and .NET developer for a while, since many of the concepts are not really app-specific but are good practice in general, such as MVVM and Inversion of Control.

But, most of all, it's a book you can use to kick-start your .NET MAUI development learning curve by focusing on what chapters interest you the most.

Software/Hardware covered in the book	OS requirements
Visual Studio Community Edition. A computer capable of running Windows 10 or later for UWP and Android. A Mac that is capable of running macOS Mojave 10.14 to use the iOS simulator	Windows 10 or later, macOS Sierra 10.12 or later
Xcode. A Mac that is capable of running macOS Sierra 10.14	macOS Mojave 10.14

If you are using the digital version of this book, we advise you to type the code yourself or access the code from the book's GitHub repository (a link is available in the next section). Doing so will help you avoid any potential errors related to the copying and pasting of code.

Download the example code files

You can download the example code files for this book from GitHub at `https://github.com/PacktPublishing/MAUI-Projects-3rd-Edition`. If there's an update to the code, it will be updated in the GitHub repository.

We also have other code bundles from our rich catalog of books and videos available at `https://github.com/PacktPublishing/`. Check them out!

Conventions used

There are a number of text conventions used throughout this book.

`Code in text`: Indicates code words in text, database table names, folder names, filenames, file extensions, pathnames, dummy URLs, user input, and Twitter handles. Here is an example: "Since all .NET MAUI programs start with the `MauiProgram.cs` file, that seems like a good place to start."

A block of code is set as follows:

```
.keypad {
    width: 300px;
    margin: auto;
    margin-top: -1.1em;
}
```

When we wish to draw your attention to a particular part of a code block, the relevant lines or items are set in bold:

```
builder.Services.AddMauiBlazorWebView();

#if DEBUG
    builder.Services.AddBlazorWebViewDeveloperTools();
    builder.Logging.AddDebug();
#endif
```

Any command-line input or output is written as follows:

```
$ mkdir ViewModels
$ cd ViewModels
```

Bold: Indicates a new term, an important word, or words that you see onscreen. For instance, words in menus or dialog boxes appear in **bold**. Here is an example: "Open Visual Studio 2022 and select **Create a new project.**"

> **Tips or important notes**
> Appear like this.

Get in touch

Feedback from our readers is always welcome.

General feedback: If you have questions about any aspect of this book, email us at `customercare@packtpub.com` and mention the book title in the subject of your message.

Errata: Although we have taken every care to ensure the accuracy of our content, mistakes do happen. If you have found a mistake in this book, we would be grateful if you would report this to us. Please visit `www.packtpub.com/support/errata` and fill in the form.

Piracy: If you come across any illegal copies of our works in any form on the internet, we would be grateful if you would provide us with the location address or website name. Please contact us at `copyright@packt.com` with a link to the material.

If you are interested in becoming an author: If there is a topic that you have expertise in and you are interested in either writing or contributing to a book, please visit `authors.packtpub.com`.

Share your thoughts

Once you've read *.NET MAUI Projects*, we'd love to hear your thoughts! Scan the QR code below to go straight to the Amazon review page for this book and share your feedback.

`https://packt.link/r/1837634912`

Your review is important to us and the tech community and will help us make sure we're delivering excellent quality content.

Download a free PDF copy of this book

Thanks for purchasing this book!

Do you like to read on the go but are unable to carry your print books everywhere?

Is your eBook purchase not compatible with the device of your choice?

Don't worry, now with every Packt book you get a DRM-free PDF version of that book at no cost.

Read anywhere, any place, on any device. Search, copy, and paste code from your favorite technical books directly into your application.

The perks don't stop there, you can get exclusive access to discounts, newsletters, and great free content in your inbox daily

Follow these simple steps to get the benefits:

1. Scan the QR code or visit the link below

https://packt.link/free-ebook/9781837634910

2. Submit your proof of purchase
3. That's it! We'll send your free PDF and other benefits to your email directly

Part 1:
Introduction

In this part, you will get an overview of .NET MAUI and learn how to install .NET MAUI in order to create your first .NET MAUI project. As you create your first projects, you will learn about the Model-View-ViewModel design pattern and how to use it in .NET MAUI apps. Xamarin users will also learn how to upgrade their projects to .NET MAUI either manually or by using the .NET Upgrade Assistant.

This part has the following chapters:

- *Chapter 1, Introduction to .NET MAUI*
- *Chapter 2, Building Your First .NET MAUI App*
- *Chapter 3, Converting a Xamarin.Forms App to .NET MAUI*

1

Introduction to .NET MAUI

This chapter is all about getting to know **.NET Multi-platform App UI (.NET MAUI)** and what to expect from it. .NET MAUI enables you to build native cross-platform mobile and desktop apps for Android, iOS, macOS, and Windows using .NET and C#. This is the only chapter that is purely theoretical; all the others cover hands-on projects. You are not expected to write any code at this point, but instead, simply read through this chapter to develop a high-level understanding of what .NET MAUI is, how .NET MAUI relates to .NET, and how to set up a development machine.

We will start by defining what a native app is and what .NET as a technology brings to the table. After that, we will look at how .NET MAUI fits into the bigger picture and learn when it is appropriate to use the traditional .NET mobile and .NET MAUI apps. We often use the term **traditional .NET mobile app** to describe apps that don't use .NET MAUI, even though .NET MAUI apps are bootstrapped through a traditional .NET mobile app.

In this chapter, we will cover the following topics:

- Defining native applications
- .NET mobile
- Exploring the .NET MAUI framework
- Setting up our development machine
- .NET mobile productivity tooling

Let's get started!

Defining native applications

The term **native application** means different things to different people. For some people, it refers to an app that is developed using the tools specified by the creator of the platform, such as an app developed for iOS with Objective-C or Swift, an Android app developed with Java or Kotlin, or a Windows app developed with C/C++ or the .NET Framework. Others use the term native application to refer to apps that are compiled into machine code that is native to the platform architecture, for example, x86, x64, or ARM. In this book, we will define a native application as one that has a native UI, performance, and API access. The following list explains these three concepts in greater detail:

- **Native UI**: Apps built with .NET MAUI use the standard controls for each platform. This means, for example, that an iOS app built with .NET MAUI will look and behave as an iOS user would expect and an Android app built with .NET MAUI will look and behave as an Android user would expect.

- **Native performance**: Apps built with .NET MAUI are compiled for native performance, meaning that they execute at nearly the same levels as apps built with the tools designed for the platform, that is, Java or Swift, and can use platform-specific hardware acceleration.

- **Native API access**: Native API access means that apps built with .NET MAUI can use everything that the target platforms and devices offer to developers. For example, .NET MAUI applications can use hardware-specific features such as the camera or maps.

.NET mobile

.NET mobile (formerly known as Xamarin) is a set of extensions to .NET that is used to develop native applications for iOS (**.NET for iOS/tvOS/Mac Catalyst**), Android (**.NET for Android**), and macOS (**.NET for macOS**). .NET is the evolution of the .NET Framework, designed for cross-platform development. .NET mobile was introduced in .NET Core 5 as optional workloads. It is technically a binding layer on top of these platforms. Using bindings to platform APIs enables .NET developers to use C# (and F#) to develop native applications with the full capacity of each platform.

The C# APIs we use when we develop apps with .NET mobile match the platform APIs, but they are modified to adhere to conventions used in .NET Core. For example, APIs are often customized to follow .NET naming conventions, and the Android `set` and `get` methods are often replaced by properties. This makes using the APIs easier and more familiar for .NET developers.

Mono (`https://www.mono-project.com`) is an open source implementation of the Microsoft .NET Framework, which is based on the **European Computer Manufacturers Association (ECMA)** standards for C# and the **Common Language Runtime (CLR)**. Mono was created to bring the .NET Framework to platforms other than Windows. It is part of the .NET Foundation (`http://www.dotnetfoundation.org`), an independent organization that supports open development and collaboration involving the .NET ecosystem. Since .NET 5, Mono is now a supported runtime for applications built on .NET. No separate installer is needed to use Mono with .NET; it is included in the installer for .NET. The Mono runtime is used for iOS, tvOS, Mac Catalyst, and Android applications, while the .NET Core CLR is used for all other supported platforms.

With a combination of the .NET mobile platforms, .NET, and Mono, we can use both the platform-specific APIs and the platform-independent parts of .NET, including namespaces such as `System`, `System.Linq`, `System.IO`, `System.Net`, and `System.Threading.Tasks`.

There are several reasons for using .NET mobile for mobile app development, which we will cover in the following sections.

Code sharing

If we use one common programming language for multiple mobile platforms (and even server platforms), then we can share a lot of code between our target platforms, as illustrated in the following diagram. All code that isn't related to the target platform can be shared with other .NET platforms. Code that is typically shared in this way includes business logic, network calls, and data models:

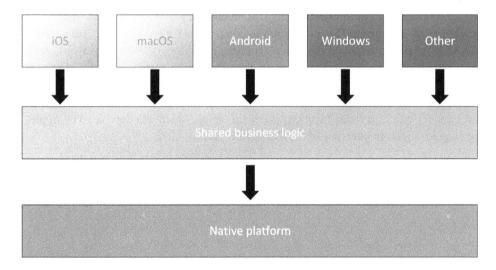

Figure 1.1 – .NET MAUI code sharing

There is also a large community based around the .NET platforms, providing a different form of code sharing. There is a wide range of third-party libraries and components that can be downloaded from NuGet (`https:// nuget.org`) that can provide you with additional features or capabilities that work across all supported .NET MAUI platforms. For example, you can find NuGet packages that provide databases, graphs, or barcode reading to include in your apps.

Code sharing across platforms leads to shorter development times. It also produces apps of a higher quality because, for example, we only need to write the code for business logic once. There is a lower risk of bugs, and it is also able to guarantee that a calculation returns the same result, regardless of what platform our users use.

Use of existing knowledge

For .NET developers who want to start building native mobile apps, it is easier to just learn the APIs for the new platforms than it is to learn programming languages and APIs for both old and new platforms.

Similarly, organizations that want to build native mobile apps can use existing developers with their knowledge of .NET to develop apps. Because there are more .NET developers than Objective-C and Swift developers, it's easier to find new developers for mobile app development projects.

.NET mobile platforms

The different .NET mobile platforms available are .NET for iOS/tvOS/Mac Catalyst, .NET for Android, and .NET for macOS. In this section, we will take a look at each of them.

.NET for iOS/tvOS/Mac Catalyst

.NET for iOS/tvOS/Mac Catalyst is used to build apps for iOS, tvOS, or Mac Catalyst, respectively, with .NET and contains the bindings to the iOS APIs mentioned previously. .NET for iOS/tvOS/Mac Catalyst uses **ahead-of-time** (AOT) compilation to compile the C# code into the **Advanced RISC Machine** (ARM) assembly language. The Mono runtime runs alongside the Objective-C runtime. Code that uses .NET namespaces, such as `System.Linq` or `System.Net`, is executed by the Mono runtime, while code that uses iOS-specific namespaces is executed by the Objective-C runtime. Both the Mono runtime and the Objective-C runtime run on top of the **X is Not Unix** (**XNU**) Unix-like kernel (`https://github.com/apple/darwin-xnu`), which was developed by Apple. The following diagram shows an overview of the iOS architecture:

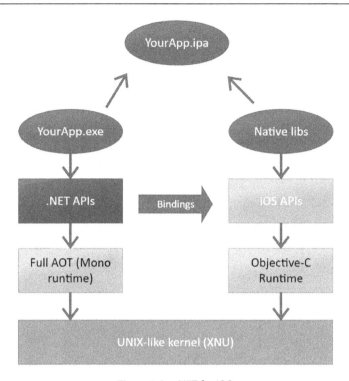

Figure 1.2 – .NET for iOS

.NET for macOS

.NET for macOS is used to build apps for macOS with .NET and contains the bindings to the macOS APIs. .NET for macOS has the same architecture as .NET for iOS—the only difference is, .NET for macOS apps are **just-in-time** (**JIT**)-compiled, unlike .NET for iOS apps, which are AOT-compiled. This is shown in the following diagram:

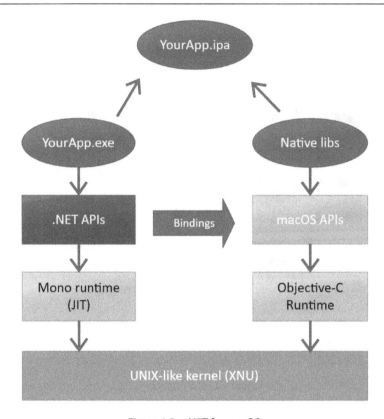

Figure 1.3 – .NET for macOS

.NET for Android

.NET for Android is used to build apps for Android with .NET and contains bindings to the Android APIs. The Mono runtime and the **Android Runtime** (**ART**) run side by side on top of a Linux kernel. .NET for Android apps could either be JIT-compiled or AOT-compiled, but to AOT-compile them, we need to use Visual Studio Enterprise.

Communication between the Mono runtime and ART occurs via a **Java Native Interface** (**JNI**) bridge. There are two types of JNI bridges—**Manage Callable Wrapper** (**MCW**) and **Android Callable Wrapper** (**ACW**). An MCW is used when code needs to run in ART and an ACW is used when ART needs to run code in the Mono runtime, as shown:

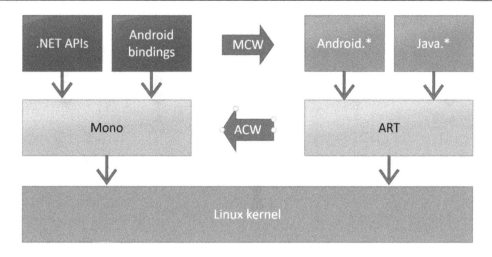

Figure 1.4 – .NET for Android

> **.NET for Tizen**
>
> .NET MAUI has additional support for the Tizen platform from Samsung. Samsung provides the binding layer and runtime to allow .NET MAUI to run on the Tizen platform. To learn more about how to install and develop for the Tizen platform, visit `https://github.com/Samsung/Tizen.NET` in your browser.

Now that we understand what .NET mobile is and how each platform works, we can explore .NET MAUI in detail.

Exploring the .NET MAUI framework

.NET MAUI is a cross-platform framework that is built on top of .NET mobile (for iOS and Android) and the **Windows UI (WinUI)** library. .NET MAUI allows developers to create a UI for iOS, Android, and WinUI in XAML. .NET MAUI improves on Xamarin.Forms by placing all platform-specific functionality in the same project as cross-platform functionality, making it easier to find and edit your code. .NET MAUI also includes all of what used to be in Xamarin.Essentials, which provides cross-platform capabilities, such as permissions, location, photos and camera, contacts, and maps, and leverages that cross-platform functionality with one shared code base, as illustrated in the following diagram:

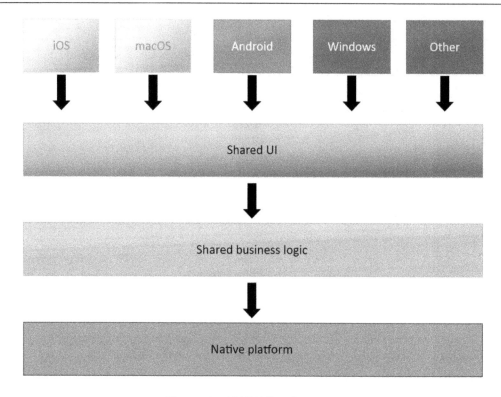

Figure 1.5 – .NET MAUI architecture

If we build an app with .NET MAUI, we can use XAML, C#, or a combination of both to create the UI.

The architecture of .NET MAUI

.NET MAUI is an abstraction layer on top of each platform. .NET MAUI has a shared layer that is used by all platforms, as well as a platform-specific layer. The platform-specific layer contains **handlers**. A handler is a class that maps a .NET MAUI control to a platform-specific native control. Each .NET MAUI control has a platform-specific handler.

The following diagram illustrates how the entry control in .NET MAUI is mapped to the correct native control for each platform. The entry control is mapped to a UITextField control from the UIKit namespace when the shared .NET MAUI code is used in an iOS app. On Android, the entry control is mapped to an EditText control from the AndroidX.AppCompat.Widget namespace. Finally, for Windows, .NET MAUI Entry handlers map to TextBox from the Microsoft.UI.Xaml.Controls namespace.

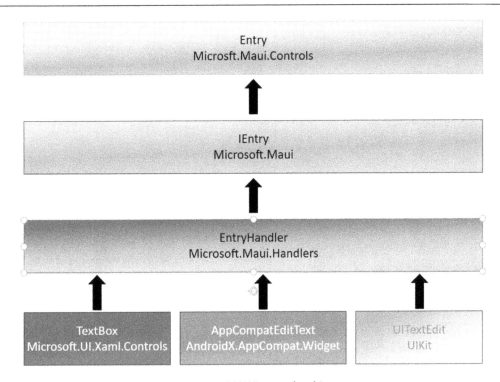

Figure 1.6 – .NET MAUI control architecture

With a firm grasp of .NET MAUI architecture and .NET mobile platforms, it is time to explore how to create UIs in .NET MAUI.

Defining a UI using XAML

The most common way to declare our UI in .NET MAUI is by defining it in a XAML document. It is also possible to create the GUI in C#, since XAML is a markup language for instantiating objects. We could, in theory, use XAML to create any type of object, provided it has a parameterless constructor. A XAML document is an **Extensible Markup Language** (**XML**) document with a specific schema.

Over the next few sections, we are going to learn about a few controls in .NET MAUI to get us started. Then, we will compare different ways that you can construct the UI using .NET MAUI.

Defining a Label control

As a simple example, let's look at the following snippet of a XAML document:

```
<Label Text="Hello World!" />
```

When the XAML parser encounters this snippet, it creates an instance of a `Label` object and then sets the properties of the object that correspond to the attributes in the XAML. This means that if we set a `Text` property in XAML, it sets the `Text` property on the instance of the `Label` object that is created. The XAML in the preceding example has the same effect as the following:

```
var obj = new Label()
{
    Text = "Hello World!"
};
```

XAML exists to make it easier to view the object hierarchy that we need to create in order to make a GUI. An object model for a GUI is also hierarchical by design, so XAML supports adding child objects. We can simply add them as child nodes, as follows:

```
<StackLayout>
    <Label Text="Hello World" />
    <Entry Text="Ducks are us" />
</StackLayout>
```

`StackLayout` is a container control that organizes the children vertically or horizontally within a container. Vertical organization is the default value and is used unless we specify otherwise. There are other containers, such as `Grid` and `FlexLayout`.

These will be used in many of the projects in the following chapters.

Creating a page in XAML

A single control is no use unless it has a container that hosts it. Let's see what an entire page would look like. A fully valid `ContentPage` object defined in XAML is an XML document. This means that we must start with an XML declaration. After that, we must have one—and only one—root node, as shown:

```
<?xml version="1.0" encoding="UTF-8"?>
<ContentPage xmlns="http://schemas.microsoft.com/dotnet/2021/maui"
xmlns:x="http://schemas.microsoft.com/winfx/2009/xaml" x:Class="MyApp.
MainPage">
    <StackLayout>
        <Label Text="Hello world!" />
    </StackLayout>
</ContentPage>
```

In the preceding example, we defined a `ContentPage` object that translates into a single view on each platform. In order to make it valid XAML, we need to specify a default namespace (`xmlns="http://schemas.microsoft.com/dotnet/2021/maui"`) and then add the x namespace (`xmlns:x="http://schemas.microsoft.com/winfx/2009/xaml"`).

The default namespace lets us create objects without prefixing them, such as the `StackLayout` object. The x namespace lets us access properties such as `x:Class`, which tells the XAML parser which class to instantiate to control the page when the `ContentPage` object is created.

A `ContentPage` object can have only one child. In this case, it's a `StackLayout` control. Unless we specify otherwise, the default layout orientation is vertical. A `StackLayout` object can, therefore, have multiple children. Later in the book, we will touch on more advanced layout controls, such as the `Grid` and `FlexLayout` controls.

As the first child of `StackLayout`, we will create a `Label` control.

Creating a page in C#

For clarity, the following code shows you how the previous example would look in C#:

```
public class MainPage : ContentPage
{
}
```

`MainPage` is a class that inherits from .NET MAUI's `ContentPage`. This class is automatically generated for us if we create a XAML page, but if we just use code, we will need to define it ourselves.

Let's create the same control hierarchy as the XAML page we defined earlier using the following code:

```
var page = new MainPage();
var stacklayout = new StackLayout();
stacklayout.Children.Add(
    new Label()
    {
        Text = "Welcome to .NET MAUI"
    });
page.Content = stacklayout;
```

The first statement creates a `page` object. We could, in theory, create a new `ContentPage` page directly, but this would prohibit us from writing any code behind it. For this reason, it's good practice to subclass each page that we plan to create.

The block following this first statement creates the `StackLayout` control, which contains the `Label` control that is added to the `Children` collection.

Finally, we need to assign `StackLayout` to the `Content` property of the page.

Since XAML is a markup language that mainly instantiates objects for us, we can see how easy it is to replicate that in C#. Next, we will take a look at some extensions that make developing your UI in C# a little better.

Using the .NET MAUI Markup Community Toolkit

The Community Toolkit organization on GitHub has a project to add **Fluent** extensions to MAUI for creating the UI in C#. The project **.NET MAUI Markup Community Toolkit** (`https://github.com/CommunityToolkit/Maui.Markup`), or **MAUI.Markup** for short, is described on the website as follows:

> *The .NET MAUI Markup Community Toolkit is a collection of Fluent C# Extension Methods that allows developers to continue architecting their apps using MVVM, Bindings, Resource Dictionaries, etc., without the need for XAML.*

Using MAUI Markup to create the same page we did in the previous two sections would look something like the following:

```
public class MainPage: ContentPage
{
    public MainPage()
    {
        Build();
    }

    public void Build() {
        Content = new StackLayout()
        {
            Children =
            {
                new Label()
                {
                    Text = "Welcome to .NET MAUI"
                }
            }
        };
    }
}
```

For more information on how to use MAUI Markup in your applications, visit `https://github.com/CommunityToolkit/Maui.Markup` using your favorite web browser.

So, what is better for creating our UI, XAML or C#?

XAML or C#?

Generally, using XAML provides a much better overview, since the page is a hierarchical structure of objects, and XAML is a very nice way of defining that structure. In code, the structure is flipped around as we need to define the innermost object first, making it harder to read the structure of our page. This was demonstrated in the *Creating a page in XAML* section of this chapter. Having said that, it is generally a matter of preference as to how we decide to define the GUI. This book will use XAML rather than C# in the projects to come.

Now that we've explored how to create our pages using .NET MAUI, it is time to review how .NET MAUI and .NET mobile compare.

.NET MAUI versus traditional .NET mobile

While this book is about .NET MAUI, we will also highlight the differences between using traditional .NET mobile and .NET MAUI. Traditional .NET mobile is used when developing UIs that use the iOS or Android **software development kit** (**SDK**) without any means of abstraction. For example, we can create an iOS app that defines its UI in a storyboard or in the code directly. This code would not be reusable for other platforms, such as Android. Apps built using this approach can still share non-platform-specific code by simply referencing a .NET standard library. This relationship is shown in the following diagram:

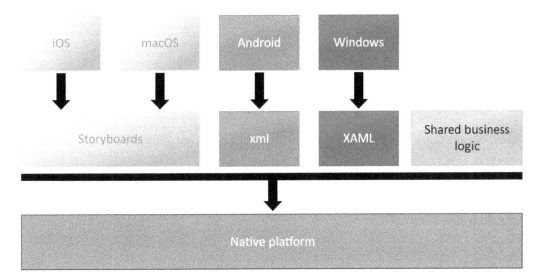

Figure 1.7 – Traditional .NET UI

.NET MAUI, on the other hand, is an abstraction of the GUI, which allows us to define UIs in a platform-agnostic way. It still builds on top of .NET for iOS, .NET for Android, and all the other supported platforms. The .NET MAUI app is created as a .NET standard library where the shared source files and platform-specific source files are all built within the same project for the platform we are currently building for. This relationship is shown in the following diagram:

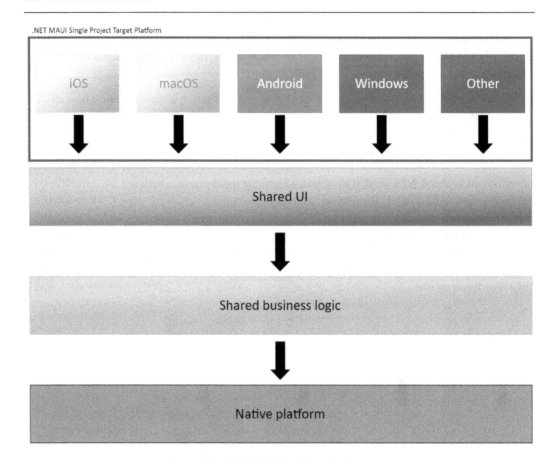

Figure 1.8 – .NET MAUI UI with a single project

Having said that, .NET MAUI cannot exist without traditional .NET mobile since it's bootstrapped through an app for each platform. This gives us the ability to extend .NET MAUI on each platform using custom renderers and platform-specific code that can be exposed to our shared code base through interfaces. We'll look at these concepts in more detail later in this chapter.

When to use .NET MAUI

We can use .NET MAUI in most cases and for most types of apps. If we need to use controls that are not available in .NET MAUI, we can always use the platform-specific APIs. There are, however, cases where .NET MAUI is not useful. The most common situation where we might want to avoid using .NET MAUI is if we build an app that should look very different across our different target platforms.

Enough theory for now; let's get our development machines ready to develop using .NET MAUI.

Setting up our development machine

Developing an app for multiple platforms imposes higher demands on our development machine. One reason for this is that we often want to run one or multiple simulators or emulators on our development machine. Different platforms also have different requirements for what is needed to begin development. Regardless of whether we use macOS or Windows, Visual Studio will be our **integrated development environment (IDE)**. There are several versions of Visual Studio, including the free community edition. Go to `https://visualstudio.microsoft.com/` to compare the available versions, and select the version that is right for you; .NET MAUI is included in all versions of Visual Studio for Windows and macOS. The following list is a summary of what we need to begin development for each platform:

- **iOS**: To develop an app for iOS, we need a **Macintosh (Mac)** device. This could either be the machine that we are developing on or a machine on our network, if we are using one. The reason we need to connect to a Mac is we need Xcode to build the app package. Xcode also provides various simulators to run and debug your app. It is possible to do some iOS development on Windows without a connected Mac; you can read more about this in the *Xamarin Hot Restart* section of this chapter.
- **Android**: Android apps can be developed on either macOS or Windows. Everything we need, including SDKs and simulators, is installed with Visual Studio.
- **WinUI**: WinUI apps can only be developed in Visual Studio on a Windows machine.

We'll start with setting up a Mac first, and then later cover Windows. If you do not own a Mac, you can skip this section and head straight to the *Setting up a Windows machine* section.

Setting up a Mac

There are two main tools that are required to develop apps for iOS and Android with .NET mobile on a Mac. These are Visual Studio for Mac (if we are only developing Android apps, this is the only tool we need) and Xcode. In the following sections, we will look at how to set up a Mac for app development.

Installing Xcode

Before we install Visual Studio, we need to download and install Xcode. Xcode is the official development IDE from Apple and contains all the tools available for iOS development, including SDKs for iOS, macOS, Mac Catalyst, and tvOS.

We can download Xcode from the Apple developer portal (`https://developer.apple.com`) or the Apple App Store. I recommend that you download it from the App Store because this guarantees you have the latest stable version. The only reason to download Xcode from the developer portal is if you want to use a prerelease version of Xcode to develop it for a prerelease of iOS.

When using prerelease versions of macOS and its accompanying version of Xcode, it is possible that .NET for iOS/tvOS/Mac Catalyst/macOS has not been updated to work with the latest Xcode changes. It is recommended to check the compatibility before installing the latest Xcode to ensure a working environment.

After the first installation, and after each update of Xcode, it is important that you open it. Xcode often needs to install additional components after an installation or an update. We also need to open Xcode to accept the license agreement with Apple.

Installing Visual Studio

To install Visual Studio, we first need to download it from `https://visualstudio.microsoft.com`.

> **Visual Studio for Mac deprecation**
>
> On August 31, 2023, Microsoft announced the deprecation of Visual Studio for Mac in accordance with their Modern Lifecycle Policy, which will mean the end of support on August 31, 2024. Visual Studio for Mac will remain supported until that date. If you have a Visual Studio subscription, you can always download the latest version of Visual Studio for Mac from `my.visualstudio.com`. Microsoft has released the C# Dev Kit and .NET MAUI Dev Kit extensions for Visual Studio Code. The extensions work on Windows, macOS, and Linux. You can use Visual Studio Code and these extensions to complete the projects in this book, although the instructions pertaining to the UI for Visual Studio for Mac will not match Visual Studio Code. To learn more, visit `https://learn.microsoft.com/en-us/visualstudio/mac/what-happened-to-vs-for-mac?view=vsmac-2022`.

When we start the Visual Studio installer via the file we downloaded, it will start to check what we already have installed on our machine. When the check is finished, we can select which platforms and tools we would like to install.

Once we have selected the platforms that we want to install, Visual Studio downloads and installs everything that we need to get started with app development using .NET mobile, as shown:

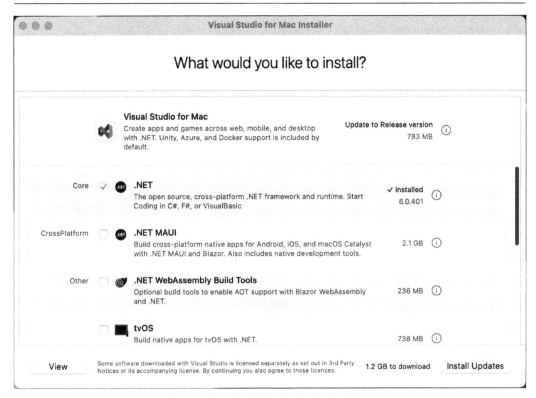

Figure 1.9 – Visual Studio for Mac installer

Configuring the Android Emulator

Visual Studio uses the Android emulators provided by Google. They are installed and configured through SDK Manager.

> **Special note for Intel-based Mac device**
>
> If we want our emulator to be fast, then we need to ensure that it is hardware-accelerated. To hardware-accelerate the Android emulator, we need to install the Intel **Hardware Accelerated Execution Manager** (**HAXM**), which can be downloaded from `https://software.intel.com/en-us/articles/intel-hardware-accelerated-execution-manager-intel-haxm`.

The next step is to create the Android emulator. First, we need to ensure that the Android emulator and the Android OS images are installed. To do this, take the following steps:

1. Go to the **Tools** tab to install the Android emulator:

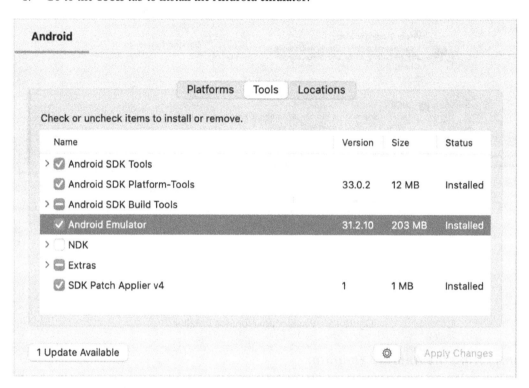

Figure 1.10 – Installing Android Emulator in Visual Studio for Mac

2. We also need to install one or multiple images to use with the emulator. We can install multiple images if, for example, we want to run our app on different versions of Android. We can select emulators with Google Play (as in the following screenshot) so that we can use Google Play services in our app, even when we are running it in an emulator. This is required if, for example, we want to use Google Maps in our app:

Figure 1.11 – Installing emulator images in Visual Studio for Mac

> **Intel versus Apple M1**
>
> If you have an Apple Mac that uses the M1 chipset, then you should use emulator images that have **ARM 64** in the name; otherwise, if you are using older Mac devices with the Intel chipset, then use images with **Intel x86** in the name.

3. Then, to create and configure an emulator, go to **Device Manager** in the **Tools** menu in Visual Studio. From **Android Device Manager**, we can start an emulator if we already have one created, or we can create new emulators, as shown:

Figure 1.12 – Android Device Manager in Visual Studio for Mac

4. If we click on the **New Device** button, we can create a new emulator with the specifications that we need. The easiest way to create a new emulator here is to select a base device that matches our needs. These base devices are preconfigured, which is often enough. However, it is also possible to edit the properties of the device so that we have an emulator that matches our specific needs.

The processor dropdown will be preselected with the correct architecture of your device. If you change this, for example, from ARM to x86 or x86 to ARM, then the emulator will be slower; always try to use the architecture that matches your device.

Figure 1.13 – Creating a new Android device

If you only have a Mac, then you are done and can skip to the *.NET mobile productivity tooling* section. If you have a Windows device, then the next section, *Setting up a Windows machine*, is for you.

Setting up a Windows machine

We can use either a virtual or physical Windows machine for development with .NET mobile. We can, for example, run a virtual Windows machine on our Mac. The only tool we need for app development on our Windows machine is Visual Studio.

Installing .NET mobile for Visual Studio 2022 or later

If we already have Visual Studio 2022 or later installed, we first need to open **Visual Studio Installer**; otherwise, we need to go to `https://visualstudio.microsoft.com` to download the installation files. Under the banner in the **Meet the Visual Studio Family** section, you can find the links to download Visual Studio 2022 for Windows or Visual Studio 2022 for Mac.

Before the installation starts, we need to select which workloads we want to install.

For .NET MAUI development, we need to install **.NET Multi-platform App UI development**. Select the **ASP.NET and web development** workload to be able to develop MAUI/Blazor hybrid apps.

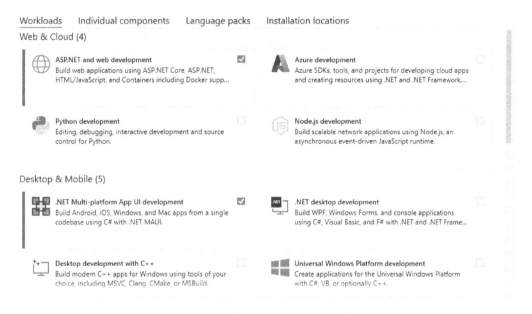

Figure 1.14 – Visual Studio 2022 installer

Hyper-V is the default hardware acceleration method when using .NET MAUI. If you want to use Intel HAXM, you will need to check the checkbox for Intel HAXM in the **Individual components** tab, as in the following screenshot:

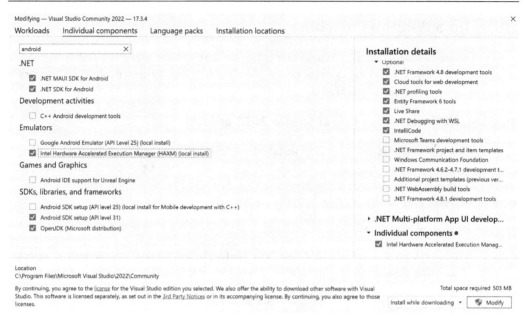

Figure 1.15 – Adding Intel HAXM

When we first start Visual Studio, we will be asked whether we want to sign in. It is not necessary for us to sign in unless we want to use Visual Studio Professional or Enterprise, in which case we will need to sign in so that our license can be verified.

Now that Visual Studio is installed, we can finish the configuration needed to run and debug apps for iOS and Android.

Pairing Visual Studio with a Mac

If we want to run, debug, and compile our iOS app from a Windows development machine, then we need to connect it to a Mac. We can set up our Mac manually, as described earlier in this chapter, or we can use **Automatic Mac Provisioning** from within Visual Studio. This installs Mono and .NET for iOS on the Mac that we are connecting to. It will not install the Visual Studio IDE, but this isn't necessary if we just want to use it as a build machine. We do, however, need to install Xcode manually.

To connect to the Mac from Visual Studio, use the **Pair to Mac** button in the toolbar (as in the following screenshot), or, in the top menu, go to **Tools | iOS | Pair to Mac**:

Figure 1.16 – Pair to Mac button

If this is the first time you have attempted to pair to a Mac, Visual Studio will open a wizard that will guide you through the steps you need to take on your Mac to enable Visual Studio to connect.

Figure 1.17 – Pair to Mac wizard

To be able to connect to a Mac—either manually or using **Automatic Mac Provisioning**—we need to be able to access the Mac via our network, and we need to enable **Remote Login** on the Mac.

To do this, go to **Settings | Sharing** and select the checkbox for **Remote Login**. To the left of the window, we can select which users are allowed to connect with **Remote Login**, as shown:

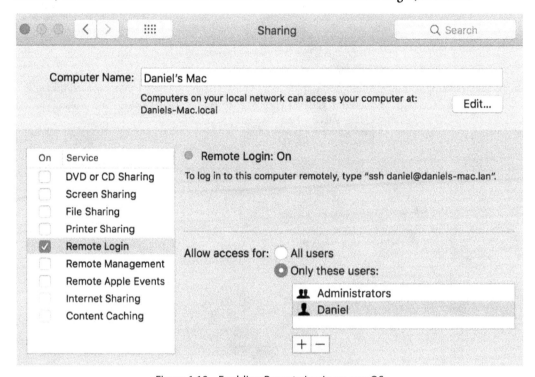

Figure 1.18 – Enabling Remote Login on macOS

A dialog box will appear showing all the Macs that can be found on the network. If your Mac doesn't appear in the list of available Macs, you can use the **Add Mac…** button in the bottom-left corner of the window to enter an IP address, as shown:

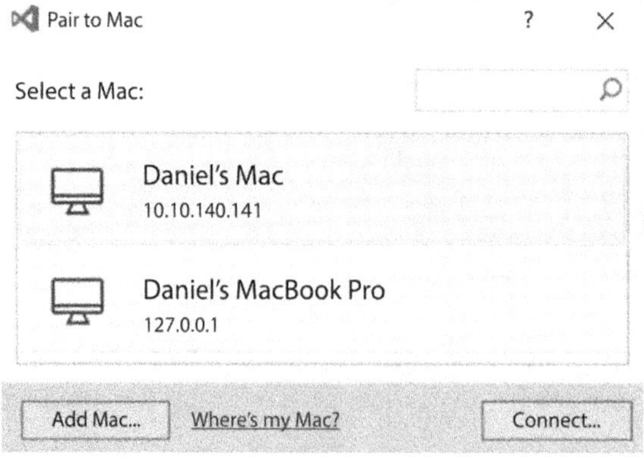

Figure 1.19 – Pair to Mac

If everything that we need is installed on the Mac, then Visual Studio will connect, and we can start building and debugging our iOS app. If Mono is missing on the Mac, a warning will appear. This warning will also give us the option to install it, as shown:

Missing Mono installation

Mono could not be found on the Mac. This is required for building iOS apps with the Visual Studio Tools for Xamarin. Would you like us to install it for you?

Install   Cancel

Figure 1.20 – Missing Mono installation dialog

Now that we have our Mac paired, we can get the Android emulator configured.

Configuring an Android emulator and hardware acceleration

If we want a fast Android emulator that works smoothly, we need to enable hardware acceleration. This can be done using either Intel HAXM or Hyper-V. The disadvantage of Intel HAXM is that it can't be used on machines with an **Advanced Micro Devices** (**AMD**) processor; we must use a machine with an Intel processor. We can't use Intel HAXM in parallel with Hyper-V.

Because of this, Hyper-V is the preferred way to hardware-accelerate an Android emulator on a Windows machine. To use Hyper-V with our Android emulator, we need to have Windows 11 or Windows 10 with the April 2018 update (or later), and Visual Studio 2017 version 15.8 (or later) installed.

Find your version of Visual Studio

To determine the version of Visual Studio you are using, when Visual Studio is open, use the **Help | About Visual Studio** menu, and you should be presented with a dialog similar to the following:

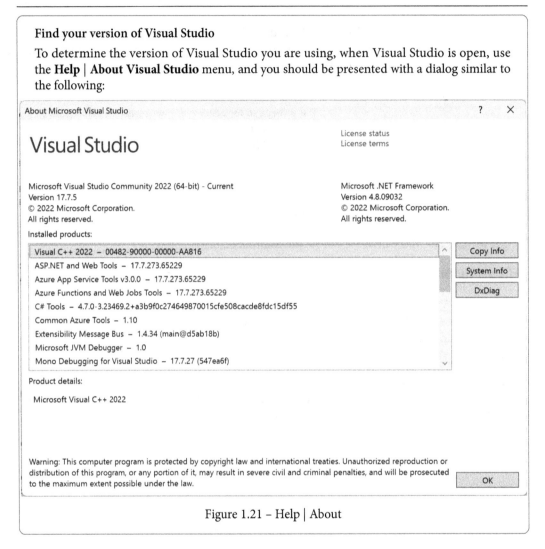

Figure 1.21 – Help | About

To enable Hyper-V, we need to take the following steps:

1. Open the search menu and type in `Turn Windows features on or off`. Click the option that appears to open it, as shown:

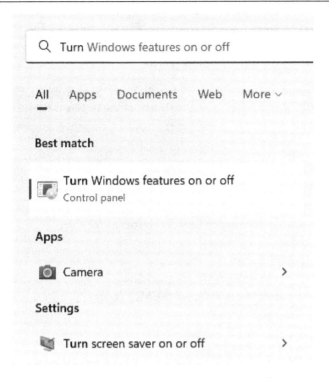

Figure 1.22 – Turn Windows features on or off

2. To enable Hyper-V, select the **Hyper-V** checkbox. Also, expand the **Hyper-V** option and check the **Hyper-V Platform** checkbox. We also need to select the **Windows Hypervisor Platform** checkbox, as shown:

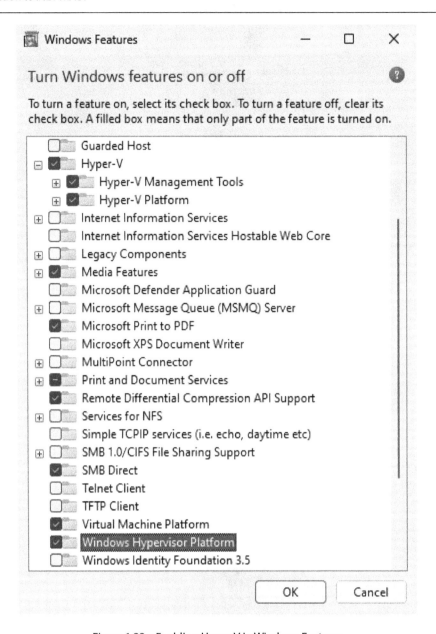

Figure 1.23 – Enabling Hyper-V in Windows Features

3. Restart the machine when Windows prompts you to do so.

4. Because we didn't install an Android emulator during the installation of Visual Studio, we need to install it now. Go to the **Tools** menu in Visual Studio, then click on **Android** and then **Android SDK Manager**.

5. Under **Tools** in **Android SDK Manager**, we can install the emulator by selecting **Android Emulator**, as in the following screenshot. Also, we should ensure that the latest version of **Android SDK Build Tools** is installed:

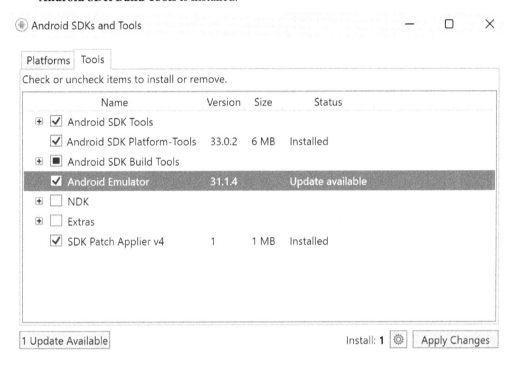

Figure 1.24 – Installing Android Emulator in Android SDK Manager

6. The Android SDK allows multiple emulator images to be installed simultaneously. We can install multiple images if, for example, we want to run our app on different versions of Android. Select emulators with **Google Play** (as in the following screenshot) so that we can use Google Play services in our app, even when we are running it in an emulator.

This is required if, for example, we want to use Google Maps in our app:

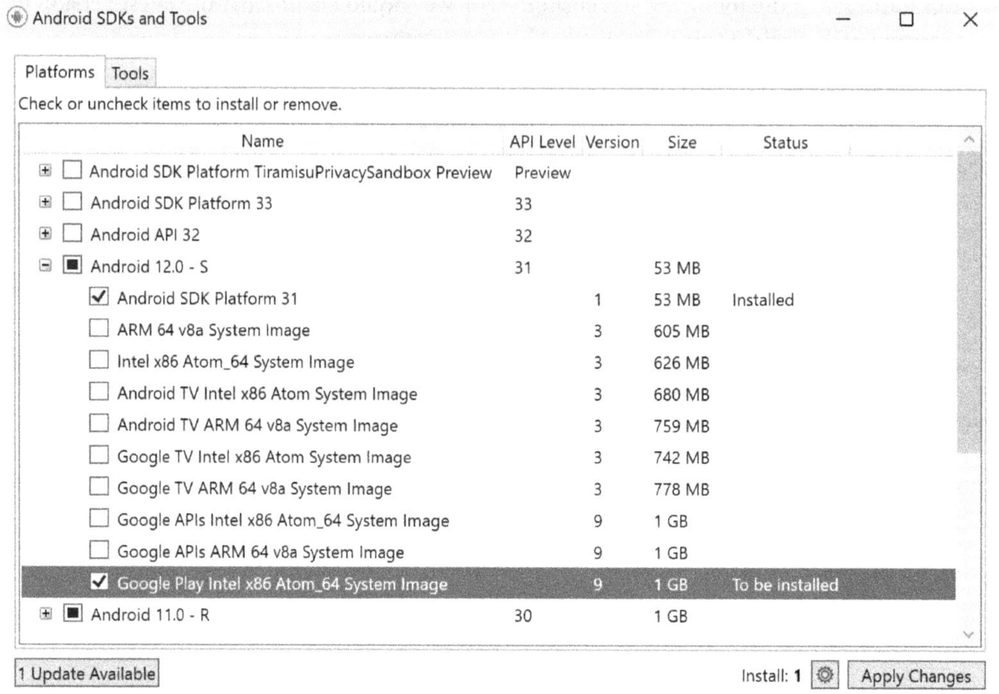

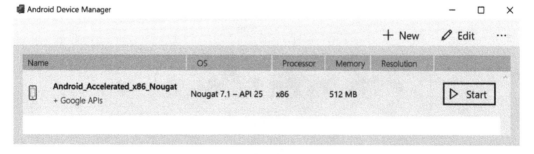

Figure 1.25 – Installing Android Emulator images in Android SDK Manager

7. Be sure to click the **Apply Changes** button to install any components you selected before closing the window.

8. The next step is to create a virtual device to use the emulator image. To create and configure an emulator, go to **Android Device Manager**, which we can open from the **Tools** tab in Visual Studio. From the device manager, we can either start an emulator—if we already have one created—or we can create new emulators, as shown:

Figure 1.26 – Android Device Manager

If we click on the **New** button, we can create a new emulator with the specifications that we need. The easiest way to create a new emulator here is to select a base device that matches our needs. These base devices are preconfigured, which is often enough. However, it is possible to edit the properties of the device so that we have an emulator that matches our specific needs.

We must select the correct processor for the emulator to match the processor in our Windows development machine. If it doesn't match, then the emulator will be slower than it needs to be. Select the **x86_64** processor (as in the following screenshot) if you are using Intel or AMD x86-based hardware, or **arm64-v*** if you have an ARM device running Windows.

Figure 1.27 – Creating a new device in Android Device Manager

Configuring Developer Mode

If we want to develop desktop apps for Windows, we need to activate Developer Mode on our development machine. To do this, go to **Settings** | **Privacy & security** | **For developers**. Then, select **Developer Mode**, as in the following screenshot. This makes it possible for us to sideload and debug apps via Visual Studio:

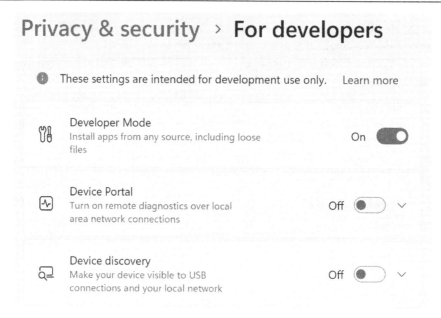

Figure 1.28 – Enabling Developer Mode

At this point, our Windows machine is ready for development. Before we dive into creating our first project, there are a few more optional features that we should review. These will help your development process as you build your apps.

.NET mobile productivity tooling

Xamarin Hot Restart and **Hot Reload** are two tools that increase productivity for .NET MAUI developers. To get even better performance from your Android emulators, you can use the **Windows Subsystem for Android (WSA)**.

Xamarin Hot Restart

Hot Restart is a Visual Studio feature to make developers more productive. It also gives us a way of running and debugging iOS apps on an iPhone without having to use a Mac connected to Visual Studio. Microsoft describes Hot Restart as follows:

> *Xamarin Hot Restart enables you to quickly test changes to your app during development, including multi-file code edits, resources, and references. It pushes the new changes to the existing app bundle on the debug target which results in a much faster build and deploy cycle.*

To use Hot Restart, you need the following:

- Visual Studio 2019 version 16.5
- iTunes (Microsoft Store or 64-bit versions)
- An Apple Developer account and paid Apple Developer Program (`https://developer.apple.com/programs/`) enrollment

Hot Restart can currently only be used with .NET for iOS apps.

Read more about the current state of Hot Restart at `https://docs.microsoft.com/en-us/xamarin/xamarin-forms/deploy-test/hot-restart`.

Hot Reload

Hot Reload is a runtime technology that allows us to update our running app with the changes we are making in the IDE. There are two major flavors of Hot Reload today: **XAML Hot Reload** and **C# Hot Reload**.

XAML Hot Reload allows us to make changes to our XAML without having to redeploy our app. When we have carried out changes to the XAML, we just save the file, and it updates the page on the simulator/emulator or on a device. XAML Hot Reload is currently supported by all .NET MAUI platforms.

C# Hot Reload allows us to make changes to our code without having to redeploy our app. C# Hot Reload is like *Edit & Continue*; however, you do not have to be in break mode in order to apply the changes to the app. Once you have made changes to your code, you can click the **Hot Reload** button in the toolbar of Visual Studio and Hot Reload will update the running app. If, for some reason, the changes cannot be applied, Hot Reload will display a dialog either asking you to fix any compilation errors or, in some cases, requiring you to restart the app.

To enable XAML Hot Reload for Visual Studio on Windows, go to **Tools | Options | Xamarin | Hot Reload**.

To enable XAML Hot Reload for Visual Studio on Mac, go to **Visual Studio | Preferences | Tools for Xamarin | XAML Hot Reload**.

C# Hot Reload is only available in Visual Studio for Windows; to enable it, go to **Tools | Options | Debugger | .NET / C++ Hot Reload**.

Windows Subsystem for Android

If you are using Windows 11 in a supported region, you can use WSA as your debugging target instead of the Android emulators. To learn more about WSA and how to set up your machine to use it, visit `https://learn.microsoft.com/en-us/windows/android/wsa/`.

If you want to use WSA to debug your .NET MAUI apps, it will help if you install the WSA Barista Visual Studio extension (`https://marketplace.visualstudio.com/items?itemName=Redth.WindowsSubsystemForAndroidVisualStudioExtension`). This will add the **Windows Subsystem for Android** menu item under **Tools**, which will prompt you to install WSA from the Windows Store, and then automatically configure WSA and set up Visual Studio to use WSA as a device.

Summary

You should now feel a bit more comfortable with what .NET mobile is and how .NET MAUI relates to .NET mobile.

In this chapter, we established a definition of what a native app is and saw how it has a native UI, performance, and API access. We talked about how .NET mobile uses Mono, which is an open source implementation of the .NET Framework, and discussed how, at its core, .NET mobile is a set of bindings to platform-specific APIs. We then looked at how .NET for iOS and .NET for Android work under the hood.

After that, we began to touch on the core topic of this book, which is .NET MAUI. We started with an overview of how platform-agnostic controls are rendered to platform-specific controls and how to use XAML to define a hierarchy of controls to assemble a page. We then spent some time looking at the difference between a .NET MAUI app and a traditional .NET mobile app.

A traditional .NET mobile app uses platform-specific APIs directly, without any abstraction, other than what .NET adds as a platform. .NET MAUI is an API that is built on top of the traditional .NET APIs and allows us to define platform-agnostic GUIs in XAML or in code that is rendered to platform-specific controls. There's more to .NET MAUI than this, but this is what it does at its core.

In the last part of this chapter, we discussed how to set up a development machine on Windows or macOS. Finally, we looked at three optional features that you can use to help improve your development cycle: Hot Restart, Hot Reload, and WSA.

Now, it's time to put our newly acquired knowledge to use! We will start by creating a *to-do* app from the ground up in the next chapter. We will look at concepts such as **Model–View–ViewModel** (**MVVM**) for a clean separation between business logic and the UI and SQLite.NET to persist data to a local database on our device. We will do this for three platforms at the same time—so, read on!

2

Building Our
First .NET MAUI App

In this chapter, we will create a to-do list app and, in doing so, explore all the bits and pieces of what makes up an app. We will look at creating pages, adding content to pages, navigating between pages, and creating a stunning layout. Well, *stunning* might be a bit of a stretch, but we will be sure to design the app so that you can tweak it to your needs once it is complete!

The following topics will be covered in this chapter:

- Setting up the project
- Persisting data locally on a device using the repository pattern
- What MVVM is and why it's a great fit for .NET MAUI
- Using .NET MAUI pages (as views) and navigating between them using .NET MAUI controls in XAML
- Using data binding
- Using styling in .NET MAUI

Technical requirements

To complete this project, you need to have Visual Studio installed on your **Macintosh (Mac)** or PC, as well as the .NET mobile components. Refer to *Chapter 1, Introduction to .NET MAUI*, for more details on how to set up your environment. This chapter provides screenshots and instructions for Visual Studio on Windows.

This chapter will be a classic **File | New | Project** chapter, guiding you step by step through the process of creating your first to-do list app. No downloads will be required whatsoever, apart from a few NuGet packages.

You can find the full source for the code in this chapter at `https://github.com/Packt Publishing/MAUI-Projects-3rd-Edition` in the `Chapter02` folder.

Project overview

Everyone needs a way of keeping track of things. To kick-start our .NET MAUI development learning curve, we've decided that a to-do list app is the best way to get started and to help you keep track of things. A simple, classic win-win scenario.

We will start by creating a project and defining a repository to store the items of a to-do list. We will render these items in list form and allow the user to edit them using a detailed user interface. We will also look at how to store the to-do list items locally on a device through **SQLite.NET** so that they don't get lost when we exit the app.

The build time for this project is about 2 hours.

Setting up the project

.NET MAUI introduces a new code-sharing paradigm called single project. Previously, in Xamarin. Forms, you would have had a separate project for each platform your app would be deployed to. In .NET MAUI, all platforms are in a single project that is multi-targeted to all the supported platforms. By default, all code is considered shared, unless it is in one of the platform-specific subfolders. We will explore this further as we progress through this and future chapters.

Let's get started!

Creating the new project

The first step is to create a new .NET MAUI project. Open Visual Studio 2022 and select **Create a new project**:

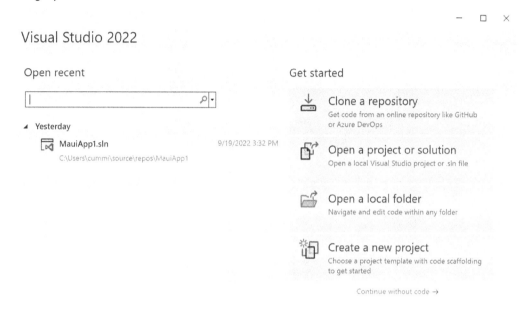

Figure 2.1 – Visual Studio 2022

This will open the **Create a new project** wizard. In the search field, type in `maui` and select the **.NET MAUI App** item from the list:

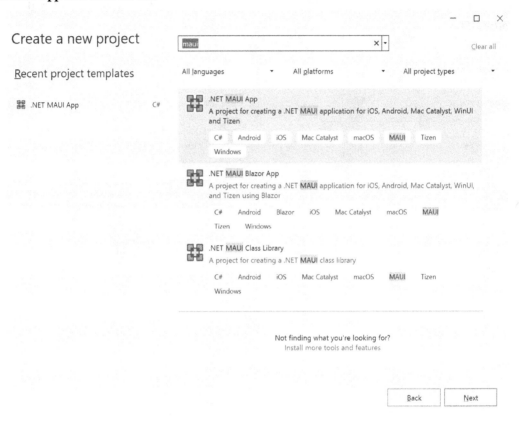

Figure 2.2 – Create a new project

Complete the next page of the wizard by naming your project DoToo, then click **Next**:

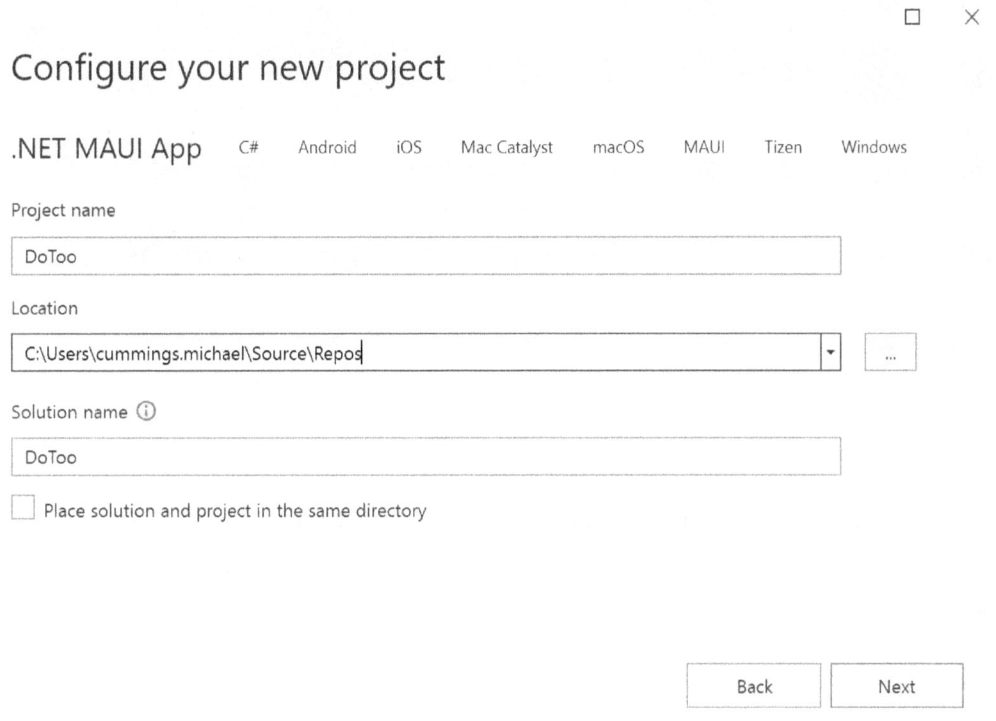

□   ✕

Configure your new project

.NET MAUI App   C#   Android   iOS   Mac Catalyst   macOS   MAUI   Tizen   Windows

Project name

DoToo

Location

C:\Users\cummings.michael\Source\Repos   ▼   ...

Solution name ⓘ

DoToo

☐ Place solution and project in the same directory

Back   Next

Figure 2.3 – Configure your new project

The next step will prompt you for the version of .NET Core to support. At the time of writing, .NET 6 is available as **Long-Term Support (LTS)**, and .NET 7 is available as **Standard Term Support**. For this book, we will assume that you will be using .NET 7:

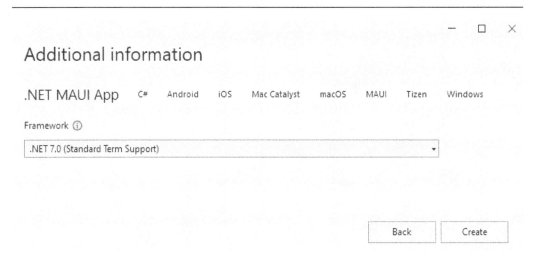

Figure 2.4 – Additional information

Finalize the setup by clicking **Create** and wait for Visual Studio to create the project.

Congratulations! We've just created our first .NET MAUI app. Let's take a look at what the template wizard generated for us.

Examining the files

The selected template has now created a single project called DoToo as a .NET library that can target iOS, Mac Catalyst (macOS), Android, and Windows platforms. You can switch the target platform using the main toolbar in Visual Studio, as shown in *Figure 2.5*:

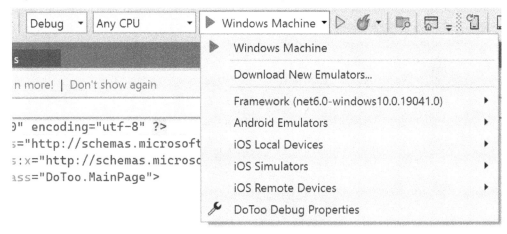

Figure 2.5 – The Debug Target drop-down menu

The Windows platform is selected by default, but you can easily switch to iOS or Android by using the **Debug Target** drop-down menu. In the dropdown under the **Framework** sub-menu, you will find all the supported target platforms.

The target framework will also change appropriately when you choose a target device. If you select an emulator under the **Android Emulators** menu item, then the Android target framework will become the current framework, whereas if you select an iOS simulator or device from one of the iOS menu items, iOS will be the current framework.

The project should now look as follows:

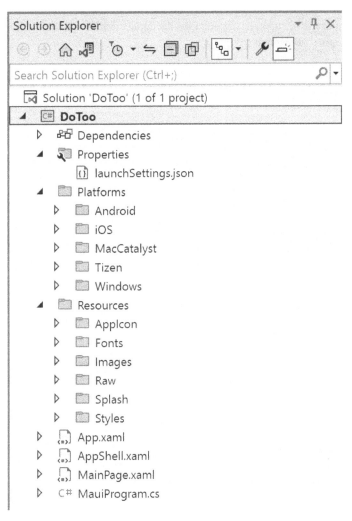

Figure 2.6 – .NET MAUI project structure

We will highlight a few important files in the project so that we have a basic understanding of what they are. First, we will look at the shared code, after which we'll look at the files/code specific to each platform (stored under the different platform folders).

Shared code

Under Dependencies, we will find references to any external dependencies, such as each referenced .NET mobile framework. Under each framework, you will find the .NET MAUI dependency under the packages folder. We will update the .NET MAUI package version in the *Updating the .NET MAUI packages* section and add more dependencies as we progress through this chapter.

The MauiProgram.cs file is the starting point for the application. The initial template will generate a MauiProgram class that looks as follows:

```
public static class MauiProgram
{
    public static MauiApp CreateMauiApp()
    {
        var builder = MauiApp.CreateBuilder();
        builder
            .UseMauiApp<App>()
            .ConfigureFonts(fonts =>
            {
                fonts.AddFont("OpenSans-Regular.ttf",
"OpenSansRegular");
                fonts.AddFont("OpenSans-Semibold.ttf",
"OpenSansSemibold");
            });

#if DEBUG
        builder.Logging.AddDebug();
#endif

        return builder.Build();
    }
}
```

The static MauiProgram class contains a single CreateMauiApp method that returns MauiApp. This instance is created by using MauiAppBuilder, which works in much the same way that the ASP. NET builders work; MauiAppBuilder uses a **Fluent API** to construct the Application instance.

> **What is a Fluent API?**
>
> A Fluent API allows method chaining wherein each method of the API returns the same context. Fluent APIs form a distinct language in themselves by using terms specific to the topic of the API. This makes the API easier to grasp and use. C#'s **Language Integrated Query (LINQ)** is a good example of a Fluent API.

Extension methods are used to add features and services to the `MauiApp` instance. The `UseMauiApp` extension method identifies the subclass of `Microsoft.Maui.Controls.Application` to use. By default, this class is defined in the `App.xaml` and `App.xaml.cs` files. Another extension method, `ConfigureFonts`, is used by the template to register custom font files in use by the application. Yet another example of an extension method that can be used is `ConfigureLifecycleEvents`, which is used to set up handlers for the cross-platform life cycle events available in .NET MAUI. We will discuss `ConfigureLifecycleEvents` more in *Chapter 3, Converting a Xamarin.Forms App into .NET MAUI.*

The `App.xaml` file is a XAML file that represents the app. This is a good place to put application-wide resources, which we will do later. We can also see the `App.xaml.cs` file, which contains the startup code.

If we open `App.xaml.cs`, we can see the starting point for our .NET MAUI application:

```
public partial class App : Application
{
    public App()
    {
        InitializeComponent();

        MainPage = new AppShell();
    }
}
```

The `MainPage` property is assigned to a page, which is particularly important as this determines which page is shown first to the user. In this template, this is the `DoToo.AppShell()` class.

The `AppShell.xaml` and `AppShell.xaml.cs` files declare the first visible UI component in the .NET MAUI app. Shell provides a form of navigation between pages. When you open `AppShell.xaml`, it should look like this:

```
<Shell
    x:Class="DoToo.AppShell"
    xmlns=http://schemas.microsoft.com/dotnet/2021/maui
    xmlns:x=http://schemas.microsoft.com/winfx/2009/xaml
    xmlns:local="clr-namespace:DoToo"
    Shell.FlyoutBehavior="Disabled">
```

```
<ShellContent
    Title="Home"
    ContentTemplate="{DataTemplate local:MainPage}"
    Route="MainPage" />

</Shell>
```

The `ShellContent` element identifies an individual page that is displayed within the shell. The `ContentTemplate` attribute is used to locate the class that implements the page – in this case, `MainPage` – while `Route` is the unique identifier for the page.

The last two files are the `MainPage.xaml` file, which contains the first page of the application, and the code-behind file, which is called `MainPage.xaml.cs`.

Next, we will work through each platform's files. Each platform has a unique folder under the `Platforms` folder. Android files are in the `Android` folder, iOS files are in the `iOS` folder, Mac Catalyst files are in the `MacCatalyst` folder, Tizen files are in the `Tizen` folder, and Windows files are in the `Windows` folder.

Android

The Android-specific platform code lives under the `Platforms/Android` folder in the project:

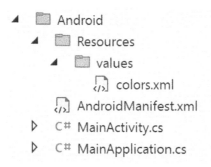

Figure 2.7 – Android-specific files

The important files here are `MainActivity.cs` and `MainApplication.cs`. These two files contain the entry point for our application when we run the app on an Android device. A standard Android app will declare **Activity** as **MainLauncher**. In .NET MAUI, this is done using the `MainLauncher` property of the `Activity` class attribute. `MauiAppCompatActivity` will search for a type decorated with `ApplicationAttribute` and instantiate it.

This attribute can be found in the `MainApplication` class in `MainApplication.cs`. During initialization, `MainApplication` will call the `CreateMauiApp` method, which, in turn, calls `MauiProgram.CreateMauiApp`, which we explored earlier in this chapter.

You don't need to understand these files in detail; just remember that they are important for initializing our app.

iOS and Mac Catalyst

The iOS and Mac Catalyst platform files are identical, but each has a folder for customizing the platform. Each platform's files are contained in their respective named folder under the `Platform` folder:

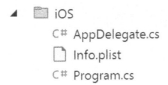

Figure 2.8 – iOS platform-specific files

`AppDelegate.cs` is the equivalent of the `MainApplication` class in the Android platform. It contains a single method called `CreateMauiApp` that has the same implementation as Android; it calls the `MauiProgram.CreateMauiApp` method.

The `Program.cs` file is the entry point for an iOS app. It contains the `Main` method, which calls `UIApplication.Main`, the launching point for an iOS application, and references the `AppDelegate` type to instantiate.

The code starts by initializing .NET MAUI and then loads the application. After that, it returns control to iOS. It must do this within 17 seconds; otherwise, the app is terminated by the OS.

The `info.plist` file is an iOS-specific file that contains information about the app, such as the bundle ID and its provisioning profiles. Visual Studio has a graphical editor for the `info.plist` file, but since it is a standard XML file, it can be edited in any text editor.

As with the Android app's startup code, we don't need to understand what is going on here in detail, other than that it's important for initializing our app.

Tizen

Tizen is Samsung's custom distribution of Android. The `Main.cs` file is the launching point and, like the Android platform, the `Program` class has a `CreateMauiApp` method. Tizen is not enabled by default. To enable it, follow the instructions in the comments in the `DoToo.csproj` file. To develop applications for Tizen, you will need to install additional software distributed by Samsung.

Windows

The last platform we will examine is the **WinUI** app. The file structure looks as follows:

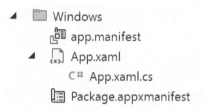

Figure 2.9 – Windows-specific files

It has an App.xaml file, which is like the one in the shared code, but specific to **Universal Windows Platform** (**UWP**) apps. It also has a related file called App.xaml.cs. This file is the Windows equivalent of Android's MauiApplication, which contains the CreateMauiApp method.

That's it for the platform project files. Next, we will look at how to keep .NET MAUI up to date.

Updating the .NET MAUI packages

> **Note – Windows users**
>
> Since .NET MAUI is distributed as part of Visual Studio, it is best to allow Visual Studio to update the packages when you update Visual Studio. If you follow these steps, you can probably get .NET MAUI into an unusable state.

.NET MAUI is distributed as a set of optional **.NET workloads**. Workloads are managed from the command line using the dotnet workload command. To see the currently installed workloads and their versions, you can use the dotnet workload list command. Visual Studio 2022 has a built-in developer PowerShell to execute commands. To access it, press *Ctrl + `* on both macOS and Windows.

Running the dotnet workload list should give you the following output. Note that your version numbers may be higher:

```
C:\Users\cummings.michael\Source\Repos\DoToo
> dotnet workload list


Installed Workload Ids     Manifest
Version                             Installation Source
-------------------------------------------------------------------
-----------------------
maui-windows               6.0.486/6.0.400
     VS 17.3.32901.215
```

```
maui-maccatalyst                 6.0.486/6.0.400
     VS 17.3.32901.215
maccatalyst                      15.4.446-ci.-release-6-0-4xx.446/6.0.400
     VS 17.3.32901.215
maui-ios                         6.0.486/6.0.400
     VS 17.3.32901.215
ios                              15.4.446-ci.-release-6-0-4xx.446/6.0.400
     VS 17.3.32901.215
maui-android                     6.0.486/6.0.400
     VS 17.3.32901.215
android                          32.0.448/6.0.400
     VS 17.3.32901.215


Use `dotnet workload search` to find additional workloads to install.
```

To update the packages in the MAUI workload, you can run `dotnet workload update`. This is a sample of the result from running that command:

```
C:\Users\cummings.michael\Source\Repos\DoToo
> dotnet workload update

No workloads installed for this feature band. To update workloads
installed with earlier SDK
versions, include the --from-previous-sdk option.
Updated advertising manifest microsoft.net.sdk.android.
Updated advertising manifest microsoft.net.sdk.tvos.
Updated advertising manifest microsoft.net.sdk.macos.
Updated advertising manifest microsoft.net.sdk.maui.
Updated advertising manifest microsoft.net.workload.emscripten.
Updated advertising manifest microsoft.net.sdk.ios.
Updated advertising manifest microsoft.net.sdk.maccatalyst.
Updated advertising manifest microsoft.net.workload.mono.toolchain.
Downloading microsoft.net.sdk.android.manifest-6.0.400.msi.x64
(32.0.465)
Installing Microsoft.NET.Sdk.Android.Manifest-6.0.400.32.0.465-x64.msi
......... Done
Downloading microsoft.net.sdk.ios.manifest-6.0.400.msi.x64 (15.4.454)
Downloading microsoft.net.sdk.maccatalyst.manifest-6.0.400.msi.x64
(15.4.454)
Downloading microsoft.net.sdk.macos.manifest-6.0.400.msi.x64
(12.3.454)
Installing Microsoft.NET.Sdk.macOS.Manifest-6.0.400.12.3.454-x64.msi
...... Done
Downloading microsoft.net.sdk.maui.manifest-6.0.400.msi.x64 (6.0.540)
```

```
Installing Microsoft.NET.Sdk.Maui.Manifest-6.0.400.6.0.540-x64.msi
...... Done
Downloading microsoft.net.sdk.tvos.manifest-6.0.400.msi.x64 (15.4.454)
Downloading microsoft.net.workload.mono.toolchain.manifest-6.0.400.
msi.x64 (6.0.9)
Installing Microsoft.NET.Workload.Mono.ToolChain.Manifest-
6.0.400.6.0.9-x64.msi ....... Done
Downloading microsoft.net.workload.emscripten.manifest-6.0.400.msi.x64
(6.0.9)
Installing Microsoft.NET.Workload.Emscripten.Manifest-
6.0.400.6.0.9-x64.msi ...... Done
No workloads installed for this feature band. To update workloads
installed
with earlier SDK versions, include the --from-previous-sdk option.

Successfully updated workload(s): .
```

To see the result, just run the `dotnet workload list` command again:

```
C:\Users\cummings.michael\Source\Repos\DoToo
> dotnet workload list

Installed Workload Ids          Manifest Version
Installation Source
-------------------------------------------------------------------
-----------------------
maui-windows                    6.0.540/6.0.400
    VS 17.3.32901.215
maui-maccatalyst                6.0.540/6.0.400
    VS 17.3.32901.215
maccatalyst                     15.4.446-ci.-release-6-0-4xx.446/6.0.400
    VS 17.3.32901.215
maui-ios                        6.0.540/6.0.400
    VS 17.3.32901.215
ios                             15.4.446-ci.-release-6-0-4xx.446/6.0.400
    VS 17.3.32901.215
maui-android                    6.0.540/6.0.400
    VS 17.3.32901.215
android                         32.0.465/6.0.400
    VS 17.3.32901.215

Use `dotnet workload search` to find additional workloads to install.
```

Now that we have a basic understanding of how .NET MAUI projects are structured, we can start building our first app!

Creating a repository and a TodoItem model

Any good architecture always involves abstraction. In this app, we need something to store and retrieve the items of our to-do list. Later, these will be stored in a SQLite database, but adding a reference to the database directly in the code that is responsible for the GUI is generally a bad idea as it tightly couples your data storage implementation to the UI layer, making it harder to test your UI code independently from the database.

So, what we need is something to abstract our database from the GUI. For this app, we've chosen to use a simple repository pattern. This repository is simply a class that sits between the SQLite database and our upcoming `ViewModel` class. This is the class that handles the interaction with the view, which, in turn, handles the GUI.

The repository will expose methods for getting, adding, and updating items, as well as events that allow other parts of the app to react to changes in the repository. It will be hidden behind an interface so that we can replace the entire implementation later without modifying anything but a line of code in the initialization of the app. This is made possible by the **Microsoft.Extensions.DependencyInjection** NuGet package.

Defining a to-do list item

We will start by creating a `TodoItem` class, which represents a single item on the list. This is a simple **Plain Old CLR Object (POCO)** class, where **CLR** stands for **Common Language Runtime**. In other words, this is a .NET class without any dependencies on third-party assemblies. To create the class, follow these steps:

1. In the `DoToo` project, create a folder called `Models`.

2. Add a class called `TodoItem.cs` to that folder and enter the following code:

    ```
    namespace DoToo.Models;

    using System;

    public class TodoItem
    {
        public int Id { get; set; }
        public string Title { get; set; }
        public bool Completed { get; set; }
        public DateTime Due { get; set; }
    }
    ```

This code is self-explanatory; it's a simple POCO class that only contains properties and no logic. We have a `Title` property that describes what we want to be done, a flag named `Completed` that determines whether the to-do list item is completed, a `Due` date for when we expect it to be done, and a unique `Id` property that we will need later for the database.

Creating a repository and its interface

Now that we have the `TodoItem` class, let's define an interface that describes a repository that will store our to-do list items:

1. In the DoToo project, create a folder called `Repositories`.

2. Create an interface called `ITodoItemRepository.cs` in the `Repositories` folder and write the following code:

    ```
    namespace DoToo.Repositories;

    using DoToo.Models;

    public interface ITodoItemRepository
    {
        event EventHandler<TodoItem> OnItemAdded;
        event EventHandler<TodoItem> OnItemUpdated;

        Task<List<TodoItem>> GetItemsAsync();
        Task AddItemAsync(TodoItem item);
        Task UpdateItemAsync(TodoItem item);
        Task AddOrUpdateAsync(TodoItem item);
    }
    ```

> **Wait, what? No Delete method?**
>
> The eagle-eyed among you might have noticed that we are not defining a `Delete` method in this interface. This is something that should be in a real-world app. While the app that we are creating in this chapter does not support deleting items, we are quite sure that you could add this yourself if you want to!

This interface defines everything we need for our app. It is there to create logical insulation between your implementation of a repository and the user of that repository. If any other parts of your application want an instance of `ITodoItemRepository`, we can pass it an object that implements `ITodoItemRepository`, regardless of how it's implemented.

With that said, let's implement `ITodoItemRepository`:

1. Create a class called `TodoItemRepository.cs` in the `Repositories` folder.

2. Enter the following code:

```
namespace DoToo.Repositories;

using DoToo.Models;

public class TodoItemRepository : ITodoItemRepository
{
    public event EventHandler<TodoItem> OnItemAdded;
    public event EventHandler<TodoItem> OnItemUpdated;

    public async Task<List<TodoItem>> GetItemsAsync()
    {
        return null; // Just to make it build
    }

    public async Task AddItemAsync(TodoItem item)
    {
    }

    public async Task UpdateItemAsync(TodoItem item)
    {
    }

    public async Task AddOrUpdateAsync(TodoItem item)
    {
        if (item.Id == 0)
        {
            await AddItemAsync(item);
        }
        else
        {
            await UpdateItemAsync(item);
        }
    }
}
```

This code is the bare-bones implementation of the interface, except for the `AddOrUpdateAsync(...)` method. This handles a small piece of logic that states that if the `Id` value of an item is 0, it's a new item. Any item with an `Id` value greater than 0 is stored in the database. This is because the database assigns a value larger than 0 when we create rows in a table.

There are also two events defined in the preceding code. They will be used to notify subscribers of a list of items that have been updated or added.

Connecting SQLite to persist data

We now have an interface, as well as a skeleton to implement that interface. The last thing we need to do to finish this section is to connect SQLite in the implementation of the repository.

Adding the SQLite NuGet package

To access SQLite in this project, we need to add a NuGet package called `sqlite-net-pcl` to the DoToo project. To do this, right-click on the **Dependencies** item under the DoToo project node of the solution and click **Manage NuGet Packages...**:

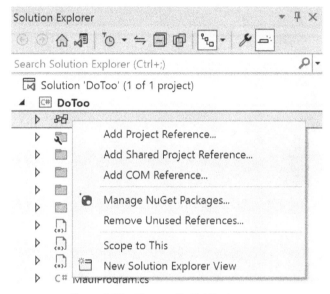

Figure 2.10 – Manage NuGet Packages...

> **Portable Class Library (PCL)**
>
> You might have noticed that the NuGet package is suffixed with `-pcl`. This is an example of what happens when naming conventions go wrong. This package supports .NET Standard 1.0, even though the name says PCL, which was the predecessor of .NET Standard.

This brings up the **NuGet Package Manager** window:

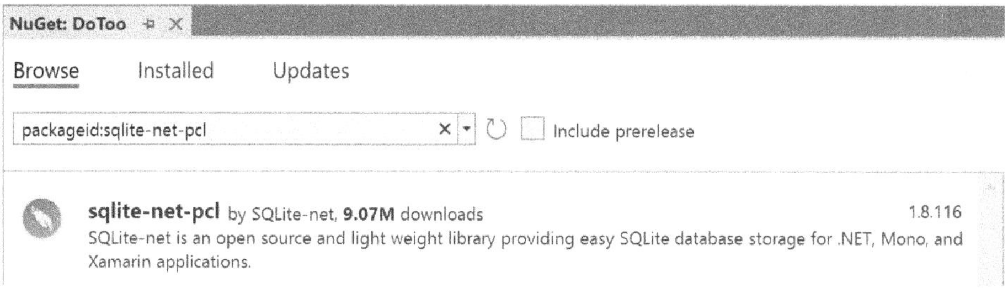

Figure 2.11 – NuGet Package Manager

To install the SQLite NuGet package, follow these steps:

1. Click **Browse** and enter `sqlite-net-pcl` in the search box.

2. Select the package by **sQLite-net** and click **Install**.

3. A dialog will be displayed showing you all the packages that will be downloaded to your system; accept the changes to complete the installation.

> Important
>
> Version 1.8.116 of the `sqlite-net-pcl` package references versions of the native library that are not fully compatible with .NET 6+ on all platforms. To work around this, you need to add additional references to the following packages manually with a version of at least 2.1:
>
> - `SQLitePCLRaw.core`
>
> - `SQLitePCLRaw.provider.sqlite3`
>
> - `SQLitePCLRaw.bundle_green`
>
> - `SQLitePCLRaw.provider.dynamic_cdecl`
>
> See `https://github.com/praeclarum/sqlite-net/issues/1102` for more details and a potential resolution.

Once the installation is complete, we can add some code to the `TodoItem` class to map the C# object to a table and create the connection to the database in the repository.

Updating the TodoItem class

Since SQLite is a relational database, it needs to know some basic information about how to create the tables that will store our objects. This is done using attributes, which are defined in the SQLite namespace:

1. Open `Models/TodoItem.cs`.

2. Add a `using SQLite` statement at the start of the file right below the `namespace` statement, as shown in the following code:

```
namespace DoToo.Models;

using SQLite;

public class TodoItem
```

3. Add the `PrimaryKey` and `AutoIncrement` attributes right before the `Id` property, as shown in the following code:

```
[PrimaryKey, AutoIncrement]
public int Id { get; set; }
```

The `PrimaryKey` attribute instructs SQLite that the `Id` property is the primary key of the table. The `AutoIncrement` attribute makes sure that the value of `Id` is increased by 1 for each new `TodoItem` class that is added to the table.

With the data object classes properly configured, it is now time to create the connection to the database.

Creating a connection to the SQLite database

We will now add all the code needed to communicate with the database. The first thing we need to do is define a connection field that will hold the connection to the database:

1. Open the `Repositories/TodoItemRepository.cs` file.

2. Add a `using SQLite` statement right below the existing `using` statements, as shown in the following code:

```
namespace DoToo.Repositories;

using DoToo.Models;
```

```
using SQLite;

public class TodoItemRepository : ITodoItemRepository
```

3. Add the following field right below the class declaration:

```
private SQLiteAsyncConnection connection;
```

The connection needs to be initialized. Once it is initialized, it can be reused throughout the lifespan of the repository. Since the method is asynchronous, it cannot be called from the constructor without introducing a locking strategy. To keep things simple, we will simply call it from each of the methods that are defined by the interface. To do so, add the following code to the `TodoItemRepository` class:

```
private async Task CreateConnectionAsync()
{
    if (connection != null)
    {
        return;
    }

    var documentPath = Environment.GetFolderPath(Environment.
SpecialFolder.MyDocuments);
    var databasePath = Path.Combine(documentPath, "TodoItems.db");
    connection = new SQLiteAsyncConnection(databasePath);
    await connection.CreateTableAsync<TodoItem>();
    if (await connection.Table<TodoItem>().CountAsync() == 0)
    {
        await connection.InsertAsync(new TodoItem()
        {
            Title = "Welcome to DoToo",
            Due = DateTime.Now
        });
    }
}
```

The method begins by checking whether we already have a connection. If we do, we can simply `return`. If we don't have a connection set up, we define a path on the disk to indicate where we want the database file to be located. In this case, we will choose the `MyDocuments` folder. .NET MAUI will find the closest match to this on each platform that we target.

Then, we create the connection and store the reference to that connection in the `connection` field. We need to make sure that SQLite has created a table that mirrors the schema of the `TodoItem` table. To make developing the app easier, we add a default to-do list item if the `TodoItem` table is empty.

Next, we'll add the implementation for the database operations.

Implementing the GetItemsAsync(), AddItemsAsync(), and UpdateItemsAsync() methods

The only thing left to do in the repository is to implement the methods for getting, adding, and updating items:

1. Locate the `GetItemsAsync()` method in the `TodoItemRepository` class.

2. Update the `GetItemsAsync()` method with the following code:

```
public async Task<List<TodoItem>> GetItemsAsync()
{
    await CreateConnectionAsync();
    return await connection.Table<TodoItem>().ToListAsync();
}
```

To ensure that the connection to the database is valid, we call the `CreateConnectionAsync()` method we created in the previous section. When this method returns, we can ensure that it is initialized and that the `TodoItem` table has been created.

Then, we use the connection to access the `TodoItem` table and return a `List<TodoItem>` item that contains all the to-do list items in the database.

> **SQLite and LINQ**
>
> SQLite supports querying data using LINQ. You can play around with this after the project is complete to get a better understanding of how to work it with databases in your app.

The code for adding items is even simpler:

1. Locate the `AddItemAsync()` method in the `TodoItemRepository` class.

2. Update the `AddItemAsync()` method with the following code:

```
public async Task AddItemAsync(TodoItem item)
{
    await CreateConnectionAsync();
```

```
        await connection.InsertAsync(item);
        OnItemAdded?.Invoke(this, item);
    }
```

The call to `CreateConnectionAsync()` makes sure that we have a connection in the same way as we did for the `GetItemsAsync()` method. After this, we insert it into the database using the `InsertAsyncAsync(...)` method on the `connection` object. After an item has been inserted into the table, we invoke the `OnItemAdded` event to notify any subscribers.

The code to update an item is the same as the `AddItemAsync()` method but also includes calls to `UpdateAsync` and `OnItemUpdated`. Let's finish up by updating the `UpdateItemAsync()` method with the following code:

1. Locate the `UpdateItemAsync()` method in the `TodoItemRepository` class.

2. Update the `UpdateItemAsync()` method with the following code:

```
public async Task UpdateItemAsync(TodoItem item)
{
    await CreateConnectionAsync();
    await connection.UpdateAsync(item);
    OnItemUpdated?.Invoke(this, item);
}
```

In the next section, we'll get started with MVVM. Grab a cup of coffee and let's get started!

Using MVVM – creating views and ViewModels

Model-View-ViewModel, or **MVVM** for short, is all about **separation of concerns**. It is an architectural pattern that defines three parts, each of which has a specific meaning:

- **Model**: This relates to anything that represents data and that can be referenced with `ViewModel`.

- **View**: This is the visual component. In .NET MAUI, this is represented by a page.

- **ViewModel**: This is the class that acts as the glue between the model and the view.

We are introducing MVVM here because the MVVM pattern was designed specifically around XAML-based GUIs. This app and the rest of the apps in this book will use XAML to define the GUI and we will use the MVVM pattern to separate the code into the three aforementioned parts.

In this app, we could say that the model is the repository and the to-do list items it returns. `ViewModel` refers to this repository and exposes properties that the view can bind to. The ground rule is that any logic should reside in `ViewModel` and no logic should reside in the view. The view should know how to present data, such as converting a Boolean value into `Yes` or `No`.

MVVM can be implemented in many ways and there are quite a few frameworks that we can use to do so, such as **Prism**, **MVVMCross**, or even **TinyMvvm**. In this chapter, we have chosen to keep things simple and implement MVVM in a vanilla way first, and then use portions of the **CommunityToolkit. Mvvm** library. CommunityTookit.Mvvm is an open source library produced by the .NET Foundation. It is a replacement for the **MVVMLight** library.

The main benefits of using MVVM as an architectural pattern are a clear separation of concerns, cleaner code, and great testability of ViewModel. If you are interested in learning more about MVVM, and how to use it with .NET MAUI, visit https://learn.microsoft.com/en-us/dotnet/ architecture/maui/mvvm.

Well, enough of that – let's write some code instead!

Defining a ViewModel base class

ViewModel is the mediator between the view and the model. We can benefit from it greatly by creating a common base class for all our ViewModel classes to inherit from. To do this, follow these steps:

1. Create a folder called ViewModels in the DoToo project.

2. Create a class called ViewModel in the ViewModels folder.

3. Add the following code:

```
using System.ComponentModel;

public abstract class ViewModel : INotifyPropertyChanged
{
    public event PropertyChangedEventHandler PropertyChanged;

    public void RaisePropertyChanged(params string[]
propertyNames)
    {
        foreach (var propertyName in propertyNames)
        {
            PropertyChanged?.Invoke(this, new
PropertyChangedEventArgs(propertyName));
        }
    }

    public INavigation Navigation { get; set; }
}
```

The ViewModel class is a base class for all ViewModel objects. It is not meant to be instantiated on its own, so we mark it as abstract. It implements INotifyPropertyChanged, which is an interface defined in System.ComponentModel in the .NET base class libraries. This interface only defines one thing – the PropertyChanged event. Our ViewModel class must raise this event whenever we want the GUI to be aware of any changes to a property. This can be done manually by adding code to a setter in a property, as we did in the current implementation, or by using the **CommunityToolkit.Mvvm** library. We will talk about this in more detail in the next section.

We will also take a little shortcut here by adding an `INavigation` property to `ViewModel`. This will help us with navigation later on. This is also something that can (and should) be abstracted since we don't want `ViewModel` to be dependent on .NET MAUI to be able to reuse the `ViewModel` classes on any platform.

Introducing the CommunityToolkit.Mvvm library's ObservableObject and ObservableProperty

The traditional way of implementing a `ViewModel` class is to inherit it from a base class (such as the `ViewModel` class that we defined previously) and then add code that might look as follows:

```
public class MyTestViewModel : ViewModel
{
    private string name;
    public string Name
    {
        get { return name; }
        set
        {
            if (name != value)
            {
                name = value;
                RaisePropertyChanged(nameof(Name));
            }
        }
    }
}
```

Each property that we want to add to a `ViewModel` class yields 13 lines of code. Not too bad, you might think. However, considering that a `ViewModel` class could potentially contain 10 to 20 properties, this rapidly turns into a lot of code. We can do better than this.

In just a few simple steps, we can use the CommunityToolkit.Mvvm library to automatically inject almost all the code during the build process:

1. In the `DoToo` project, install the CommunityToolkit.Mvvm NuGet package.

2. Update the `ViewModel` class so that it looks like this:

```
using CommunityToolkit.Mvvm.ComponentModel;

[ObservableObject]
public abstract partial class ViewModel
{
```

```
        public INavigation Navigation { get; set; }
    }
```

We have changed the base class of our `ViewModel` class so that it has an `ObservableObject` attribute and added the `partial` modifier. This attribute will add the base implementation of `INotifyPropertyChanged` that was previously in our `ViewModel` base class automatically during the build process.

Once our base class has been modified, we can use the `ObservableProperty` attribute to automatically generate the property implementation. The result is that the test class we had previously is reduced to a single line of code per property. This makes the code base more readable because everything happens behind the scenes:

```
public partial class MyTestViewModel : ViewModel
{
    [ObservableProperty]
    string name;
}
```

There are a few things to note about the previous example. First, the class must be marked as `partial` for the `ObservableProperty` attribute to work, just like the `ObservableObject` attribute. Second, when using the `ObservableProperty` attribute, you place it on a private field, not a property. The CommunityToolkit.Mvvm library uses **Source Generators**, a feature added in .NET 5, to generate the actual property implementation.

One of the great things about using Source Generators is that you can always view the generated source to see how things work. For example, to view the generated source for the `ViewModel` class, do the following:

1. Open the `ViewModel.cs` file.

2. Right-click the `ViewModel` type name.

3. Select **Goto Implementation**.

Normally, this would do nothing, as you are in the implementation of `ViewModel`. However, since there is additional generated code, Visual Studio will show you a list of locations that contain implementations for `ViewModel`, similar to what's shown here:

Figure 2.12 – Finding all implementations of ViewModel

The first item in the list is what we added to the `ViewModel.cs` file, and the second item in the list is the generated code. By double-clicking the item, it will open the generated code in a new code window. In the *Creating TodoItemViewModel* section, you can follow the same steps to see what is generated for the property implementations.

Now that we've seen how to implement properties using a sample `ViewModel`, it is time to create the concrete `ViewModel` classes.

Creating MainViewModel

So far, we have mainly prepared to write the code that will make up the app itself. `MainViewModel` is the `ViewModel` class for the first view that is displayed to the user. It is responsible for providing data and logic to a list of to-do list items. We will create the bare-bones `ViewModel` classes and add code to them as we progress through this chapter:

1. Create a class called `MainViewModel` in the `ViewModels` folder.

2. Add the following template code and resolve the references:

```
public class MainViewModel : ViewModel
{
    private readonly ITodoItemRepository repository;

    public MainViewModel(ITodoItemRepository repository)
    {
        this.repository = repository;
        Task.Run(async () => await LoadDataAsync());
    }

    private async Task LoadDataAsync()
    {

    }
}
```

The structure of this class is something that we will reuse for all the `ViewModel` classes to come.

Let's summarize the important features we want the `ViewModel` class to have:

*   We inherit from the `ViewModel` class to gain access to shared logic, such as the `INotifyPropertyChanged` interface and common navigation code.

*   All dependencies to other classes, such as repositories and services, which are passed through the constructor of `ViewModel`. This is handled by the **dependency injection** pattern and, more specifically in our case, by `Microsoft.Extensions.DependencyInjection`,

which is the implementation of the dependency injection we are using. We will add support for automatic dependency injection in the *Wiring up dependency injection* section.

- We use an asynchronous call to `LoadDataAsync()` as an entry point to initialize the `ViewModel` class. Different MVVM libraries might do this in different ways, but the basic functionality is the same.

Creating TodoItemViewModel

`TodoItemViewModel` is the `ViewModel` class that represents each item in the to-do list on `MainView`. It does not have an entire view of its own, although it could have. Instead, it is rendered by a template in `ListView`. We will get back to this when we create the controls for `MainView`.

The important thing here is that this `ViewModel` object represents a single item, regardless of where we choose to render it.

Let's create the `TodoItemViewModel` class:

1. Create a class called `TodoItemViewModel` in the `ViewModels` folder.

2. Update the class so that it matches the following code:

```
namespace DoToo.ViewModels;

using CommunityToolkit.Mvvm.ComponentModel;
using DoToo.Models;

public partial class TodoItemViewModel : ViewModel
{
    public TodoItemViewModel(TodoItem item) => Item = item;

    public event EventHandler ItemStatusChanged;

    [ObservableProperty]
    TodoItem item;

    public string StatusText => Item.Completed ? "Reactivate" :
    "Completed";
}
```

As with any other `ViewModel` class, we inherit the `TodoItemViewModel` class from `ViewModel`. We conform to the pattern of injecting all the dependencies into the constructor. In this case, we pass an instance of the `TodoItem` class to the constructor that the `ViewModel` object will use to expose the view.

The `ItemStatusChanged` event handler will be used later when we want to signal to the view that the state of the `TodoItem` class has changed. The `Item` property allows us to access the item that we passed in.

The `StatusText` property is used to make the status of the to-do item human-readable in the view.

Creating the ItemViewModel class

`ItemViewModel` represents the to-do list item in a view that can be used to create new items and edit existing ones:

1. In the `ViewModels` folder, create a class called `ItemViewModel`.

2. Add the following code:

```
namespace DoToo.ViewModels;

using DoToo.Repositories;

public class ItemViewModel : ViewModel
{
    private readonly ITodoItemRepository repository;

    public ItemViewModel(ITodoItemRepository repository)
    {
        this.repository = repository;
    }
}
```

The pattern is the same as for the previous two `ViewModel` classes:

*   We use dependency injection to pass the `TodoItemRepository` class to the `ViewModel` object

*   We use inheritance from the `ViewModel` base class to add the common features defined by the base class

Creating the MainView view

Now that we are done with the `ViewModel` classes, let's create the skeleton code and the XAML required for the views. The template created a file named `MainPage.xml`. In MVVM, the convention is to use a `-View` suffix instead. We will also want to place all our views together in a subfolder, as we did with the `ViewModel` classes. Let's deal with the `MainPage.xml` file first, which is the view that will be loaded first:

1.   Delete the `MainPage.xml` file from the root of the project.

2.   Create a folder called `Views` in the `DoToo` project.

3.   Right-click on the `Views` folder, select **Add**, and then click **New Item...**.

4.   Select **.NET MAUI** under the **C# Items** node on the left.

5.   Select **.NET MAUI ContentPage (XAML)** and name it `MainView`.

6.   Click **Add** to create the page:

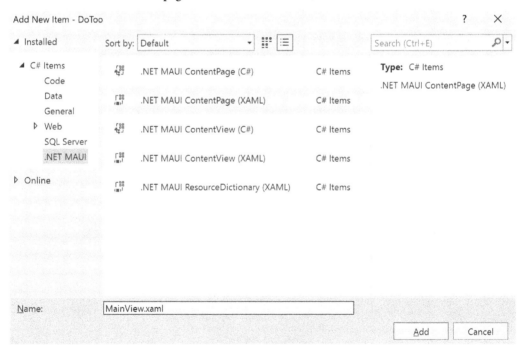

Figure 2.13 – Adding a new XAML file

Let's add some content to the newly created view:

1. Open `MainView.xaml`.

2. Remove all the template code below the `ContentPage` root node and add the XAML code highlighted in the following code:

```xml
<?xml version="1.0" encoding="utf-8"?>
<ContentPage xmlns="http://schemas.microsoft.com/dotnet/2021/
maui"
             xmlns:x="http://schemas.microsoft.com/winfx/2009/
xaml"
    x:Class="DoToo.Views.MainView"
    Title="Do Too!">

    <ContentPage.ToolbarItems>
      <ToolbarItem Text="Add" />
    </ContentPage.ToolbarItems>

    <Grid>
      <Grid.RowDefinitions>
        <RowDefinition Height="auto" />
        <RowDefinition Height="*" />
      </Grid.RowDefinitions>

      <Button Text="Toggle filter" />

      <ListView Grid.Row="1">
      </ListView>
    </Grid>
</ContentPage>
```

To be able to access custom converters, we need to add a reference to a local namespace. The `xmlns:local="clr-namespace:DoToo"` line defines this namespace for us. We will not use it directly in this case, but it's a good idea to have a local namespace defined. If we create custom controls, we can access them by writing something such as `<local:MyControl />`.

The `Title` property on the `ContentPage` page gives the page a title. Depending on the platform we are running on, the title is displayed differently. If we use a standard navigation bar, it will be displayed at the top, for example, in both iOS and Android. A page should always have a title.

The `ContentPage.ToolbarItems` node defines a toolbar item for adding new to-do items. It will also be rendered differently based on the platform, but it always follows the platform-specific UI guidelines.

A page in .NET MAUI (and in an XML document, in general) can only have one root node. The root node in a .NET MAUI page populates the Content property of the page itself. Since we want our MainView view to contain a list of items and a button at the top to toggle a filter (to switch between all items and only active items), we need to add a Layout control to position them on the page. Grid is a control that allows you to split up the available space based on rows and columns.

For our MainView view, we want to add two rows. The first row is a space calculated by the height of the button (Height="auto") and the second row takes up all the remaining space for ListView (Height="*"). Elements such as ListView are positioned in the grid using the Grid.Row and Grid.Column attributes. Both properties default to 0 if they are not specified, just like the button.

ListView is a control that presents items in a list, which is coincidently exactly what our app will do. It's worth noting that .NET MAUI does have a control called CollectionView, which can handle displaying collections of items better than ListView. Subsequent chapters will use this control, but we wanted to introduce you to the good old ListView control as well.

> **Tip**
> If you are interested in how Grid works, you can search for more information about .NET MAUI grids on the internet or check out the official documentation at https://learn.microsoft.com/en-us/dotnet/maui/user-interface/layouts/grid.

We also need to wire up ViewModel to the view. This can be done by passing the ViewModel class in the constructor of the view:

1. Open the code-behind file of MainView by expanding the MainView.xaml file in **Solution Explorer**. The code-behind file is named MainView.xaml.cs.

2. Add a using DoToo.ViewModels statement to the top of the file, adjacent to the existing using statements.

3. Modify the constructor of the class so that it looks as follows by adding the highlighted code:

```
public MainView(MainViewModel viewModel)
{
    InitializeComponent();
    viewModel.Navigation = Navigation;
    BindingContext = viewModel;
}
```

We follow the same pattern as we did with the ViewModel classes by passing any dependencies through the constructor. A view is always dependent on a ViewModel class. To simplify the project, we also assign the Navigation property of the page directly to the Navigation property defined in the ViewModel base class. In a larger project, we might want to abstract this property as well to make sure that we separate the `ViewModel` classes from .NET MAUI. For the sake of this app, however, it is OK to reference it directly.

Lastly, we assign `ViewModel` to the `BindingContext` class of the page. This tells the .NET MAUI binding engine to use our `ViewModel` object for the bindings that we will create later.

At this point, since we have removed `MainPage`, the project will no longer run. We will fix this in the *Making the app run* section later.

Creating the ItemView view

The second view we will add is `ItemView`. We will use this to add and edit the to-do list items:

1. Create a new content page (in the same way that we created the `MainView` view) and name it `ItemView`.

2. Edit the XAML file so that it appears as in the following code. The changes are highlighted:

```xml
<?xml version="1.0" encoding="UTF-8"?>
<ContentPage xmlns=http://schemas.microsoft.com/dotnet/2021/maui
             xmlns:x=http://schemas.microsoft.com/winfx/2009/
xaml
  x:Class="DoToo.Views.ItemView"
  Title="New todo item">

  <ContentPage.ToolbarItems>
    <ToolbarItem Text="Save" />
  </ContentPage.ToolbarItems>

  <StackLayout Padding="14">
    <Label Text="Title" />
    <Entry />
    <Label Text="Due" />
    <DatePicker />
    <StackLayout Orientation="Horizontal">
      <Switch />
      <Label Text="Completed" />
    </StackLayout>
  </StackLayout>
</ContentPage>
```

As with `MainView`, we need a title. We will give it a default title of `New todo item` for now, but we will change this to `Edit todo item` when we reuse this view for editing later on. The user must be able to save a new or edited item, so we have added a toolbar `Save` button. The content of the page uses `StackLayout` to structure the controls. `StackLayout` adds an element vertically (the default option) or horizontally based on the space it calculates that the element takes up. This is a CPU-intensive process, so we should only use it on small portions of our layout. In `StackLayout`, we add a `Label` control, which is a line of text over the `Entry` control that comes underneath it. The `Entry` control is a text input control that contains the name of the to-do list item. Then, we have a section for `DatePicker`, where the user can select a due date for the to-do list item. The final control is a `Switch` control, which renders a toggle button to control when an item is complete, as well as a heading next to it. Since we want these to be displayed next to each other horizontally, we use a horizontal `StackLayout` control to do this.

The last step for the views is to wire up the `ItemViewModel` model to `ItemView`:

1. Open the code-behind file of `ItemView` by expanding the `ItemView.xaml` file in **Solution Explorer**.

2. Add a using `DoToo.ViewModels` statement to the top of the file, adjacent to the existing `using` statements.

3. Modify the constructor of the class so that it looks as follows. Add the code that is marked in bold:

```
public ItemView (ItemViewModel viewmodel)
{
    InitializeComponent ();
    viewmodel.Navigation = Navigation;
    BindingContext = viewmodel;
}
```

This code is identical to the code that we added for `MainView`, except for the type of `ViewModel` class.

Wiring up dependency injection

Earlier, we discussed the dependency injection pattern, which states that all dependencies, such as the repositories and view models, must be passed through the constructor of the class. This requirement has several benefits:

- It increases the readability of the code since we can quickly determine all the external dependencies

- It makes dependency injection possible

- It makes unit testing possible by mocking classes

- We can control the lifetime of an object by specifying whether it should be a singleton or a new instance for each resolution

Dependency injection is a pattern that lets us determine, at runtime, which instance of an object should be passed to a constructor when an object is created. We do this by defining a container where we register all the types of a class. We let the framework that we are using resolve any dependencies between them. Let's say that we ask the container for a `MainView` class. The container takes care of resolving `MainViewModel` and any dependencies that the class has.

.NET MAUI uses the `Microsoft.Extensions.DependencyInjection` NuGet library internally and it is exposed for us to use in our applications. The first step is to register the classes we want to participate in dependency injection.

Registering View, ViewModels, and Services

For our classes to be available through dependency injection, they need to be registered with the dependency injection service. .NET MAUI exposes the dependency injection service using the `Services` property of the `MauiAppBuilder` class. The `Services` property will return an `IServiceCollection` object, also referred to as the container. `IServiceCollection` has two methods we are interested in, `AddSingleton`, and `AddTransient`. The "Transient" and "Singleton" in the method names refer to the lifetime of the objects. Transient objects are created every time they are requested from the container. Singleton objects are created only once, and that one instance is returned every time the class is requested from the container.

When registering classes with the container, it is recommended to use extension methods to group the types. For this app, there are three groups: `View`, `ViewModels`, and `Services`. The extension methods will take a single parameter and return a single value, the `MauiAppBuilder` instance. This is how the Builder pattern is implemented and allows us to chain the methods on the builder defined in the `CreateMauiApp` method.

To implement the methods, follow these steps:

1. Open the `MauiProgram.cs` file.

2. Make the following changes to the `MauiProgram` class. The changes are highlighted in bold:

```
using DoToo.Repositories;

public static class MauiProgram
{
    public static MauiApp CreateMauiApp()
    {
        var builder = MauiApp.CreateBuilder();
        builder
            .UseMauiApp<App>()
            .ConfigureFonts(fonts =>
            {
                fonts.AddFont("OpenSans-Regular.ttf",
"OpenSansRegular");
                fonts.AddFont("OpenSans-Semibold.ttf",
"OpenSansSemibold");
```

```
        })
        .RegisterServices()
        .RegisterViewModels()
        .RegisterViews();

        return builder.Build();
    }

    public static MauiAppBuilder RegisterServices(this
MauiAppBuilder mauiAppBuilder)
    {
    mauiAppBuilder.Services.
AddSingleton<ITodoItemRepository,TodoItemRepository>();

        return mauiAppBuilder;
    }

    public static MauiAppBuilder RegisterViewModels(this
MauiAppBuilder mauiAppBuilder)
    {
        mauiAppBuilder.Services.AddTransient<ViewModels.
MainViewModel>();

        mauiAppBuilder.Services.AddTransient<ViewModels.
ItemViewModel>();
        return mauiAppBuilder;
    }

    public static MauiAppBuilder RegisterViews(this
MauiAppBuilder mauiAppBuilder)
    {
        mauiAppBuilder.Services.AddTransient<Views.MainView>();

        mauiAppBuilder.Services.AddTransient<Views.ItemView>();
        return mauiAppBuilder;
    }
}
```

Normally, registering a type is done by using the type name as the generic argument to the registration method, as in `mauiAppBuilder.Services.AddTransient<Views.MainView>();`. But that doesn't work if you need to resolve an interface to an implementation, like what is happening in the `RegisterServices` method. There, the registration method doesn't use the generic argument; instead, it passes in the type to register as the first argument, and the second argument is the type of the instance to return.

> **Info**
>
> To learn more about how `Microsoft.Extensions.DependencyInjection` works, visit `https://learn.microsoft.com/en-us/dotnet/core/extensions/dependency-injection` in your favorite browser.

Now that dependency injection is wired up, we can get the project running again.

Making the app run

There are just a few more changes we need to make to enable the app to run:

1. Open the `App.xaml.cs` file by expanding the `App.xaml` node in the `DoToo` project.

2. Modify the following lines in bold:

   ```
   public App(Views.MainView view)
   {
       InitializeComponent();

       MainPage = new NavigationPage(view);
   }
   ```

3. `AppShell.Xaml` and `AppShell.xaml.cs` are no longer used, so they can be deleted from the project.

When .NET MAUI initializes the `App` class via the builder, it does so by using the dependency injection container, so any arguments you add to the `App` constructor are resolved from the container as well, and their dependencies too. In this case, we are importing the `MainView` class (and all its dependencies, including `MainViewModel` and `TodoItemRepository`) and wrapping it in `NavigationPage`. `NavigationPage` is a page defined in .NET MAUI that adds a navigation bar and enables the user to navigate to other views.

> **Information**
>
> .NET MAUI includes `Shell`, and we have a whole chapter about it in this book. However, to become a good .NET MAUI developer, you need to know the basics, and the basics of navigating in .NET MAUI uses the good old `NavigationPage` control.

That's it! Now, your project should start. Depending on the platform you are using, it might look as follows:

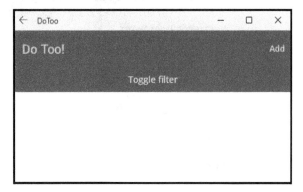

Figure 2.14 – The DoToo application in the Windows subsystem for Android

Now that we have the app running with a basic UI, let's add some functionality, starting with displaying data.

> **Tip**
> If you are debugging the app using the Windows target framework and it isn't working, and you aren't getting any error messages, try using the Android target framework. Sometimes, you can get better error reporting from a different platform.

Adding data bindings

Data binding is the heart and soul of MVVM. This is the way that the views and ViewModel communicate with each other. In .NET MAUI, we need two things to make data binding happen:

- We need an object to implement `INotifyPropertyChanged`.
- We need to set the `BindingContext` class of the page to that object. We already do this on both `ItemView` and `MainView`.

A useful feature of data binding is that it allows us to use two-way communication. For example, when data binding text to an `Entry` control, the property on the data-bound object is updated directly. Consider the following XAML:

```
<Entry Text="{Binding Title}" />
```

To make this work, we need a property named `Title` on the string object. We have to look at the documentation, define an object, and let **IntelliSense** provide us with a hint to find out what type our property should be.

Controls that perform an action, such as `Button`, usually expose a property called `Command`. This property is of the `ICommand` type, and we can either return `Microsoft.Maui.Controls.Command` or an implementation of our own. The `Command` property is explained in the next section, where we will use it to navigate to `ItemView`.

In the next few sections, we will be adding the data binding and command implementations to our views and ViewModels, starting with navigating from `MainView` to `ItemView`.

> **Information**
>
> It's also worth noting that .NET MAUI supports one-way binding in addition to two-way data binding, which comes in handy when you want to display data in a view but not allow it to update the ViewModel. From a performance perspective, it's a good idea to mark those bindings as one-way bindings.

Navigating from MainView to ItemView to add a new item

We have an `Add` toolbar button in `MainView`. When the user taps this button, we want it to take them to `ItemView`. The MVVM way to do this is to define a command and then bind that command to the button.

In .NET MAUI, to navigate to a view, you need a reference to an instance of the destination. In this case, that would be `ItemView`. Since all our views have been registered with the dependency injection container, we will need a reference to the container to request a new instance of the view when we are ready to navigate. We will use constructor injection to have the container provide us with its instance, like this:

1. Open `ViewModels/MainViewModel.cs`.

2. Add a `using` statement for `DoToo.Views`.

3. Add the following field to the class:

    ```
    private readonly IServiceProvider services;
    ```

4. Modify the constructor as follows. The changes are highlighted:

    ```
    public MainViewModel(ITodoItemRepository repository,
    IServiceProvider services)
    {
        this.repository = repository;
        this.services = services;
        Task.Run(async () => await LoadDataAsync());
    }
    ```

This will capture the instance of `ItemView` that was created by the dependency injection container in a class field. Now, let's look at the command implementation.

All commands should be exposed as a generic `ICommand` type. This abstracts the actual command implementation, which is good general practice to follow. The command must be a property; in our case, we are creating a new `Command` object that we assign to this property. The property is read-only, which is usually fine for a `Command` object. The action of the command (the code that we want to run when the command is executed) is passed to the constructor of the `Command` object.

Following those requirements, you might end up writing something like the following:

```
public ICommand AddItem => new Command(async () =>
{
    await Navigation.PushAsync(itemView);
});
```

There is a lot of boilerplate code in that implementation that you would have to repeat for each command. This boilerplate code can get in the way of what the command is doing. Like how we were able to eliminate boilerplate code with properties, we can do the same with `ICommand`, but instead, we can use the `RelayCommand` attribute. The `RelayCommand` attribute uses a source generator to wrap a method in a new `Command` instance and expose it through a property. The property name that's generated is the method name with "Command" appended to it.

Now, we can add the `Command` object's implementation:

1. Open `ViewModels/MainViewModel.cs`.

2. Add the following method to the class:

   ```
   [RelayCommand]
   public async Task AddItemAsync() => await Navigation.
   PushAsync(services.GetRequiredService<ItemView>());
   ```

3. Add using `CommunityToolkit.Mvvm.Input;` to the usings section of the file.

4. Update the class definition to allow the Source Generators to do their thing:

   ```
   public partial class MainViewModel : ViewModel
   ```

The action of the command is simply to use the `Navigation` service to push the `itemView` instance onto the stack for us.

After that, we just need to wire up the `AddItemAsync` command from `ViewModel` to the `Add` button in the view:

1. Open `Views/MainView.xaml`.

2. Update the `ContentPage` element:

```
<ContentPage xmlns=http://schemas.microsoft.com/dotnet/2021/maui
               xmlns:x=http://schemas.microsoft.com/winfx/2009/
xaml
    xmlns:viewModels="clr-namespace:DoToo.ViewModels"
    x:Class="DoToo.Views.MainView"
    x:DataType="viewModels:MainViewModel"
    Title="Do Too!">
```

3. Add the `Command` attribute to `ToolbarItem`:

```
<ContentPage.ToolbarItems>
    <ToolbarItem Text="Add" Command="{Binding
AddItemAsyncCommand}" />
</ContentPage.ToolbarItems>
```

Run the app and tap the `Add` button to navigate to the new `ItemView` view. Notice that the back button appears automatically.

Adding new items to the list

Now that we have finished adding navigation to a new item, let's add the code to create a new item and save it to the database:

1. Open `ViewModels/ItemViewModel.cs`.

2. Add the following code in bold:

```
using CommunityToolkit.Mvvm.ComponentModel;
using CommunityToolkit.Mvvm.Input;
using DoToo.Models;
using DoToo.Repositories;

public partial class ItemViewModel : ViewModel
{
    private readonly ITodoItemRepository repository;

    [ObservableProperty]
    TodoItem item;

    public ItemViewModel(ITodoItemRepository repository)
    {
        this.repository = repository;
        Item = new TodoItem() { Due = DateTime.Now.AddDays(1) };
    }

    [RelayCommand]
```

```
    public async Task SaveAsync()
    {
        await repository.AddOrUpdateAsync(Item);
        await Navigation.PopAsync();
    }

}
```

The `Item` property holds a reference to the current item that we want to add or edit. A new item is created in the constructor and when we want to edit an item, we can simply assign our own item to this property. The new item is not added to the database unless we execute the `Save` command defined at the end. Once the item has been added or updated, we remove the view from the navigation stack and return to `MainView` again.

> **Information**
>
> Since the navigation keeps pages in a stack, the framework declares methods that reflect operations that you can perform on a stack. The operation of removing the topmost item in a stack is known as *popping the stack*, so instead of `RemoveAsync()`, we have `PopAsync()`. To add a page to the navigation stack, we push it, so the method is called `PushAsync()`.

Now that we have extended `ItemViewModel` with the necessary commands and properties, it's time to data-bind them in the XAML:

1. Open `Views/ItemView.xaml`.

2. Add the code marked in bold:

```xml
<?xml version="1.0" encoding="UTF-8"?>
<ContentPage
    xmlns=http://schemas.microsoft.com/dotnet/2021/maui
        xmlns:x=http://schemas.microsoft.com/winfx/2009/xaml
    xmlns:viewModels="clr-namespace:DoToo.ViewModels"
    x:Class="DoToo.Views.ItemView"
    x:DataType="viewModels:ItemViewModel" >
    <ContentPage.ToolbarItems>
        <ToolbarItem Text="Save" Command="{Binding
SaveAsyncCommand}" />
    </ContentPage.ToolbarItems>
    <StackLayout Padding="14">
        <Label Text="Title" />
        <Entry Text="{Binding Item.Title}" />
        <Label Text="Due" />
        <DatePicker Date="{Binding Item.Due}" />
        <StackLayout Orientation="Horizontal">
```

```
        <Switch IsToggled="{Binding Item.Completed}" />
        <Label Text="Completed" />
    </StackLayout>
</ContentPage>
```

The binding to the `ToolbarItems` command attribute triggers the `SaveAsync` command exposed by `ItemViewModel` when a user taps the `Save` link. It's worth noting again that any attribute called `Command` indicates that an action will take place and we must bind it to an instance of an object implementing the `ICommand` interface.

The `Entry` control that represents the title is data-bound to the `Item.Title` property of `ItemViewModel`, and the `Datepicker` and `Switch` controls bind similarly to their respective properties.

We could have exposed `Title`, `Due`, and `Complete` as properties directly on `ItemViewModel`, but instead, we chose to reuse the already-existing `TodoItem` object as a reference. This is fine, so long as the properties of the `TodoItem` object implement the `INotifyPropertyChange` interface.

Binding ListView in MainView

A to-do list is not much use without a list of items. Let's extend `MainViewModel` with a list of items:

1. Open ViewModels/`MainViewModel.cs`.

2. Add using `System.Collections.ObjectModel` in the using section of the class.

3. Add a property for the to-do list items:

   ```
   [ObservableProperty]
   ObservableCollection<TodoItemViewModel> items;
   ```

`ObservableCollection` is like an ordinary collection, but it has a useful superpower: it can notify listeners about changes in the list, such as when items are added or deleted. The `ListView` control listens to changes in the list and updates itself automatically based on these. However, it's important to be aware that a change to an item in the list will not trigger an update. Changing the title of an item will not cause the list to re-render. Let's move on to implementing the rest of `MainViewModel`.

Now, we need some data:

1. Open ViewModels/`MainViewModel.cs`.

2. Replace (or complete) the `LoadDataAsync` method and create the `CreateTodoItemViewModel` and `ItemStatusChanged` methods:

   ```
   private async Task LoadDataAsync()
   {
       var items = await repository.GetItemsAsync();
   ```

```
        var itemViewModels = items.Select(i =>
    CreateTodoItemViewModel(i));

        Items = new ObservableCollection<TodoItemViewModel>
    (itemViewModels);
    }


    private TodoItemViewModel CreateTodoItemViewModel(TodoItem item)
    {
        var itemViewModel = new TodoItemViewModel(item);
        itemViewModel.ItemStatusChanged += ItemStatusChanged;
        return itemViewModel;
    }


    private void ItemStatusChanged(object sender, EventArgs e)
    {
    }
```

3.   Resolve all new references by adding the following using statements:

```
using CommunityToolkit.Mvvm.ComponentModel;
using DoToo.Models;
```

The LoadData method calls the repository to fetch all items. Then, we wrap each to-do list item in TodoItemViewModel. This contains more information that is specific to the view that we don't want to add to the TodoItem class. It is good practice to wrap plain objects in ViewModel; this makes it simpler to add actions or extra properties to it. ItemStatusChanged is a stub that is called when we change the status of the to-do list item from active to completed, and vice versa.

We also need to hook up some events from the repository to know when data changes:

1.   Open ViewModels/MainViewModel.cs.

2.   Add the following code in bold:

```
public MainViewModel(TodoItemRepository repository,
IServiceProvider services)
{
    repository.OnItemAdded += (sender, item) =>
        items.Add(CreateTodoItemViewModel(item));

    repository.OnItemUpdated += (sender, item) =>
        Task.Run(async () => await LoadDataAsync());

    this.repository = repository;
    this.services = services;
```

```
        Task.Run(async () => await LoadDataAsync());
    }
```

When an item is added to the repository, no matter who added it, `MainView` will add it to the `items` list. Since the items collection is an observable collection, the list updates. If an item is updated, we simply reload the list.

Let's data-bind our items to `ListView`:

1. Open `MainView.xaml` and locate the `ListView` element.

2. Modify it so that it reflects the following code:

```
<ListView Grid.Row="1"
  RowHeight="70" ItemsSource="{Binding Items}">
  <ListView.ItemTemplate>
    <DataTemplate x:DataType="viewModels:TodoItemViewModel">
      <ViewCell>
        <Grid Padding="15,10">
          <Grid.RowDefinitions>
              <RowDefinition />
              <RowDefinition />
          </Grid.RowDefinitions>
          <Grid.ColumnDefinitions>
            <ColumnDefinition Width="10" />
            <ColumnDefinition Width="*" />
          </Grid.ColumnDefinitions>

          <BoxView Grid.RowSpan="2" />
          <Label Grid.Column="1"
            Text="{Binding Item.Title}" FontSize="Medium" />
          <Label Grid.Column="1" Grid.Row="1"
            Text="{Binding Item.Due}" FontSize="Micro" />
          <Label Grid.Column="1" Grid.Row="1"
            HorizontalTextAlignment="End" Text="Completed"
            IsVisible="{Binding Item.Completed}"
            FontSize="Micro" />
        </Grid>
      </ViewCell>
    </DataTemplate>
  </ListView.ItemTemplate>
</ListView>
```

The `ItemsSource` binding tells `ListView` where to find the collection to iterate over and is local to `ViewModel`. Any bindings in the `ViewCell` node, however, are local to each item that we iterate in the list. In this case, we are binding to `TodoItemViewModel`, which contains a property named `Item`. This, in turn, has properties such as `Title`, `Due`, and `Completed`. We can navigate down the hierarchy of objects without any problem when defining a binding.

The `DataTemplate` element defines what each row will look like. We use a grid to partition the space, just as we did earlier.

You may have noticed that we didn't discuss what `BoxView` was for, and it isn't bound to any properties of `ViewModel`. The next two sections will cover how we can use the `Completed` property to color code our items with `BoxView`.

Creating a ValueConverter object for the item's status

Sometimes, we want to bind to objects that are a representation of the original value. This could be a piece of text that is based on a Boolean value. Instead of `true` and `false`, for example, we might want to write `Yes` and `No` or return a color. This is where `ValueConverter` comes in handy. It can be used to convert a value to and from another value. We are going to write a `ValueConverter` object that converts the status of a to-do list item into a color:

1. In the root of the `DoToo` project, create a folder called `Converters`.

2. Create a class called `StatusColorConverter.cs` in the `Converters` folder and add the following code:

```
using System;
using System.Globalization;

public class StatusColorConverter : IValueConverter
{
    public object Convert(object value, Type targetType, object
parameter, CultureInfo culture)
    {
        return (Color)Application.Current.Resources[
        (bool)value ? "CompletedColor" : "ActiveColor"];
    }

    public object ConvertBack(object value, Type targetType,
object parameter, CultureInfo culture)
    {
        return null;
    }
}
```

A `ValueConverter` object is a class that implements `IValueConverter`. This, in turn, only has two methods defined. The `Convert` method is called when the view reads data from `ViewModel` and the `ConvertBack` method is used when `ViewModel` gets data from the view. The `ConvertBack` method is only used for controls that return data from plain text, such as the `Entry` control.

If we look at the implementation of the `Convert` method, we'll notice that any value passed to the method is of the `object` type. This is because we don't know what type the user has bound to the property to which we are adding this `ValueConverter` class. We may also notice that we fetch colors from a resource file. We could have defined the colors in the code, but this is not recommended. So, instead, we went the extra mile and added them as a global resource to the `App.xaml` file. Resources are a good thing to take another look at once you have finished this chapter:

1. Open `App.xaml` in the DoToo project.

2. Add the following `ResourceDictionary` element:

```
<Application ...>
    <Application.Resources>
        <ResourceDictionary>
            <ResourceDictionary.MergedDictionaries>
                <ResourceDictionary Source="Resources/Styles/
Colors.xaml" />
                <ResourceDictionary Source="Resources/Styles/
Styles.xaml" />
            </ResourceDictionary.MergedDictionaries>
            <ResourceDictionary>
                <Color x:Key="CompletedColor"> #1C8859 </Color>
                <Color x:Key="ActiveColor"> #D3D3D3 </Color>
            </ResourceDictionary>
        </ResourceDictionary>
    </Application.Resources>
</Application>
```

`ResourceDictionary` can define a wide range of different objects. We only need the two colors that we want to access from `ValueConverter`. Notice that these can be accessed by the key given to them and from any other XAML file using a static resource binding.

`ValueConverter` itself is referenced as a static resource but from a local scope.

Using ValueConverter

We want to use our brand-new `StatusColorConverter` object in `MainView`. Unfortunately, we have to jump through some hoops to make this happen. We need to do three things:

- Define a namespace in XAML

- Define a local resource that represents an instance of the converter

- Declare that we want to use the converter in the binding

Let's start with the namespace:

1. Open `Views/MainView.xaml`.

2. Add the following namespace to the page:

```
<ContentPage  xmlns="http://schemas.microsoft.com/dotnet/2021/
maui"
                xmlns:x="http://schemas.microsoft.com/winfx/2009/
xaml"
    xmlns:local="clr-namespace:DoToo.ViewModels"
    xmlns:converters="clr-namespace:DoToo.Converters"
    x:Class="DoToo.Views.MainView" Title="Do Too!>
```

Add a `Resource` node to the `MainView.xaml` file:

1. Open `Views/MainView.xaml`.

2. Add the following `ResourceDictionary` element, shown in bold under the root element of the XAML file:

```
<ContentPage ...>

    <ContentPage.Resources>
        <ResourceDictionary>
            <converters:StatusColorConverter
    x:Key="statusColorConverter"/>
        </ResourceDictionary>
    </ContentPage.Resources>

    <ContentPage.ToolBarItems>
        <ToolbarItem Text="Add" Command="{Binding AddItem}" />
    </ContentPage.ToolbarItems>

    <Grid ...>
    </Grid>
</ContentPage>
```

This has the same form as the global resource dictionary, but since this one is defined in `MainView`, it can only be accessed from there. We could have defined this in the global resource dictionary, but it's usually more efficient to define objects that you only consume in one place as close to that place as possible.

The last step is to add the converter:

1. Locate the `BoxView` node in the XAML file.

2. Add the `BackgroundColor` XAML, which is marked in bold:

    ```
    <BoxView Grid.RowSpan="2"
            BackgroundColor="{Binding Item.Completed,
            Converter={StaticResource statusColorConverter}}" />
    ```

What we have done here is bind a `bool` value to a property that takes a `Color` object. Right before the data binding takes place, however, `ValueConverter` converts the `bool` value into a color. This is just one of the many cases where `ValueConverter` comes in handy. Keep this in mind when you define the GUI.

Navigating to an item using a command

We want to be able to see the details for a selected to-do list item. When we tap a row, we should navigate to the item in that row.

To do this, we need to add the following code:

1. Open `ViewModels/MainViewModel.cs`.

2. Add the `SelectedItem` property, the `OnSelectedItemChanging` event handler, and the `NavigateToItemAsync` method to the class:

    ```
    [ObservableProperty]
    TodoItemViewModel selectedItem;

    partial void OnSelectedItemChanging(TodoItemViewModel value)
    {
        if (value == null)
        {
            return;
        }

        MainThread.BeginInvokeOnMainThread(async () => {
            await NavigateToItemAsync(value);
        });
    }

    private async Task NavigateToItemAsync(TodoItemViewModel item)
    {
        var itemView = services.GetRequiredService<ItemView>();
        var vm = itemView.BindingContext as ItemViewModel;
    ```

```
            vm.Item = item.Item;
            itemView.Title = "Edit todo item";

            await Navigation.PushAsync(itemView);
    }
```

The `SelectedItem` property is a property that we will data-bind to `ListView`. When we select a row in `ListView`, this property is set to the `TodoItemViewModel` object that represents that row. We are using the `ObservableProperty` attribute here to carry out its `PropertyChanged` magic. However, since the setter is being generated through the `ObservableProperty` attribute, there is no place to add additional code to the property. Luckily, the `ObservableProperty` source generator also adds two partial methods that can be implemented. We are using `OnSelectedItemChanging` to add additional functionality to the setter. The other partial method is `OnSelectedItemChanged`. `OnSelectedItemChanging` is called before the property value has changed and `OnSelectedItemChanged` is called after the value has changed. Remember that you can always view the generated source to learn more about how these attributes are extending your code.

The `OnSelectedItemChanging` method then calls `NavigateToItem`, which creates a new `ItemView` view using the .NET MAUI dependency injection container. At this point, we change the `Title` of the view from `"Add todo item"` to `"Edit todo item"`. We extract `ViewModel` from the newly created `ItemView` view and assign the current `TodoItem` object that `TodoItemViewModel` contains. Confused? Remember that `TodoItemViewModel` wraps a `TodoItem` object, and it is that item that we want to pass to `ItemView`.

We are not done yet. Now, we need to data-bind the new `SelectedItem` property to the right place in the view:

1. Open `Views/MainView.xaml`.
2. Locate `ListView` and add the attributes in bold:

    ```
    <ListView x:Name="ItemsListView" Grid.Row="1" RowHeight="70"
        ItemsSource="{Binding Items}"
        SelectedItem="{Binding SelectedItem}">
    ```

The `SelectedItem` attribute binds the `SelectedItem` property's `ListView` view to the `ViewModel` property. When the selection of an item in `ListView` changes, the `ViewModel` property's `SelectedItem` property is called and we navigate to the new and exciting views.

The x:Name attribute is for naming ListView because we need to make a small and ugly hack to make this work. ListView stays selected after the navigation is done. When we navigate back, it cannot be selected again until we select another row. To mitigate this, we need to hook up to the ItemSelected event of ListView and reset the selected item directly on ListView. This is not recommended because we shouldn't have any logic in our views, but sometimes, we have no other choice:

1. Open Views/MainView.xaml.cs.

2. Add the following code in bold:

```
public MainView(MainViewModel viewmodel)
{
    InitializeComponent();
    viewmodel.Navigation = Navigation;
    BindingContext = viewmodel;

    ItemsListView.ItemSelected += (s, e) =>
        ItemsListView.SelectedItem = null;
}
```

We should now be able to navigate to an item in the list. Next, we will mark it as complete.

Marking an item as complete using a command

We need to add a functionality that allows us to toggle items between complete and active. It is possible to navigate to the detailed view of the to-do list item, but this is too much work for a user. Instead, we'll add a ContextAction item to ListView. In iOS, for example, this is accessed by swiping left on a row:

1. Open ViewModel/TodoItemViewModel.cs.

2. Add a using statement for CommunityToolkit.Mvvm.Input.

3. Add a command to toggle the status of the item and a piece of text that describes the status:

```
[RelayCommand]
void ToggleCompleted()
{
    Item.Completed = !Item.Completed;
    ItemStatusChanged?.Invoke(this, new EventArgs());
}
```

Here, we have added a command for toggling the state of an item. When executed, it inverses the current state and raises the ItemStatusChanged event so that subscribers are notified. To change the text of the context action button depending on the status, we added a StatusText property. This is not recommended practice because we are adding code that only exists because of a specific UI case to ViewModel. Ideally, this would be handled by the view, perhaps by using ValueConverter. To save us from having to implement these steps, however, we have left it as a string property:

1. Open Views/MainView.xaml.

2. Locate the ListView.ItemTemplate node and add the following ViewCell.ContextActions node:

```
<ListView.ItemTemplate>
  <DataTemplate>
    <ViewCell>
      <ViewCell.ContextActions>
        <MenuItem Text="{Binding StatusText}" Command="{Binding ToggleCompletedCommand}" />
      </ViewCell.ContextActions>
      <Grid Padding="15,10">
        . . .
      </Grid>
    </ViewCell>
  </DataTemplate>
</ListView.ItemTemplate>
```

Creating the filter toggle function using a command

We want to be able to toggle between viewing active items only and all the items. We will create a simple mechanism to do this.

Hook up the changes in MainViewModel as follows:

1. Open ViewModels/MainViewModel.cs and locate ItemStatusChangeMethod.

2. Add the implementation to the ItemStatusChanged method and a property called ShowAll to control the filtering:

```
private void ItemStatusChanged(object sender, EventArgs e)
{
    if (sender is TodoItemViewModel item)
    {
        if (!ShowAll && item.Item.Completed)
        {
            Items.Remove(item);
        }
    }
```

```
            Task.Run(async () => await repository.
    UpdateItemAsync(item.Item));
        }
    }

    [ObservableProperty]
    bool showAll;
```

The ItemStatusChanged event handler is triggered when we use the context action from the previous section. Since the sender is always an object, we try to cast it to TodoItemViewModel. If this is successful, we check whether we can remove it from the list if ShowAll is not true. This is a small optimization; we could have called LoadData and reloaded the entire list, but since the Items list is set to ObservableCollection, it communicates to ListView that one item has been removed from the list. We also call the repository to update the item to persist the change of status.

The ShowAll property is what controls which state our filter is in. We need to adjust the LoadData method to reflect this:

1. Locate the Load method in MainViewModel.

2. Add the following lines of code marked in bold:

```
    private async Task LoadDataAsync()
    {
        var items = await repository.GetItemsAsync();

        if (!ShowAll)
        {
            items = items.Where(x => x.Completed == false).ToList();
        }

        var itemViewModels = items.Select(i =>
    CreateTodoItemViewModel(i));
        Items = new ObservableCollection<TodoItemViewModel>
    (itemViewModels);
    }
```

If ShowAll is false, we limit the content of the list to the items that have not been completed. We can do this either by having two methods, GetAllItems() and GetActiveItems(), or by using a filter argument that can pass to GetItemsAsync(). Take a minute to think about how we could implement this.

Let's add the code that toggles the filter:

1. Open `ViewModels/MainViewModel.cs`.

2. Add the `FilterText` and `ToggleFilterAsync` properties:

```
[RelayCommand]
private async Task ToggleFilterAsync()
{
    ShowAll = !ShowAll;
    await LoadDataAsync();
}
```

The `ShowAll` property is a Boolean value, and that does not display well in a human-readable form. We will use another `ValueConverter` to change the status into a human-readable form:

1. Create a new class in the `Converters` folder named `FilterTextConverter.cs`.

2. Add the following code:

```
using System;
using System.Globalization;

internal class FilterTextConverter : IValueConverter
{
    public object Convert(object value, Type targetType, object
parameter, CultureInfo culture)
    {
        return (bool)value ? "All" : "Active";
    }

    public object ConvertBack(object value, Type targetType,
object parameter, CultureInfo culture)
    {
        return null;
    }
}
```

`FilterTextConverter` is very similar to the previous converter we created. The difference is that in the `Convert` method, we convert a `bool` value into the `"All"` or `"Active"` string. This converter will be used in the view to change the value of `ShowAll` into a value more suitable for display in the user interface.

The logic for the `ToggleFilterAsync` command is a simple inversion of the state and then a call to `LoadDataAsync`. This, in turn, causes the list to be reloaded.

Before we can filter the items, we need to hook up the filter button to `Command` and `Converter`:

1. Open `Views/MainView.xaml`.

2. Add the following highlighted entry to `ResourceDictionary`:

```
<ResourceDictionary>
    <converters:StatusColorConverter   x:Key=
"statusColorConverter"/>
    <converters:FilterTextConverter
x:Key="filterTextConverter"/>
</ResourceDictionary>
```

3. Locate the button that controls the filter (the only button in the file).

4. Adjust your code to reflect the following code:

```
<Button Text="{Binding ShowAll,Converter={StaticResource
filterTextConverter}, StringFormat='Filter: {0}'}"
    Command="{Binding ToggleFilterAsyncCommand}" />
```

We have now finished with this feature! However, our app isn't very attractive; we'll deal with this in the following section.

Laying out the contents

This last section is about making the app look a bit nicer. We will just scratch the surface of the possibilities here, but this should give you some ideas about how styling works.

Setting an application-wide background color

Styles are a great way of applying styling to elements. They can be applied either to all elements of a type or the elements referenced by a key if you add an `x:Key` attribute:

1. Open `App.xaml`.

2. Add the following XAML, which is in bold, to the file:

```
<ResourceDictionary>
  <Style TargetType="NavigationPage">
    <Setter Property="BarBackgroundColor" Value="#A25EBB" />
    <Setter Property="BarTextColor" Value="#FFFFFF" />
  </Style>
  <Style x:Key="FilterButton" TargetType="Button">
    <Setter Property="Margin" Value="15" />
    <Setter Property="BorderWidth" Value="1" />
    <Setter Property="BorderColor" Value="Silver" />
    <Setter Property="TextColor" Value="Black" />
```

```
    </Style>
    <Color x:Key="CompletedColor">#1C8859</Color>
    <Color x:Key="ActiveColor">#D3D3D3</Color>
</ResourceDictionary>
```

The first style we will apply is a new background color and text color to the navigation bar. The second style will be applied to the filter button. We can define a style by setting `TargetType`, which tells .NET MAUI which type of object this style can be applied to. We can then add one or more properties that we want to set. The result will be the same as if we had added these properties directly to the element in the XAML code.

Styles that lack the `x:Key` attribute are applied to all instances of the type defined in `TargetType`. Styles that have a key must be explicitly assigned in the XAML of the user interface. We will see examples of this when we define the filter button in the next section.

Laying out the MainView and ListView items

In this section, we will improve the appearance of `MainView` and `ListView`. Open `Views/MainView.xaml` and apply the changes in bold in the XAML code in each of the following sections.

The filter button

The filter button allows us to toggle the state of the list to show only the active to-do items or all the to-do items. Let's style it to make it stand out a bit in the layout:

1. Find the filter button.

2. Make the following changes:

```xml
<Button Style="{DynamicResource FilterButton}"
        Text="{Binding ShowAll,Converter={StaticResource
filterTextConverter}, StringFormat='Filter: {0}'}"
        BackgroundColor="{DynamicResource ActiveColor}"
        TextColor="Black"
        Command="{Binding ToggleFilterCommand}">
    <Button.Triggers>
      <DataTrigger TargetType="Button" Binding="{Binding ShowAll}"
Value="True">
        <Setter Property="BackgroundColor" Value="{DynamicResource
CompletedColor}" />
        <Setter Property="TextColor" Value="White" />
      </DataTrigger>
    </Button.Triggers>
  </Button>
```

The style is applied using `DynamicResource`. Anything defined in a resource dictionary, either in the `App.xaml` file or in the local XAML file, is accessible through it. Then, we set `BackgroundColor`, again setting `DynamicResource` to `ActiveColor` and `TextColor` to `Black`.

The `Button.Triggers` node is a useful feature. We can define several types of triggers that fire when certain criteria are met. In this case, we use a data trigger that checks whether the value of `ShowAll` changes to `true`. If it does, we set `TextColor` to white and `BackgroundColor` to `CompletedColor`. The coolest part is that when `ShowAll` becomes `false` again, it switches back to whichever value it was before.

Touching up ListView

`ListView` could use a couple of minor changes. The first change is formatting the due date string to a more human-readable format and the second is changing the color of the `Completed` label to a nice green tint:

1. Open `Views/MainView.xaml`.

2. Locate the labels that bind `Item.Due` and `Item.Completed` in `ListView`:

    ```
    <Label Grid.Column="1" Grid.Row="1"
           Text="{Binding Item.Due, StringFormat='{0:MMMM d, yyyy}'}"
           FontSize="Micro" />
    <Label Grid.Column="1" Grid.Row="1"
           HorizontalTextAlignment="End"
           Text="Completed"
           IsVisible="{Binding Item.Completed}"
           FontSize="Micro"
           TextColor="{StaticResource CompletedColor}" />
    ```

Here, we added a formatting string to the binding to format the date using a specific format. In this case, we used the `0:MMMM d, yyyy` format, which will display the date as a string in the format of, for example, May 5, 2020.

We also added a text color to the `Completed` label that is only visible if an item is completed. We did this by referencing our dictionary in `App.xaml`.

Now that all the code changes are complete, run our application. Here is a small gallery of screenshots that should match your application:

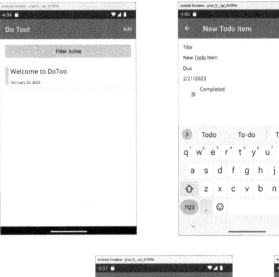

Figure 2.15 – DoToo on Android

Summary

You should now have a good grasp of all the steps involved in creating a .NET MAUI app from scratch. In this chapter, we learned about the project structure and the important files in a newly created project. We talked about dependency injection and learned the basics of MVVM by creating all the views and the `ViewModel` classes needed. We also covered data storage in SQLite to persist data on our device in a fast and secure way. Using the knowledge you've gained from this chapter, you should now be able to create the backbone of any app you'd like.

The next chapter will focus on upgrading an existing Xamarin.Forms application to .NET MAUI.

3

Converting a Xamarin.Forms App into .NET MAUI

Before we dive too far into .NET MAUI, we will look back at an existing Xamarin.Forms app and convert it into .NET MAUI. This chapter will guide you through the steps to convert an existing Xamarin. Forms app running on Mono into a .NET MAUI app running on .NET 7. We will discuss two different methods for converting your Xamarin.Forms application into .NET MAUI. The first method will use a new .NET MAUI project and move our old Xamarin.Forms code into the new project. The second method will use the **.NET Upgrade Assistant** tool to do some of the upgrades for us.

If you are new to .NET MAUI and are not coming from Xamarin.Forms app development, feel free to skip this chapter and go straight to the next project.

The following topics will be covered in this chapter:

- Moving code to a new .NET MAUI project

- An overview of upgrading a Xamarin.Forms app

- Installing and running .NET Upgrade Assistant

Technical requirements

To complete this project, you need to have Visual Studio installed on your **Macintosh (Mac)** or PC, as well as the .NET Mobile components. Refer to *Chapter 1, Introduction to .NET MAUI*, for more details on how to set up your environment. There are additional components that will be installed in this chapter, so you will need an internet connection to download and install .NET Upgrade Assistant. This chapter provides screenshots and instructions for Visual Studio on Windows.

This chapter will both be a classic **File | New | Project** chapter and use an existing app, guiding you step by step through the process of migrating an app from Xamarin.Forms to .NET MAUI. For the second app, you will need to download the source from this book's GitHub repository.

You can find the full source for the code in this chapter at `https://github.com/PacktPublishing/MAUI-Projects-3rd-Edition` under the `Chapter03` folder.

Project overview

This chapter is not meant to be an exhaustive tome of all the things you need to be aware of when converting your Xamarin.Forms app into .NET MAUI. Rather, it is an overview of what you need to consider when migrating your app, along with two walk-throughs of methods to accomplish that task. There are too many variations in application styles, versions, frameworks, custom controls, and so on for this one chapter to cover every scenario. That could take an entire book and would most likely be outdated by the time it was published. So, in this chapter, we are going to focus on a simple migration method that has the benefit of using the .NET MAUI Single Project feature and .NET Migration Assistant, which will automate much of the manual stuff and is constantly being updated.

The first part of this chapter will use a new Xamarin.Forms app created from the Shell template. The second part of this chapter will use .NET Upgrade Assistant to upgrade the open source app, BuildChat, which is available on GitHub at `https://github.com/mindofai/Build2019Chat`.

Every development book needs to have a chat app; this book is no different. For .NET Migration Assistant, we will use an existing chat app that was built using Xamarin.Forms. The app can be debugged and tested locally, so there is no need to set up and configure any cloud services.

The build time for this project is about one hour.

Migrating into a blank .NET MAUI template

This method of moving your existing app code into a new .NET MAUI app is used mainly for smaller, simpler apps that do not have a lot of external dependencies, such as NuGet, or native libraries. The largest benefit of this method is that your migrated app will be in a single project, targeting all the .NET MAUI-supported platforms. If your original app only targeted Android and iOS, using this method could get you Mac Catalyst and Windows targets for free. Using .NET Upgrade Assistant will not add any platforms that you didn't already target.

To illustrate these steps, we will create a new project, just like we did in *Chapter 2*. This time, however, we will have to create a Xamarin.Forms project first.

Creating a new Xamarin.Forms app

The following steps will guide you through creating a new Xamarin.Forms project:

1. Open Visual Studio 2022 and select **Create a new project**:

Figure 3.1 – Visual Studio 2022

This will open the **Create a new project** wizard.

2. In the search field, type Xamarin.Forms and select the **Mobile App (Xamarin.Forms)** item from the list:

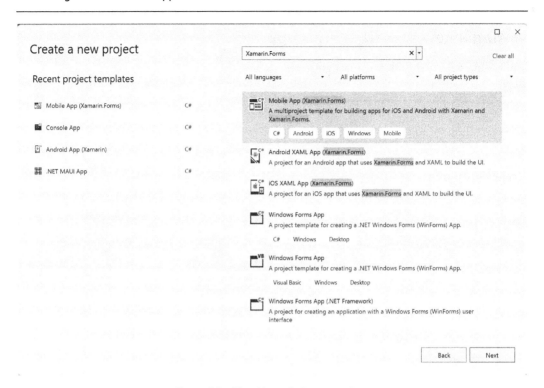

Figure 3.2 – New Xamarin.Forms project

3.   Complete the next page of the wizard by naming your project MauiMigration, and then click **Next**:

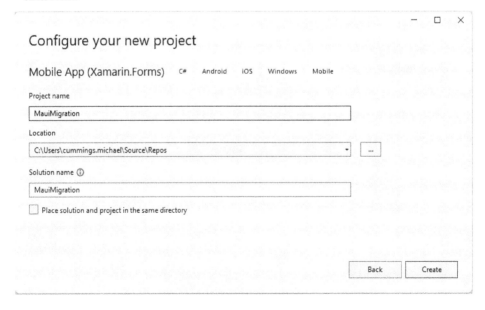

Figure 3.3 – Configuring the Xamarin.Forms project

4. Select the **Flyout** template and ensure that all three Xamarin.Forms platforms are checked:

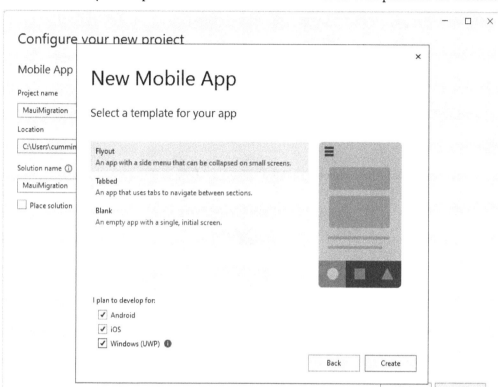

Figure 3.4 – Selecting the Xamarin.Forms template and platforms

You may receive a few messages about out-of-support components. These are expected if this is the first time you are creating a Xamarin.Forms app and can be safely ignored.

5. Finalize the setup by clicking **Create** and wait for Visual Studio to create the project.

Before we begin to migrate this app, it's a good idea to make sure it works properly.

6. Run the app and test out all the buttons, flyout options, and menus.

    On the **About** page, there is a **Learn more** button that will open a browser and navigate you to the Xamarin.Forms quickstart web page:

    Learn more at **https://aka.ms/xamarin-quickstart**

    Figure 3.5 – The Learn more button

7. The flyout menu has three options: **About**, **Browse**, and **Logout**. Ensure you click on each one and explore all their functionality:

    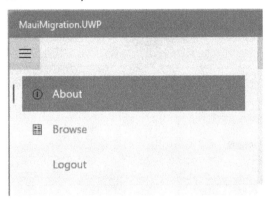

    Figure 3.6 – Flyout menu options

Now that we have explored this Xamarin.Forms app, let's move on to creating the new .NET MAUI project, which will act as our new app.

Creating a new .NET MAUI app

We will add a new project to the current solution to make things easier. To create a new .NET MAUI project, follow these steps:

1. In Visual Studio, right-click the `MauiMigration` solution item in **Solution Explorer** and select **File | Add | New Project**:

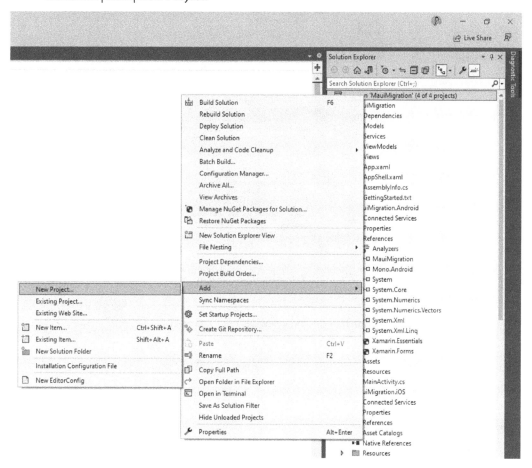

Figure 3.7 – Adding a new project to solution

2.   In the **Add a new project** dialog, select **.NET MAUI App** and click **Next**:

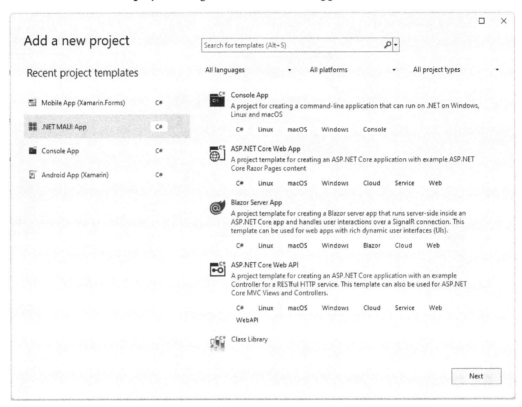

Figure 3.8 – Adding a new .NET MAUI project

3. In the **Configure your new project** dialog, set the project name to MyMauiApp and click **Next**:

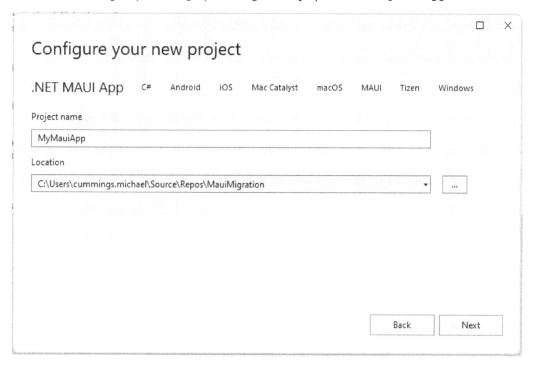

Figure 3.9 – Configuring the .NET MAUI project

4. In the **Additional Information** dialog, ensure that the selected framework is **.NET 7.0 (Standard Term Support)** and click **Create**.

Now that we have the shell of the .NET MAUI app, we can start moving the important parts of the Xamarin.Forms app over to the .NET MAUI project.

Migrating the MauiMigration app to MyMauiApp

Migrating a Xamarin.Forms app into a blank new .NET MAUI template will involve the following high-level steps:

1. Copy your app files into the new template.

2. Change the Xamarin.Forms namespace to its .NET MAUI equivalent.

3. Update the app startup so that it uses your views.

The next few sections will explain how to accomplish these steps.

Copying files to a new project

To begin, let's copy the XAML and C# files and folders from the Xamarin.Forms project to the .NET MAUI project. In the `MauiMigration` project, select the `Models`, `Services`, `ViewModels`, and `Views` folders. We will copy these files instead of moving them so that we don't destroy the original project. Right-click any of the selected folders and select **Copy**. Right-click the `MauiApp` project and click **Paste**.

A few images are also needed; we will copy them from the `MauiMigration.UWP` project. Under the `MauiMigration.UWP` project, you will find three image files called `icon_about.png`, `icon_feed.png`, and `xamarin_logo.png`. Select all three files and copy them just like you did for the previous files. Paste the files into the `MyMauiApp/Resources/Images` folder.

Visual Studio may make some changes to your project file when you copy/paste files, such as adding new item groups that remove and add the same file to the project. You can safely remove these changes as the single project system knows how to handle XAML and `.png` files. If you get compilation errors related to missing images or errors in the XAML files, check `MyMAuiApp.csproj` for any additional `ItemGroups` that reference either the `.png` files or XAML files and remove them.

The following screenshot shows an example of the `MyMauiApp.csproj` file after copying the images from `MauiMigration.UWP`. *Figure 3.10* shows the changes that Visual Studio added; these can be removed:

```
46
47        <!-- Raw Assets (also remove the "Resources\Raw" prefix) -->
48        <MauiAsset Include="Resources\Raw\**" LogicalName="%(RecursiveDir)%(Filename)%(Extension)" />
49      </ItemGroup>
50
51      <ItemGroup>
52        <MauiImage Remove="Resources\Images\icon_about.png" />
53        <MauiImage Remove="Resources\Images\icon_feed.png" />
54        <MauiImage Remove="Resources\Images\xamarin_logo.png" />
55      </ItemGroup>
56
57      <ItemGroup>
58        <None Remove="Resources\Images\icon_about.png" />
59        <None Remove="Resources\Images\icon_feed.png" />
60        <None Remove="Resources\Images\xamarin_logo.png" />
61      </ItemGroup>
62
63      <ItemGroup>
64        <AndroidResource Include="Resources\Images\icon_about.png" />
65        <AndroidResource Include="Resources\Images\icon_feed.png" />
66        <AndroidResource Include="Resources\Images\xamarin_logo.png" />
67      </ItemGroup>
68
69      <ItemGroup>
70        <PackageReference Include="Microsoft.Extensions.Logging.Debug" Version="7.0.0" />
71      </ItemGroup>
72
73    </Project>
74
```

Figure 3.10 – Visual Studio added unnecessary items

Next, we will need to update the XAML files so that they reference .NET MAUI controls.

Updating namespaces

Currently, the XAML files are still using the Xamarin.Forms namespace. To update these files, we need to change the following:

```
xmlns="http://xamarin.com/schemas/2014/forms"
```

We must amend this like so:

```
xmlns="http://schemas.microsoft.com/dotnet/2021/maui"
```

You can use the **Find and Replace** feature in Visual Studio to make these changes or manually edit each file. To use the **Find and Replace** dialog, follow these steps:

1. In **Solution Explorer**, select the MyMauiApp project.

2. From the Visual Studio menu select **Edit | Find and Replace | Replace in Files** (or *Ctrl + Shift + H*) to open the **Find and Replace** dialog:

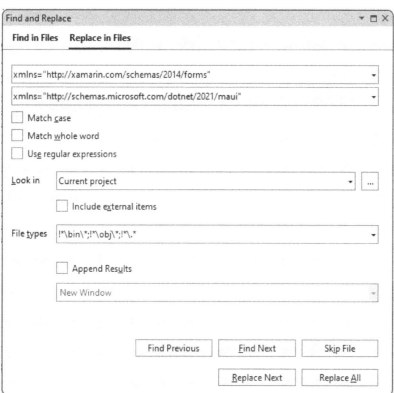

Figure 3.11 – The Find and Replace dialog

3. In the **Look in** field, select **Current project**.

4. Click the **Replace All** button; there should only be five places to make that change.

Once completed, there are a few other changes that need to be made.

In the `NewItemPage.xaml` file, in the `Views` folder of the `MyMauiApp` project, remove the following highlighted text:

```
Title="New Item"
```

```
xmlns:ios="clr-namespace:Xamarin.Forms.PlatformConfiguration.
iOSSpecific;assembly=Xamarin.Forms.Core"
```

```
ios:Page.UseSafeArea="true">
```

Now that the XAML file namespace changes are complete, we can move on to the C# namespace changes.

Using the **Find and Replace** dialog again, we can remove all the Xamarin.Forms namespace references. This time, by using a regular expression, we can remove multiple lines. Use the following expression in the **Find** entry box:

```
^using Xamarin\.[Forms,Essentials].*;
```

Then, check the **Use regular expressions** checkbox and select **Current project** for **Look in**, as shown here:

Figure 3.12 – Find and Replace – C# namespaces

Most of the changes have been made, and the app should compile at this point. However, it is not using any of the code we have copied and modified as .NET MAUI Shell does not reference the copied pages. The next step is to hook Shell into our code.

Modifying the app startup

The MyMauiApp project is still using the default MainPage.xaml file as its startup page. The next step in our migration is to make the AppShell.xaml file the same as it is for our Xamarin.Forms app.

What we need to copy is in the MauiMigration project, in the AppShell.xaml file, starting at line 76 until line 101, as shown in the following code snippet:

```
<!--
        When the Flyout is visible this defines the content to display
in the flyout.
        FlyoutDisplayOptions="AsMultipleItems" will create a separate
flyout item for each child element              https://docs.microsoft.
com/dotnet/api/xamarin.forms.shellgroupitem.flyoutdisplayoptions?view
=xamarin-forms
    -->
    <FlyoutItem Title="About" Icon="icon_about.png">
        <ShellContent Route="AboutPage" ContentTemplate="{DataTemplate
local:AboutPage}" />
    </FlyoutItem>
    <FlyoutItem Title="Browse" Icon="icon_feed.png">
        <ShellContent Route="ItemsPage" ContentTemplate="{DataTemplate
local:ItemsPage}" />
    </FlyoutItem>


    <!-- When the Flyout is visible this will be a menu item you can
tie a click behavior to  -->
    <MenuItem Text="Logout" StyleClass="MenuItemLayoutStyle"
Clicked="OnMenuItemClicked">

    </MenuItem>


    <!--
        TabBar lets you define content that won't show up in a flyout
menu. When this content is active
        the flyout menu won't be available. This is useful for
creating areas of the application where
        you don't want users to be able to navigate away from. If you
would like to navigate to this
        content you can do so by calling
        await Shell.Current.GoToAsync("//LoginPage");
    -->

    <TabBar>
        <ShellContent Route="LoginPage" ContentTemplate="{DataTemplate
local:LoginPage}" />
    </TabBar>
```

Copy the preceding code and replace the following lines in the MyMauiApp project's AppShell. xaml file:

```
<ShellContent
    Title="Home"
    ContentTemplate="{DataTemplate local:MainPage}"
    Route="MainPage" />
```

To enable the flyout, you will need to remove the following highlighted text in AppShell.xaml:

```
xmlns:local="clr-namespace:MyApp"
Shell.FlyoutBehavior="Disabled">
```

The final change that we need to make will register the correct instance for the **Browse** page to use for the list of items. Open the MauiProgram.cs file and make the changes highlighted in the following code block:

```
using Microsoft.Extensions.Logging;
using MauiMigration.Services;
using MauiMigration.Models;
namespace MyMauiApp
{
    public static class MauiProgram
    {
        public static MauiApp CreateMauiApp()
        {
            var builder = MauiApp.CreateBuilder();
            builder
                .UseMauiApp<App>()
                .ConfigureFonts(fonts =>
                {
                    fonts.AddFont("OpenSans-Regular.ttf",
"OpenSansRegular");
                    fonts.AddFont("OpenSans-Semibold.ttf",
"OpenSansSemibold");
                });
            DependencyService.RegisterSingleton<IDataStore<Item>>(new
MockDataStore());
#if DEBUG
            builder.Logging.AddDebug();
#endif
            return builder.Build();
        }
    }
}
```

At this point, you should be able to compile and run the converted project. Make sure you set `MyMauiApp` as the startup project before running. Play around with the app and make sure it is all working before moving on to the next section on manually migrating your apps to .NET MAUI.

Manual migration overview

In the previous section, we converted a simple Xamarin.Forms app into .NET MAUI using the Single Project system. This project did not use any advanced features of Xamarin.Forms, such as external NuGets, custom controls, or any commercial controls. These are additional items that you will have to consider when migrating your apps from Xamarin.Forms to .NET MAUI.

In this section, we are going to discuss the basic flow that you should go through to migrate your Xamarin.Forms app to .NET MAUI. This is by no means an all-inclusive list; the .NET MAUI team is updating a wiki page that details all their knowledge in one location.

> **Official migration guide**
>
> The guidelines for migrating your Xamarin.Forms apps to .NET MAUI are continually evolving since the release of .NET MAUI, based on usage and feedback. To review the latest guidelines, please visit the following URL in your favorite browser: `https://learn.microsoft.com/en-us/dotnet/maui/get-started/migrate?view=net-maui-7.0`.

When migrating your apps from Xamarin.Forms to .NET MAUI, you will need to follow the overall steps outlined here:

1. Convert the Xamarin.Forms projects from .NET Framework into .NET SDK style.
2. Update the code from Xamarin.Forms to .NET MAUI.
3. Update any incompatible dependencies with .NET 6+ versions.
4. Address any breaking API changes.
5. Run the converted app and verify its functionality.

.NET Upgrade Assistant is a tool that will attempt to perform the first four steps for you. However, before we dive into using .NET Upgrade Assistant, we are going to look at what each of these steps entails so that we have a firm understanding of how to migrate our app if .NET Upgrade Assistant is unable to operate on an app project.

Converting the Xamarin.Forms projects from .NET Framework into .NET SDK style

Some Xamarin.Forms projects are based on the .NET Framework project template. It is a verbose project format that has been updated for .NET projects. The new format, typically referred to as the

SDK style, is a much more concise format with better defaults. Xamarin.Forms projects created with Visual Studio 16.5 or later use the newer SDK format.

To convert an old format project file, such as the one we will be using with .NET Upgrade Assistant, in the *Installing and running .NET Upgrade Assistant* section later in this chapter, you would need to change the `<Project />` element to the new SDK style, like this:

1. Find a line like this in your project file:

   ```
   <Project ToolsVersion="4.0" DefaultTargets="Build"
   xmlns="http://schemas.microsoft.com/developer/msbuild/2003">
   ```

2. Replace it with the following:

   ```
   <Project Sdk="Microsoft.NET.Sdk">
   ```

Remember that you will have to make this change for all your Xamarin.Forms projects, platform-specific projects, and any shared library projects.

Let's look at the specific changes that you might have to make for a typical Xamarin.Forms project that targets both Android and iOS.

To convert the shared library project of a Xamarin.Forms app into a .NET MAUI project, we need to replace the contents of the `csproj` file with the following:

```
<Project Sdk="Microsoft.NET.Sdk">
  <PropertyGroup>
    <TargetFrameworks>net7.0-ios;net7.0-android; </TargetFrameworks>
    <TargetFrameworks Condition="$([MSBuild]::IsOSPlatform
('windows'))">$(TargetFrameworks);net7.0-windows10.0.19041.0
</TargetFrameworks>
    <UseMaui>True</UseMaui>
    <OutputType>Library</OutputType>
    <ImplicitUsings>enable</ImplicitUsings>
    <!-- Required for C# Hot Reload -->
    <UseInterpreter Condition="'$(Configuration)' == 'Debug'">True</
UseInterpreter>
    <SupportedOSPlatformVersion
Condition="$([MSBuild]::GetTargetPlatformIdentifier('$
(TargetFramework)')) == 'ios'">14.2</SupportedOSPlatformVersion>
    <SupportedOSPlatformVersion
Condition="$([MSBuild]::GetTargetPlatformIdentifier('$
(TargetFramework)')) == 'android'">21.0</SupportedOSPlatformVersion>
    <SupportedOSPlatformVersion
Condition="$([MSBuild]::GetTargetPlatformIdentifier('$
(TargetFramework)')) == 'windows'">10.0.17763.0</SupportedOS
PlatformVersion>
```

```
    <TargetPlatformMinVersion
Condition="$([MSBuild]::GetTargetPlatformIdentifier('$
(TargetFramework)')) == 'windows'">10.0.17763.0</Target
PlatformMinVersion>
  </PropertyGroup>
</Project>
```

The important bits are highlighted. `<UseMaui>True</UseMaui>` will enable the project system to automatically add the correct references to the project for .NET MAUI libraries. The `<TargetFrameworks>` elements are updated for the correct **Target Framework Monikers (TFMs)** for .NET 6 or 7.

.NET MAUI also requires an additional `<SupportedOSPlatformVersion>` property, which is set conditionally based on `TargetPlatformIdentifier`.

That is a significant amount of reduction from the .NET Framework `csproj` files. In addition to the property reduction, you can remove most of the `<ItemGroup>...</ItemGroup>` entries since all source files are now included by default. The groups to leave are the groups containing `<ProjectReferences />` entries.

Now that we've converted the shared project, let's review the changes needed for an Android project.

As with all .NET Framework projects, we will need to change the `<Project ...>` element to the following:

```
<Project Sdk="Microsoft.NET.Sdk">
```

Now, remove all the `<PropertyGroup>...</PropertyGroup>` elements since they are default values, and replace them with the following:

```
<PropertyGroup>
  <UseMaui>True</UseMaui>
  <TargetFramework>net7.0-android</TargetFramework>
  <OutputType>Exe</OutputType>
  <ImplicitUsings>enable</ImplicitUsings>
  <SupportedOSPlatformVersion Condition="'$(TargetFramework)' ==
'net7.0-android'">31.0</SupportedOSPlatformVersion>
</PropertyGroup>
<PropertyGroup>
  <UseInterpreter Condition="$(TargetFramework.Contains('-
android'))">True</UseInterpreter>
</PropertyGroup>
```

Again, in the Android project, we set the `<UseMaui>` and `<TargetFramework>` properties.

To finish migrating this project file, remove all the `<ItemGroup>` elements and their contents, except for the one group that contains the `<AndroidResource>` elements. Finally, remove the `<Import>` element at the bottom of the project file; it is no longer needed.

The next project type to look at is an iOS project. The changes here will be very similar to the changes in an Android project but with an iOS twist. Xamarin.Forms iOS projects are also a .NET Framework style project, so we will need to change the `<Project ...>` element to the following:

```
<Project Sdk="Microsoft.NET.Sdk">
```

Then, we will need to remove all the existing `<PropertyGroup>` elements and add the following:

```
<PropertyGroup>
  <UseMaui>true</UseMaui>
  <TargetFramework>net7.0-ios</TargetFramework>
  <OutputType>Exe</OutputType>
  <ImplicitUsings>enable</ImplicitUsings>
  <SupportedOSPlatformVersion
Condition="$([MSBuild]::GetTargetPlatformIdentifier('$
(TargetFramework)'))== 'ios'">14.2</SupportedOSPlatformVersion>
</PropertyGroup>
```

Again, you will notice the `<UseMaui>` property being set, and that the `<TargetFramwork>` value is `net7.0-ios`.

To finish converting an iOS project, remove the `<ItemGroup>` and `<Import>` elements except for the one containing the `<ProjectReference>` items; those are still needed.

Updating code from Xamarin.Forms to .NET MAUI

Updating the code from Xamarin.Forms to .NET MAUI consists of several steps. First, we need to add some new code that is required to initialize .NET MAUI. Refer to the *Examining the files section* of *Chapter 2* for more details on the files that are needed.

The first change you will need is in the shared project. Add a new class file called `MauiProgram.cs` with the following contents, minimally:

```
public static class MauiProgram
{
    public static MauiApp CreateMauiApp()
    {
        var builder = MauiApp.CreateBuilder();
        builder
            .UseMauiApp<App>()
            .ConfigureFonts(fonts =>
            {
                fonts.AddFont("OpenSans-Regular.ttf",
"OpenSansRegular");
            });
```

```
        return builder.Build();
    }
}
```

You may have to add additional code to this file so that it can handle additional needs for your application, such as registering types for dependency injection or registering additional library features, such as logging.

In the Platforms/Android folder, add a new class file named MainApplication.cs and update the class so that it looks like this:

```
[Application]
public class MainApplication : MauiApplication
{
    public MainApplication(IntPtr handle, JniHandleOwnership
ownership)
    : base(handle, ownership)
    {
    }
    protected override MauiApp CreateMauiApp() => MauiProgram.
CreateMauiApp();
}
```

We must also make a few adjustments to the MainActivity class. Open the MainActivity.cs file and make the changes highlighted in the following code:

```
[Activity(Label = "ManualMigration", Icon = "@mipmap/icon",
Theme = "@style/Theme.MaterialComponents", MainLauncher = true,
ConfigurationChanges = ConfigChanges.ScreenSize | ConfigChanges.
Orientation | ConfigChanges.UiMode | ConfigChanges.ScreenLayout |
ConfigChanges.SmallestScreenSize )]

public class MainActivity : MauiAppCompatActivity
{
    protected override void OnCreate(Bundle savedInstanceState)
    {
        base.OnCreate(savedInstanceState);
    }
}
```

The final change we would need to make in the Android project is to update targetSdkVersion in the AndroidManifest.xml file to version 33, like this:

```
<uses-sdk android:minSdkVersion="21" android:targetSdkVersion="33" />
```

Android SDK versions

Since the Android SDK is updated yearly, it may be the case that the version of .NET for Android and .NET MAUI that you are using is also using a version of the Android SDK that is greater than 33. The good news is that you will get an error in Visual Studio if `targetSdkVersion` is too low or is not installed. Just follow the instructions in the build error to set the SDK version correctly.

That is all the changes we need to make in the Android project for now. Moving on to the iOS project, the `AppDelegate.cs` file can be updated so that it matches the following:

```
public partial class AppDelegate : MauiUIApplicationDelegate
{
    protected override MauiApp CreateMauiApp() => MauiProgram.
CreateMauiApp();
}
```

The final change for the iOS project is to open the `Info.plist` file and change the `MinimumOSVersion` property to `15.2`.

With the base changes needed to start your app as a .NET MAUI app done, the next changes are much broader brush strokes:

1. Remove all `Xamarin.*` namespaces from `.cs` files.

2. Change all `xaml` namespace declarations from the following:

    ```
    xmlns="http://xamarin.com/schemas/2014/forms"
    ```

    You will need to amend them like so:

    ```
    xmlns="http://schemas.microsoft.com/dotnet/2021/maui"
    ```

You may notice that these are the same changes we made in the previous section when using a .NET MAUI Single Project.

About images

.NET MAUI has improved image handling for the various platforms that it targets. You can provide a single SVG image file, and it will resize the image correctly for all platforms for you. Since the SVG format is based on vectors, it will render much better than other formats such as JPG and PNG after resizing. It is recommended that you convert your images into SVG format, if possible, to take advantage of this feature in .NET MAUI.

Updating any incompatible NuGet packages

There are a lot of NuGet packages out there and there is no way we can cover them all. But, in general, for each of the NuGet packages that are in use in your app, be sure to look for a version that specifically supports .NET MAUI or the version of .NET that you are targeting. You can use the NuGet Gallery page to determine whether a package supports .NET MAUI. Using a popular package such as PCLCrypto version 2.0.147 (https://www.nuget.org/packages/PCLCrypto/2.0.147#support edframeworks-body-tab) targets classic Xamarin projects but not .NET 6 or .NET 7. You can find the compatible frameworks under the **Frameworks** tab:

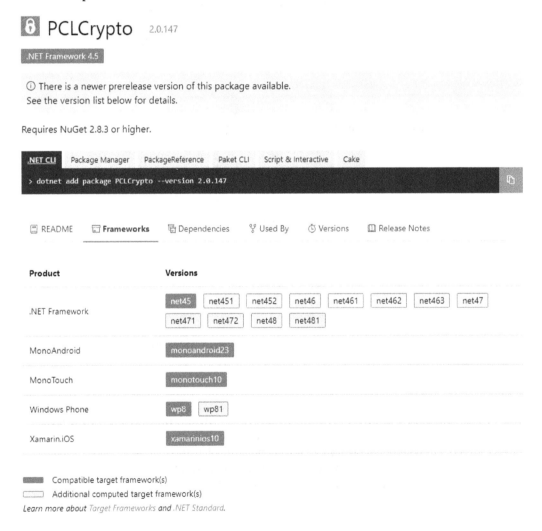

Figure 3.13 – NuGet Gallery page for PCLCrypto v2.0.147

However, version 2.1.40-alpha (`https://www.nuget.org/packages/PCLCrypto/2.1.40-alpha#supportedframeworks-body-tab`) lists .NET 6 and .NET 7 as compatible frameworks:

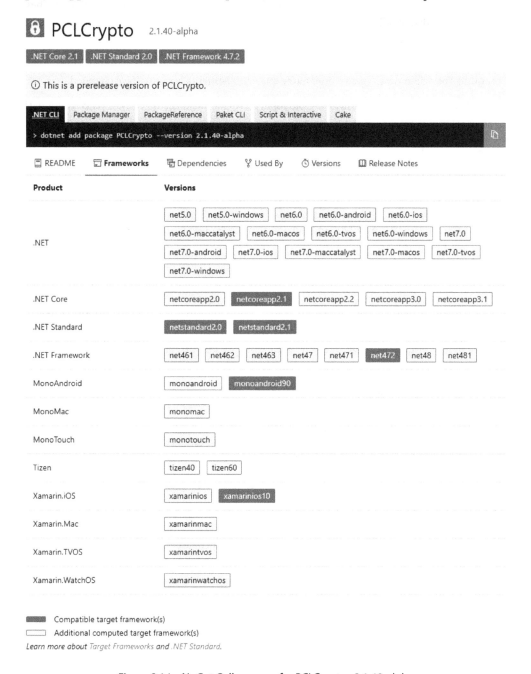

Figure 3.14 – NuGet Gallery page for PCLCrypto v2.1.40-alpha

Currently, we know of the following NuGet changes:

- Remove all Xamarin.Forms and Xamarin.Essentials NuGet references from your projects. These are now included in .NET MAUI directly. You will have to make some namespace adjustments as those have changed.

- Replace Xamarin.Community Toolkit with the latest preview of .NET MAUI Community Toolkit. You will have to make some namespace adjustments as those have changed.

- If you reference any of the following SkiaSharp NuGet packages directly, replace them with the latest previews:

  - **SkiaSharp.Views.Maui.Controls**

  - **SkiaSharp.Views.Maui.Core**

  - **SkiaSharp.Views.Maui.Controls.Compatibility**

You can find the latest version of NuGet packages on the NuGet Gallery website at `https://nuget.org/packages`.

Addressing any breaking API changes

Unfortunately, there is no magic bullet for any of these types of changes. You will simply have to start from the top of your error list and work your way through them. You can review the release notes linked in the official migration guide, available at `https://learn.microsoft.com/en-us/dotnet/maui/migration/`, as helpful hints.

For example, a common type of error that's seen is `error CS0104: 'ViewExtensions' is an ambiguous reference between 'Microsoft.Maui.Controls.ViewExtensions' and 'Microsoft.Maui.ViewExtensions'`. This can be fixed by explicitly using the full namespace when referencing the type or by using a type alias – for example, `using ViewExtensions = Microsoft.Maui.Controls.ViewExtensions`.

Custom renderers and effects

Your application may use custom renderers or effects to provide a unique user experience. Covering how to upgrade these components is beyond the scope of this chapter. To learn more about how to upgrade renderers and effects, visit the Microsoft Learn site for .NET MAUI migration at `https://learn.microsoft.com/en-us/dotnet/maui/migration/`.

Running the converted app and verifying its functionality

This is not the last step – I recommend that you attempt to do this after each change. Building your app as you make changes ensures that you are moving in the right direction. I recommend that you **rebuild** over **building** and even go so far as to delete the `obj` and `bin` folders beforehand. This will ensure that you are building with the latest changes and dependencies, by forcing a NuGet restore.

Now that we know the basics of how to convert a Xamarin.Forms app, let's use .NET Upgrade Assistant to migrate a project for us.

Installing and running .NET Upgrade Assistant

As stated previously, .NET Upgrade Assistant will attempt to perform the first four steps of the migration of your Xamarin.Forms app to .NET MAUI outlined in the previous section. The tool is under active development and as the team discovers new improvements, they are added. This is mostly due to feedback they receive from developers like you.

At the time of writing, .NET Upgrade Assistant did not work on all projects and has the following limitations:

- Xamarin.Forms must be version 5.0 and higher
- Only Android and iOS projects are converted
- .NET MAUI must be properly installed with the appropriate workloads

If you have followed the steps from *Chapter 1*, then the last should should already be satisfied.

If your Xamarin.Forms app meets these criteria, then we can get started by installing the tool.

Installing .NET Upgrade Assistant

.NET Upgrade Assistant is a Visual Studio extension on Windows and a command-line tool on Windows and macOS. You can use the integrated developer PowerShell in Visual Studio or any command-line prompt to install the tool. Follow these steps to install the tool using Visual Studio on Windows:

1. In Visual Studio, select the **Extensions** menu, then the **Manage Extensions** item. This will open the **Manage Extensions** dialog.
2. In the **Manage Extensions** dialog, use the search box to search for `upgrade`.

3.   Select **.NET Upgrade Assistant** and click **Download**:

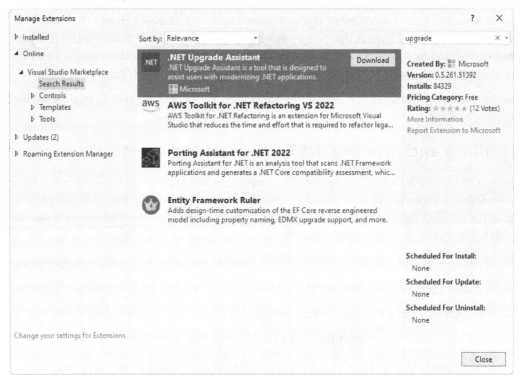

Figure 3.15 – Visual Studio – the Manage Extensions dialog

4.   Once the extension has been downloaded, you will need to close and reopen Visual Studio to install the extension:

Figure 3.16 – Installing the .NET Upgrade Assistant VSIX

5. Click **Modify** and then follow the instructions to complete the installation.

6. Once the installation is complete, reopen Visual Studio.

Now that the tool has been installed, we can use it to convert a Xamarin.Forms project!

Preparing to run .NET Upgrade Assistant

For the remainder of this chapter, we will be using Visual Studio on Windows to migrate a Xamarin. Forms app to .NET MAUI.

To run .NET Upgrade Assistant, we will need a Xamarin.Forms project to upgrade. For this portion of the chapter, we will use the Xamarin.Forms app that was demoed at the Microsoft Build conference in 2019. The source can be found at `https://github.com/mindofai/Build2019Chat`, though you can find it in this book's GitHub repository at `https://github.com/PacktPublishing/MAUI-Projects-3rd-Edition` under the `Chapter03/Build2019Chat` folder.

Once you have downloaded the source, open the `BuildChat.sln` file in Visual Studio. Once the project has finished loading, make sure your configuration is correct by running the app first. You should see a screen that looks like this on Android:

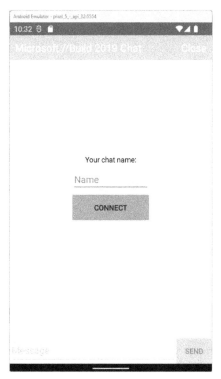

Figure 3.17 – The original app on Android

Now that we have confirmed that the original app runs, we can follow these steps to prepare for running .NET Upgrade Assistant:

1. Right-click the `BuildChat` solution node in **Solution Explorer** and select **Manage NuGet Packages for Solution…**:

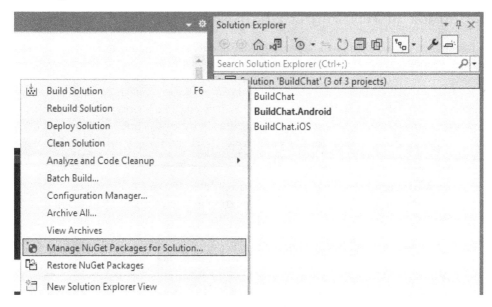

Figure 3.18 – Solution context menu

2. In the **NuGet – Solution** window that opens, select the **Updates** tab:

Figure 3.19 – The NuGet – Solution window

3. Click the **Select all packages** checkbox, then click **Update**.

4. Visual Studio will prompt you with a preview of all the changes that will be made. Click **OK** once you have reviewed them.

5. Visual Studio will then prompt you to accept the license terms for packages that have them. Once you have reviewed the license terms, click **I Accept**.

6. After updating, you may still have a **gold bar** indicator in the Visual Studio window from running the application earlier. You can safely dismiss the message by clicking the **X** button on the right:

XAML Hot Reload is disabled because it requires Xamarin.Forms 5.0.0.2012 or newer.   Update Xamarin.Forms Packages   Don't show this message again for this project   ✕

Figure 3.20 – Xamarin.Forms version gold bar

7. Once the packages have been updated, let's make sure the app is still working by running it again. You should get a build error like the following:

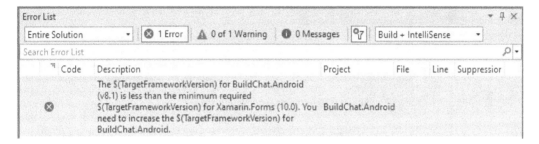

Figure 3.21 – Error after upgrading packages

8. To resolve this error, in **Solution Explorer**, select the `BuildChat.Android` project, then press *Alt* and *Enter* at the same time to open the project properties page.

9. Use the **Target Framework** dropdown to change the value from `Android 8.1 (Oreo)` to `Android 10.0`.

> **Google Play support**
>
> You may get a warning about Google Play requiring new apps and updates to support a specific version of Android. To remove that warning, just set **Target Framework** to the version indicated in the warning message.

10. Visual Studio will prompt you to confirm the change as it has to close and re-open the project. Select **Yes**.

11. Visual Studio may also prompt you to install the Android version if you haven't installed it. Follow the prompts to install the Android version.

12. Attempting to run the project again yields a new set of errors:

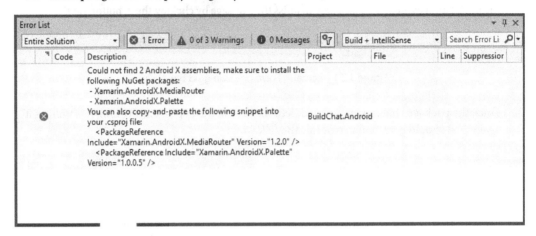

Figure 3.22 – Missing packages error

13. To resolve this error, right-click the BuildChat.Android project and select **Unload Project**. The project file should open in Visual Studio automatically.

14. Locate <ItemGroup> in the file with <PackageReference> items and make the changes highlighted in the following snippet:

```
<ItemGroup>
    <PackageReference Include="Microsoft.AspNetCore.SignalR.
Client">
        <Version>7.0.9</Version>
    </PackageReference>
    <PackageReference Include="Xamarin.Forms"
Version="5.0.0.2578" />
    <PackageReference Include="Xamarin.Android.Support.Design"
Version="28.0.0.3" />
    <PackageReference Include="Xamarin.Android.Support.
v7.AppCompat" Version="28.0.0.3" />
    <PackageReference Include="Xamarin.Android.Support.v4"
Version="28.0.0.3" />
    <PackageReference Include="Xamarin.Android.Support.
v7.CardView" Version="28.0.0.3" />
    <PackageReference Include="Xamarin.AndroidX.MediaRouter"
Version="1.2.0" />
    <PackageReference Include="Xamarin.AndroidX.Palette"
Version="1.0.0.5" />
</ItemGroup>
```

You could also use Visual Studio's NuGet Package Manager to add these packages.

15. Save and reload the project before trying to run it again. Since the project was unloaded, you will need to set the `BuildChat.Android` project as the startup project again.

You should be able to run the application at this time since some warnings can be ignored. If not, review the previous steps to make sure you made all the changes correctly. At this point, we are ready to run the upgrade assistant to convert from Xamarin.Forms into .NET MAUI.

> **Treat warnings as errors**
>
> If you have the project option to treat warnings as errors set to anything other than none, then the warnings will prevent you from running the app. Set the option to none to allow the app to run. The option defaults to none.

Running .NET Upgrade Assistant

Running .NET Upgrade Assistant from within Visual Studio is a straightforward process. We will upgrade each project individually; there isn't any method to upgrade all the projects in one go.

Upgrading the BuildChat project

Let's start with the shared project, `BuildChat`, by following these steps:

1. Select the `BuildChat` project in **Solution Explorer**.

2. Use the context menu to select the **Upgrade** menu item:

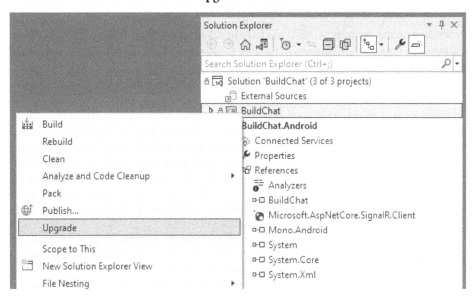

Figure 3.23 – Upgrading the BuildChat project

This will open the **Upgrade** assistant in a document window:

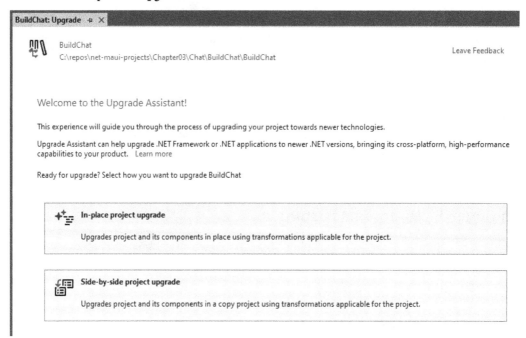

Figure 3.24 – Upgrading the BuildChat project

3. Select the **In-place project upgrade** option.

4. Depending on the versions of .NET you have installed, you will be prompted to choose one. If you followed the setup instructions in *Chapter 1*, you should have the .NET 7.0 option available. Select **.NET 7.0** and select **Next**:

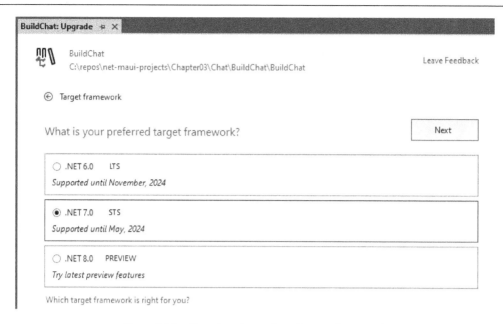

Figure 3.25 – Choosing the preferred target framework

5.   At this point, you are allowed to review the changes that will be made by expanding each node in the list. You can also choose to not upgrade certain items by removing the check in the checkbox next to that item. When you have inspected all the changes, make sure all items are checked again, then click **Upgrade selection**:

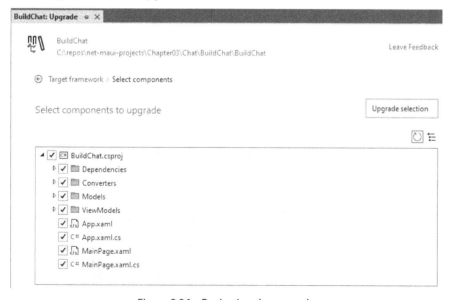

Figure 3.26 – Reviewing the upgrade

6.   Visual Studio will start the upgrade process. You can monitor it as it completes each item:

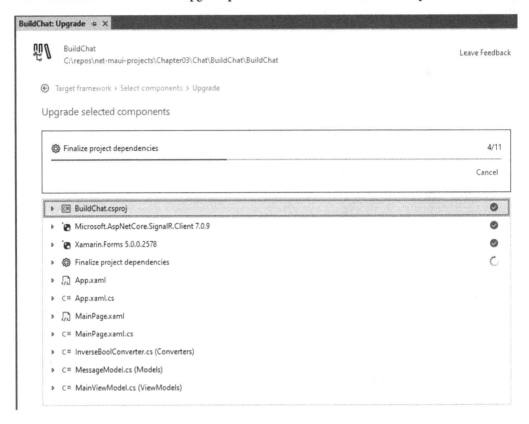

Figure 3.27 – Upgrade in progress

7.   When it's finished, you can inspect each item to see what the result of the upgrade was:

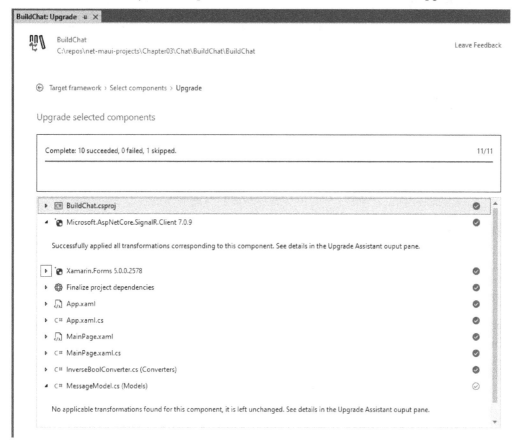

Figure 3.28 – Upgrade complete

A white check in a green circle indicates some transformation was completed and successful, a green check with a white background means the step was skipped since nothing was needed, and a red cross (not shown) means the transformation failed. You can view the complete output from the tool by inspecting the **Upgrade Assistant** log in the output pane:

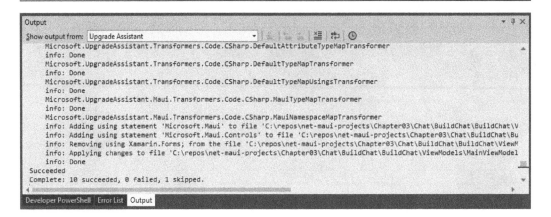

Figure 3.29 – Upgrade Assistant log output

Do not be concerned with the errors in the error window at this point. There will be errors until we finish upgrading the remaining projects. Now that the `BuildChat` project has been upgraded, we can upgrade the `BuildChat.Android` project.

Upgrading the BuildChat.Android project

The steps for upgrading the remaining projects are largely the same – the only difference will be the steps involved in upgrading each project. The next two sections will skip the screenshots and just provide the steps. To complete the upgrade for the `BuildChat.Android` project, follow these steps:

1. Select the **BuildChat.Android** project in **Solution Explorer**.
2. Use the context menu to select the **Upgrade** menu item.
3. This will open the **Upgrade** assistant in a document window.
4. Select the **In-place project upgrade** option.
5. Select the **.NET 7.0** option, then select **Next**.
6. Review the changes that will be made by expanding each node in the list. Make sure all items are checked, then click **Upgrade Selection**.
7. Visual Studio will complete the upgrade process.

Now that .NET Upgrade Assistant has completed the `BuildChat.Android` project, we can upgrade the `BuildChat.iOS` project.

Upgrading the BuildChat.iOS project

The steps for upgrading the iOS project are largely the same – the only difference will be the steps involved in upgrading each project. To complete the upgrade for the `BuildChat.iOS` project, follow these steps:

1. Select the `BuildChat.iOS` project in **Solution Explorer**.

2. Use the context menu to select the **Upgrade** menu item.

3. This will open the **Upgrade** assistant in a document window:

4. Select the **In-place project upgrade** option.

5. Select the **.NET 7.0** option, then select **Next**.

6. Review the changes that will be made by expanding each node in the list. Make sure all items are checked, then click **Upgrade Selection**.

7. Visual Studio will complete the upgrade process.

Now that .NET Upgrade Assistant has completed the `BuildChat.iOS` project, we can see how well it worked.

Completing the upgrade to .NET MAUI

With .NET Upgrade Assistant having done all the work it can to upgrade the projects, we can now see what is left for us to complete the upgrade to .NET MAUI.

The first thing we want to do is make sure that the project is clean of all the previous build artifacts. This will ensure we are referencing all the right dependencies in our build output by forcing a restore and build. The best way to accomplish this is to remove the `bin` and `obj` folders from each project folder.

Use **File Explorer** to remove the `bin` and `obj` folders from the `BuildChat`, `BuildChat.Android` and `BuildChat.iOS` folders, then build the solution.

We'll end up with a few build errors for each project, as shown in the following figure:

Figure 3.30 – Package issues

To resolve these errors, either use Visual Studio's NuGet Package Manager to add a reference to version 7.0.1 of the `Microsoft.Extensions.Logging.Abstractions` package to all the projects, or follow these steps to update the project files manually:

1. Select the `BuildChat` project in **Solution Explorer**.

    The project file will open in a document window automatically.

2. Locate the `ItemGroup` element that contains the `PackageReference` items.

3. Make the changes highlighted in the following snippet:

    ```
    <ItemGroup>
        <PackageReference Include="Microsoft.AspNetCore.SignalR.
    Client" Version="7.0.9" />
        <PackageReference Include="Microsoft.Extensions.Logging.
    Abstractions" Version="7.0.1" />
    </ItemGroup>
    ```

4. Select the `BuildChat.Android` project in **Solution Explorer**.

    The project file will open in a document window automatically.

5. Locate the `ItemGroup` element that contains the `PackageReference` items.

6. Make the changes highlighted in the following snippet:

    ```
    <ItemGroup>
        <PackageReference Include="Xamarin.AndroidX.MediaRouter"
    Version="1.2.0" />
        <PackageReference Include="Xamarin.AndroidX.Palette"
    Version="1.0.0.5" />
        <PackageReference Include="Microsoft.AspNetCore.SignalR.
    Client" Version="7.0.9" />
        <PackageReference Include="Microsoft.Extensions.Logging.
    Abstractions" Version="7.0.1" />
    </ItemGroup>
    ```

7. Select the `BuildChat.iOS` project in **Solution Explorer**.

    The project file will open in a document window automatically.

8. Locate the `ItemGroup` element that contains the `PackageReference` items.

9. Make the changes highlighted in the following snippet:

    ```
    <ItemGroup>
        <PackageReference Include="Microsoft.AspNetCore.SignalR.
    Client" Version="7.0.9" />
        <PackageReference Include="Microsoft.Extensions.Logging.
    Abstractions" Version="7.0.1" />
    </ItemGroup>
    ```

Now that we have added the required package references, we can try building the app again. After this build, we'll get two new errors:

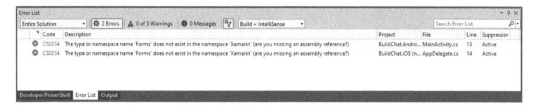

Figure 3.31 – Namespace does not exist errors

These two errors show two areas that the upgrade assistant did not upgrade. Luckily, we covered how to upgrade these two files easier in this chapter. Let's upgrade them again, starting with the BuildChat.Android project.

Open the MainActivity.cs file and make the changes highlighted in the following code:

```
using Microsoft.Maui;
namespace BuildChat.Droid
{
    [Activity(Label = "BuildChat", Icon = "@mipmap/icon", Theme
= "@style/Theme.MaterialComponents", MainLauncher = true,
ConfigurationChanges = ConfigChanges.ScreenSize | ConfigChanges.
Orientation)]
    public class MainActivity : MauiAppCompatActivity
    {
        protected override void OnCreate(Bundle savedInstanceState)
        {
            base.OnCreate(savedInstanceState);
        }
    }
}
```

That should complete the changes needed for the Android project. Now, to upgrade the iOS project, open the AppDelegate.cs file in BuildChat.iOS and update it so that it matches the following:

```
using Foundation;
using Microsoft.Maui;
using Microsoft.Maui.Hosting;
namespace BuildChat.iOS
{
    // The UIApplicationDelegate for the application. This class is
responsible for launching the
    // User Interface of the application, as well as listening (and
optionally responding) to
```

```
    // application events from iOS.
    [Register("AppDelegate")]
    public partial class AppDelegate : MauiUIApplicationDelegate
    {
        protected override MauiApp CreateMauiApp() => MauiProgram.
CreateMauiApp();
    }
}
```

The final change for the iOS project is to open the Info.plist file and change the MinimumOSVersion property to 15.2. To make this change, use **Generic PList Editor**. To open the Info.plist file and make this change, follow these steps:

1.   Select the Info.plist file in the BuildChat.iOS project.

2.   Use the context menu (right-click) and select **Open With….**

3.   In the **Open With** dialog, select **Generic Plist Editor**, then select **OK**:

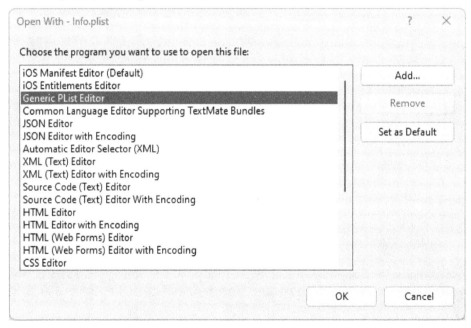

Figure 3.32 – Opening the Info.plist file

4. Find the entry labeled Minimum system version and change the value from 8.0 to 15.1:

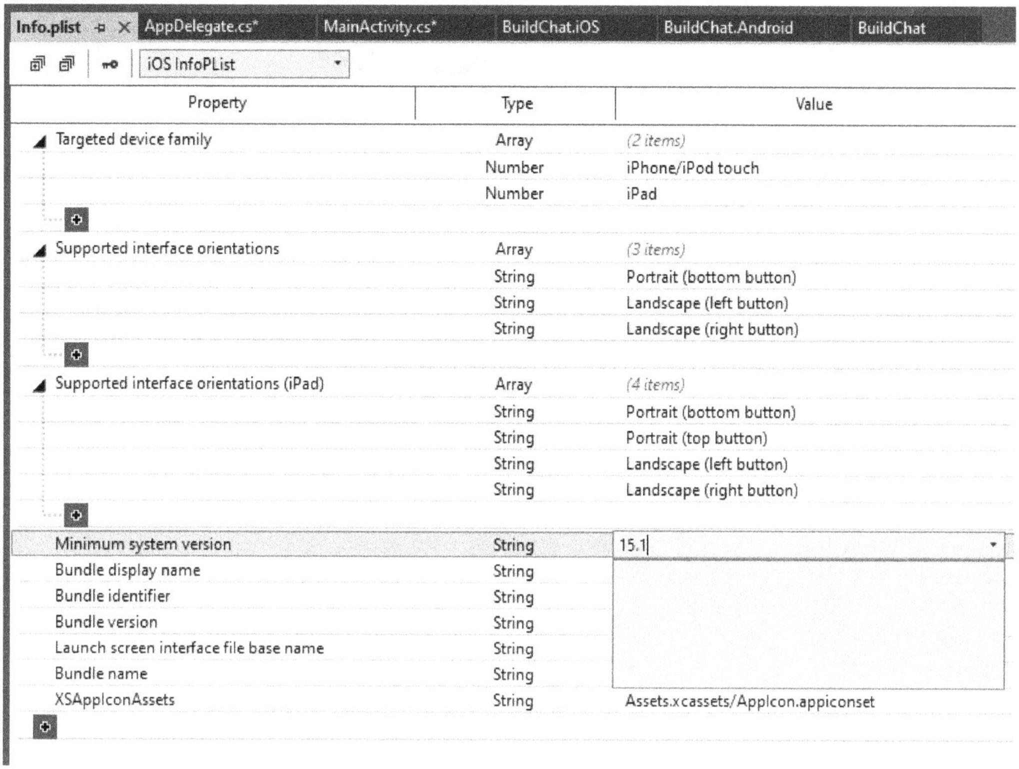

Figure 3.33 – Changing the minimum system version

Great – that should complete the changes needed to get the app running as a .NET MAUI application! The following are the before and after screenshots of the application; *before* is on the left and *after* is on the right:

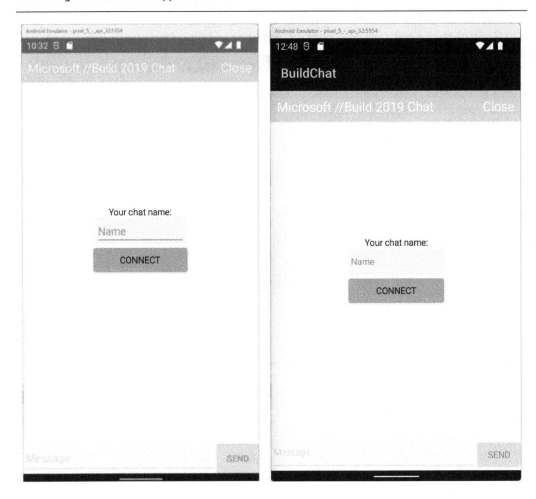

Figure 3.34 – Xamarin.Forms versus .NET MAUI

There are some visual changes between Xamarin.Forms and .NET MAUI, and you can tweak the .NET MAUI settings for the layouts and controls to get a very similar output.

Summary

In this chapter, we focused on upgrading a Xamarin.Forms app to .NET MAUI. We learned how to upgrade the project files from .NET Framework to SDK-style projects, application startup files, XAML views, and C# files needed for .NET MAUI. We started by doing this manually to learn about all the required steps and changes. We ended this chapter by using .NET Upgrade Assistant to make many of the changes for us. We also learned how to upgrade to the Single Project format, which is the default for .NET MAUI. Now, we can pick the best method for the project we are upgrading.

While we covered a lot in this chapter, it was not exhaustive. There is a lot of variation in different projects, from NuGet package dependencies and vendor-provided controls to customizations using renderers and effects. The app you are upgrading may use one or all of these, and you can find additional help on the Microsoft Learn site for upgrading Xamarin.Forms to .NET MAUI: `https://learn.microsoft.com/en-us/dotnet/maui/migration/`.

If you are interested in seeing the `BuildChat` app fully functional, try using .NET Upgrade Assistant on the service that was also built for the 2019 Microsoft Build conference. You can find the source on GitHub at `https://github.com/mindofai/SignalRChat/tree/master`. You could also use ChatGPT to help you build the service yourself using Azure Functions and SignalR.

In the next chapter, we will build an app that displays news articles using the new .NET MAUI Shell.

Part 2:
Basic Projects

In this part, you will learn about .NET MAUI features such as Shell, CollectionView, Image, Button, Label, CarouselView, Grid, Custom Controls, and Gestures. You will explore using location services, calling custom web APIs, and designing your XAML for different form factors.

This part has the following chapters:

- *Chapter 4, Building a News App Using .NET MAUI Shell*
- *Chapter 5, Building a Matchmaking App with Rich UX Using Animations*
- *Chapter 6, Building a Photo Gallery App Using CollectionView and CarouselView*
- *Chapter 7, Building a Location Tracking App Using GPS and Maps*
- *Chapter 8, Building a Weather App for Multiple Form Factors*

4

Building a News App Using .NET MAUI Shell

In this chapter, we will create a news app that leverages the **Shell** navigation functionality provided to us by the .NET MAUI team at Microsoft. The old way of doing this, which involved using **ContentPage**, **FlyoutPage**, **TabbedPage**, or **NavigationPage** as the main page, as we did in *Chapter 2*, still works, but we are sure that you will enjoy the new way of defining the structure of your app. Also, you can mix and match the old and new.

By the end of this chapter, you will have learned how to define an app structure using Shell, consume data from a REST API, configure navigation, and pass data between views using query-style routes.

So, what is Shell, then? In Shell, you define the structure of your app using **Extensible Application Markup Language** (**XAML**) instead of hiding it in spread-out pieces of code in your app. You can also navigate using routes, just like those fancy web developers are doing.

The following topics will be covered in this chapter:

- Defining a Shell navigation page
- Creating a flyout
- Creating a navigation bar
- Navigating using routes and passing data in query strings
- Consuming data from a public **representational state transfer** (**REST**) **application programming interface** (**API**)
- Adding content in the form of a `CollectionView` control

Technical requirements

To be able to complete this project, you will need to have Visual Studio for Mac or Windows installed, as well as the necessary .NET MAUI workload components. See *Chapter 1, Introduction to .NET MAUI*, for more details on how to set up your environment.

You can find the source code for this chapter at `https://github.com/PacktPublishing/MAUI-Projects-3rd-Edition`.

Project overview

We will create a .NET MAUI project using the **single-project** feature as the code-sharing strategy. It will contain two parts, detailed as follows:

- In the first part, we will create views and make them navigable using Shell
- In the second part, we will add some content by consuming a **REST API** for news

The second part is not needed to learn about Shell, but it will take you a bit further down the road to a complete app.

The build time for this project is about 1.5 hours.

Building the news app

This chapter will be all about building a news app from the beginning. It will guide you through every step, but it will not go into every detail. For that, we recommend *Chapter 2, Building Our First .NET MAUI App*, which goes into more detail.

Happy coding!

Setting up the project

This project, like all the rest, is a **File | New | Project...**-style project. This means that we will not be importing any code at all. So, this first section is all about creating the project and setting up the basic project structure.

Creating the new project

The first step is to create a new .NET MAUI project:

1. Open Visual Studio 2022 and select **Create a new project**:

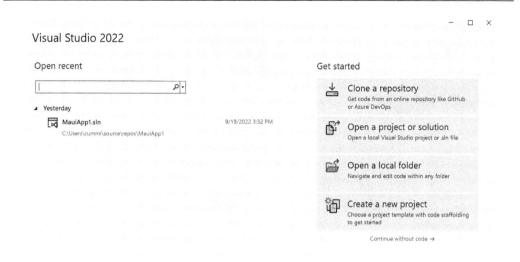

Figure 4.1 – Visual Studio 2022

This will open the **Create a new project** wizard.

2.   In the search field, type in `maui` and select the **.NET MAUI App** item from the list:

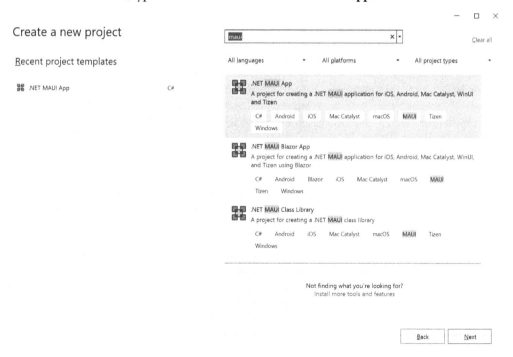

Figure 4.2 – Create a new project

3. Click **Next**.

4. Enter News as the name of the app, as shown in the following screenshot:

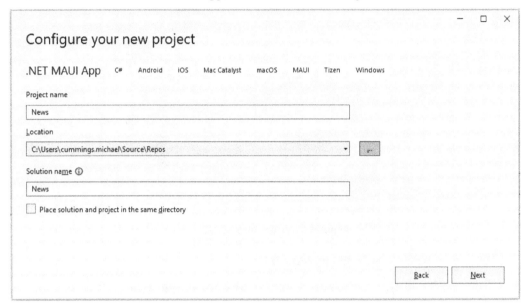

Figure 4.3 – Configure your new project

5. Click **Next**.

6. The last step will prompt you for the version of .NET Core to support. At the time of writing, .NET 6 is available as **Long-Term Support** (**LTS**), and .NET 7 is available as **Standard Term Support**. For this book, we will assume that you will be using .NET 7:

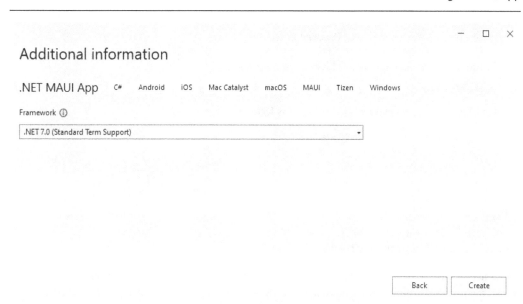

Figure 4.4 – Additional information

7.   Finalize the setup by clicking **Create** and wait for Visual Studio to create the project.

That's it for project creation.

Let's continue by setting up the structure of the app.

Creating the structure of the app

In this section, we will start to build the **Views** and **ViewModels** of the app. The *Using MVVM – creating views and ViewModels* section in *Chapter 2* contains more details on **Model-View-ViewModel** (**MVVM**) as a design pattern. It's recommended that you read that first if you don't know what MVVM is.

Creating the ViewModel base class

ViewModel is the mediator between View and Model. Let's create a base class for ViewModels with common functionality that we can reuse. In practice, ViewModel must implement an interface called INotifyPropertyChanged for MVVM to function. We will do so in the base class and will also add a little handy helper tool called **CommunityToolkit.Mvvm** that will save us a lot of time. Again, please check out *Chapter 2, Building Our First .NET MAUI App*, if you are feeling unsure about MVVM.

The first step is to create a base class. Proceed as follows:

1. In the `News` project, create a folder called `ViewModels`.

2. In the `ViewModels` folder, create a class called `ViewModel`.

3. Change the existing class so that it looks as follows:

    ```
    namespace News.ViewModels;
    public abstract class ViewModel
    {
    }
    ```

Excellent! Let's implement `INotifyPropertyChanged` in the base `ViewModel` class.

A quick recap of CommunityToolkit.Mvvm

CommunityToolkit.Mvvm is a NuGet package that contains a few source generators that can automatically generate the necessary implementation details for `INotifyPropertyChanged`. More specifically, it will inject a call that will raise the `PropertyChanged` event whenever a setter is called. It also takes care of property dependencies; if I change the `FirstName` property, the `FullName` read-only property will also get a `PropertyChanged` event. Before CommunityToolkit.Mvvm, you would have had to write this code manually.

It's all explained in more detail in *Chapter 2, Building Our First .NET MAUI App*. Have you read it yet?

Adding a reference to CommunityToolkit.Mvvm

CommunityToolkit.Mvvm and its dependencies are installed using NuGet. So, let's install the NuGet package:

1. In the `News` project, install the CommunityToolkit.Mvvm NuGet package, version 8.0.0.

2. Accept any license dialog boxes.

This will install the relevant NuGet packages.

Implementing INotifyPropertyChanged

`ViewModel` sits between `View` and `Model`. When a change in `ViewModel` occurs, `View` must be notified. The mechanism for this is the `INotifyPropertyChanged` interface, which defines an event that the controls in `View` subscribe to. The `ObservableObject` attribute is the magic that will generate the `INotifyPropertyChanged` implementation for us. Follow these steps:

1. In the `News` project, open up `ViewModels.cs`.

2. Add the following code in bold:

    ```
    using CommunityToolkit.Mvvm.ComponentModel;
    ```

```
[ObservableObject]
public abstract partial class ViewModel
{
}
```

This instructs CommunityToolkit.Mvvm to implement the `INotifyPropertyChanged` interface. The next step is all about reducing the number of lines of code that we will have to write. Normally, you would have to manually raise the `PropertyChanged` event from your code, but thanks to source generators, which write code at build time, we simply have to create normal properties and let CommunityToolkit.Mvvm do the magic.

Let's move on and create our first `ViewModel`.

Creating the HeadlinesViewModel class

We will now start to create some `View` and `ViewModel` placeholders that we will expand on during this chapter. We will not directly implement all graphical features; instead, we'll keep it simple and think of all these pages as placeholders for what's to come next.

The first one is the `HeadlinesViewModel` class, which will serve as the `ViewModel` for `HeadlinesView`. Proceed as follows:

1. In the `News` project, under the `ViewModels` folder, create a new class called `Head linesViewModel`.

2. Edit the class so that it inherits from the `ViewModel` base class, as shown in bold in the following code snippet:

    ```
    namespace News.ViewModels;
    public class HeadlinesViewModel : ViewModel
    {
        public HeadlinesViewModel()
        {
        }
    }
    ```

OK – not bad. It doesn't do much yet, but we'll just leave it for now. Let's create the matching view.

Creating HeadlinesView

This view will eventually show a list of news, but for now, it will be kept simple. Follow these steps to create the page:

1. In the `News` project, create a folder named `Views`.

2. Right-click on the `Views` folder, select **Add**, and then click **New Item...**.

If you are using Visual Studio 17.7 or later, click the **Show all Templates** button in the dialog that pops up. Otherwise, move on to the next step.

3. Under the **C# Items** node on the left, select **.NET MAUI**.

4. Select **.NET MAUI ContentPage (XAML)** and name it `HeadlinesView`.

5. Click **Add** to create the page.

Refer to the following screenshot to view the preceding information:

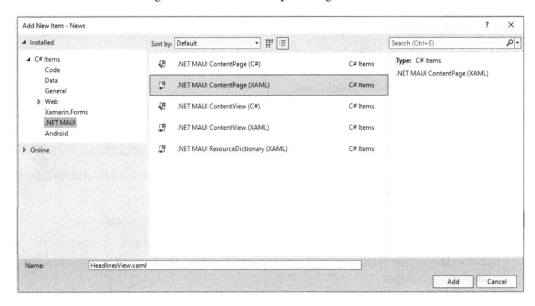

Figure 4.5 – Add New Item

Let's add some placeholder code to `HeadlinesView`, just to have something to navigate to and from. We will replace it with something hotter later on in this chapter, but to keep things simple, let's add a label. To do so, proceed as follows:

1. In the `News` project, under the `Views` folder, open `HeadlinesView.xaml`.

2. Edit the XAML code by adding the following code marked in bold:

```
<?xml version="1.0" encoding="utf-8" ?>
<ContentPage xmlns="http://schemas.microsoft.com/
dotnet/2021/maui"
             xmlns:x="http://schemas.microsoft.com/
winfx/2009/xaml"
    x:Class="News.Views.HeadlinesView"
```

```
        Title="Home">
        <VerticalStackLayout>
            <Label
                Text="HeadLinesView!"
                VerticalOptions="Center"
                HorizontalOptions="Center" />
        </VerticalStackLayout>
    </ContentPage>
```

This will set the title of the page and add a label with the text `HeadlinesView` centered in the middle of the page. Let's move on and create some additional view placeholders.

Creating ArticleItem

The app will eventually display a list of articles where each article will be rendered using a reusable component. We will call this reusable component `ArticleItem`. In .NET MAUI, a reusable component is called a **ContentView**. Please don't confuse this with an MVVM View, which is represented by a page in .NET MAUI. We know that this is confusing, but the rule is that a .NET MAUI page is an MVVM View and a .NET MAUI ContentView is essentially a reusable control.

That said, let's create the `ArticleItem` class, as follows:

1. In the `News` project, right-click the `Views` folder, select **Add**, and then click **New Item....**

    If you are using Visual Studio 17.7 or later, click the **Show all Templates** button in the dialog that pops up. Otherwise, move on to the next step.

2. Under the **C# Items** node on the left, select **.NET MAUI**.

    Important: Make sure that you select the **ContentView** template in the next step and *not* the **ContentPage** template.

3. Select **.NET MAUI ContentView (XAML)** (important, remember?) and name it `ArticleItem`.

4. Click **Add** to create the view.

Refer to the following screenshot to view the preceding information:

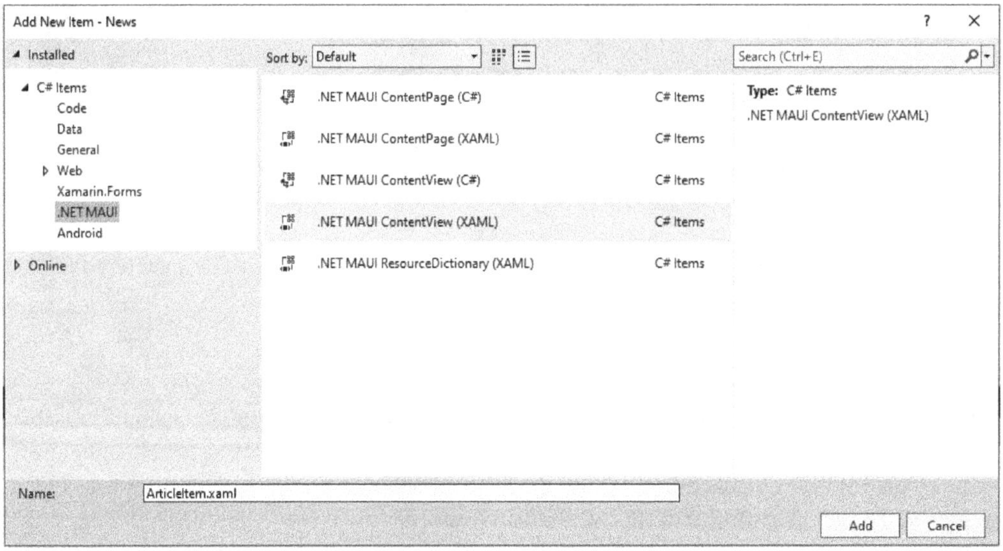

Figure 4.6 – Add New Item – ArticleItem.xaml

We don't need to alter the XAML code that's been generated at this point, so we'll simply leave it as is.

Creating ArticleView

In the previous section, we created the `ArticleItem` content view. This view (`ArticleView`) will contain `WebView` to display each article. But for the moment, let's just add `ArticleView` as a placeholder. Follow these steps to do so:

1. In the `News` project, right-click the `Views` folder, select **Add**, and then click **New Item....**

    If you are using Visual Studio 17.7 or later, click the **Show all Templates** button in the dialog that pops up. Otherwise, move on to the next step.

2. Select **.NET MAUI** under the **C# Items** node on the left.

3. Select **.NET MAUI ContentPage (XAML)** and name it `ArticleView`.

4. Click **Add** to create the page.

Since this view is also a placeholder view at the moment, we'll just add a label to indicate the type of page. Edit the content by following these steps:

1. In the `News` project, open `ArticleView.xaml`.

2. Edit the XAML code by adding the following code marked in bold:

```xml
<?xml version="1.0" encoding="utf-8" ?>
<ContentPage xmlns="http://schemas.microsoft.com/dotnet/2021/maui"
             xmlns:x="http://schemas.microsoft.com/winfx/2009/xaml"
    x:Class="News.Views.ArticleView"
    Title="ArticleView">
    <VerticalStackLayout>
        <Label
            Text="ArticleView!"
            VerticalOptions="Center"
            HorizontalOptions="Center" />
    </VerticalStackLayout>
</ContentPage>
```

Alright – just one more view to mock before we can start to wire things up.

Creating AboutView

The last view will be created in the same way as all the others. Proceed as follows:

1. In the News project, right-click the Views folder, select **Add**, and then click **New Item...**.

2. Under the **C# Items** node on the left, select **.NET MAUI**.

3. Select **.NET MAUI ContentPage (XAML)** and name it AboutView.

4. Click **Add** to create the page.

This view is the only view that will stay a placeholder view. It's up to you to do something with it if you choose to build something cool out of this project later. So, we will only add a label that states that this is AboutView:

1. In the News project, open AboutView.xaml.

2. Edit the XAML code by adding the following code marked in bold:

```xml
<?xml version="1.0" encoding="utf-8" ?>
<ContentPage xmlns="http://schemas.microsoft.com/dotnet/2021/maui"
             xmlns:x="http://schemas.microsoft.com/winfx/2009/xaml"
    x:Class="News.Views.AboutView"
    Title="AboutView">
    <VerticalStackLayout>
        <Label
            Text="AboutView!"
```

```
                    VerticalOptions="Center"
                    HorizontalOptions="Center" />
        </VerticalStackLayout>
    </ContentPage>
```

With that, we have all the views we need to start wiring the app up. The first step is to configure dependency injection.

Wiring up dependency injection

By using dependency injection as a pattern, we can keep our code cleaner and more testable. This app will use constructor injection, which means that all dependencies that a class has must be passed through its constructor. The container then constructs objects for you, so you don't have to care too much about the dependency chain. Since .NET MAUI already includes a dependency injection framework, called Microsoft.Extensions.DependencyInjection, there is nothing extra to install.

> **Confused about dependency injection?**
>
> Check out the *Wiring up a dependency injection* section in *Chapter 2, Building Our First .NET MAUI App*, for more details on dependency injection.

Registering Views and ViewModels with dependency injection

When registering classes with the container, it is recommended to use extension methods to group the types. The extension methods will take a single parameter and return a single value, the MauiAppBuilder instance. This is how the **Builder** pattern is implemented and it allows us to chain the methods on the builder defined in the CreateMauiApp method. For this app, we have Views and ViewModels to register for now. Let's create the method:

1. In the News project, open the MauiProgram.cs file.

2. Make the following changes to the MauiProgram class; the changes are highlighted in the code:

```
public static class MauiProgram
{
    public static MauiApp CreateMauiApp()
    {
        var builder = MauiApp.CreateBuilder();
        builder
            .UseMauiApp<App>()
            .ConfigureFonts(fonts =>
            {
                fonts.AddFont("OpenSans-Regular.ttf",
"OpenSansRegular");
                fonts.AddFont("OpenSans-Semibold.ttf",
```

```
    "OpenSansSemibold");
            })
        .RegisterAppTypes();
        return builder.Build();
    }
    public static MauiAppBuilder RegisterAppTypes(this
MauiAppBuilder mauiAppBuilder)
    {
        // ViewModels
        mauiAppBuilder.Services.AddTransient<ViewModels.
        HeadlinesViewModel>();

        // Views
        mauiAppBuilder.Services.AddTransient<Views.AboutView>();
        mauiAppBuilder.Services.AddTransient<Views.
ArticleView>();
        mauiAppBuilder.Services.AddTransient<Views.
HeadlinesView>();
        return mauiAppBuilder;
    }
}
```

The .NET MAUI `MauiAppBuilder` class exposes the `Services` property, which is the dependency injection container. We simply need to add the types we want dependency injection to know about; the container will do the rest for us. Think of a builder as something that collects a lot of information on what needs to be done, and then finally builds the object we need. It's a very useful pattern on its own, by the way.

We're only using the builder for one thing at the moment. Later on, we will use it to register any class in the assembly that inherits from our abstract `ViewModel` class. The container is now prepared for us to ask for these types.

Now, we need to make some graphical touches to our app. We will rely on **Font Awesome** to do the magic.

Downloading and configuring Font Awesome

Font Awesome is a free collection of images packaged into a font. .NET MAUI has excellent support for using Font Awesome in toolbars, navigation bars, and all over the place. It's not strictly needed to make this app, but we think that it's worth the extra round trip since you are most likely going to need something like this in your new killer app.

The first step is to download the font.

Downloading Font Awesome

Downloading the font is straightforward. Please note the renaming of the file – it's not needed but it's easier to edit configuration files and such if they have simpler names. Follow these steps to acquire and copy the font to each project:

1. Browse to `https://fontawesome.com/download`.

2. Click the **Free for Desktop** button to download Font Awesome.

3. Unzip the downloaded file, then locate the `otfs` folder.

4. Rename the `Font Awesome 5 Free-Solid-900.otf` file to `FontAwesome.otf` (you can keep the original name, but it's just less to type if you rename it). Your filename may be different since Font Awesome is continually updating but it should be similar.

5. Copy `FontAwesome.otf` to the `Resources/Fonts` folder in the `News` project.

Alright – now, we need to register Font Awesome with .NET MAUI.

Configuring .NET MAUI to use Font Awesome

It would be nice if all we needed to do was to copy the font file into the project folders. A lot does happen with just that action. The default .NET MAUI template includes all the fonts in the `Resources/Fonts` folder with the following item definition in the `News.csproj` file:

```
<!-- Custom Fonts -->
<MauiFont Include="Resources\Fonts\*" />
```

This ensures that the font files are processed and included in the app package automatically. What is left is to register the font with the .NET MAUI runtime so that it is available to our XAML resources. To do that, add the following highlighted line to the `MauiProgram.cs` file:

```
.ConfigureFonts(fonts =>
{
    fonts.AddFont("FontAwesome.otf", "FontAwesome");
    fonts.AddFont("OpenSans-Regular.ttf", "OpenSansRegular");
    fonts.AddFont("OpenSans-Semibold.ttf", "OpenSansSemibold");
})
```

This line adds an alias that we can use in the next section to create static resources. The first parameter is the filename for the font file, while the second is the alias for the font that we can use in the `FontFamily` attribute.

All that is left is to define some icons in a resource dictionary.

Defining some icons in the resource dictionary

Now that we have defined the font, we will put that to use and define five icons to use in our app. We'll add the XAML first; then, we'll examine one of the `FontImage` tags.

Proceed as follows:

1. In the `News` project, open `App.xaml`.

2. Add the following code marked in bold under the existing `ResourceDictionary.MergedDictionaries` tag:

```
<ResourceDictionary>
  <ResourceDictionary.MergedDictionaries>
    <ResourceDictionary Source="Resources/Styles/Colors.xaml" />
    <ResourceDictionary Source="Resources/Styles/Styles.xaml" />
  </ResourceDictionary.MergedDictionaries>
  <FontImage x:Key="HomeIcon" FontFamily="FontAwesome"
Glyph="&#xf015;" Size="22" Color="Black" />
  <FontImage x:Key="HeadlinesIcon" FontFamily="FontAwesome"
Glyph="&#xf70e;" Size="22" />
  <FontImage x:Key="NewsIcon" FontFamily=" FontAwesome"
Glyph="&#xf1ea;" Size="22" />
  <FontImage x:Key="SettingsIcon" FontFamily="FontAwesome"
Glyph="&#xf013;" Size="22" Color="Black" />
  <FontImage x:Key="AboutIcon" FontFamily="FontAwesome"
Glyph="&#xf05a;" Size="22" Color="Black" />
</ResourceDictionary>
```

`FontImage` is a class that can be used anywhere in .NET MAUI that expects an `ImageSource` object. It's designed to render one character (or glyph) into an image. The `FontImage` class needs a few attributes to work, detailed as follows:

*   A key for reference that will be used by other views in the application.

*   A `FontFamily` resource that uses the alias reference back to the `Font` resource that we defined in the preceding section.

*   The `Glyph` object, which represents the image to be shown. To find which image these cryptic values refer to, check out `fontawesome.com`, click on **Icons**, select **free and open-source icons**, and start browsing.

*   `Size` and `Color`. These are not strictly needed but they're nice to define. They are used for a few of the icons in this app so that they render properly for light themes.

Font Awesome is now installed and configured. We've done a lot of work to get to the actual topic of this chapter. It's time to define the shell!

Defining the shell

As stated earlier, .NET MAUI Shell is the newest way of defining the structure of your app. The different projects in this book use alternate ways of defining the overall structure of the app, but .NET MAUI Shell is, in our opinion, the best way of defining the UI structure. We hope you find it as exciting as we do!

Defining the basic structure

We're going to start by defining a basic structure for the app without really adding any of our defined views to it. After that, we'll add the actual views one by one. But let's start by adding some content and creating `ContentPage` objects using XAML directly. Follow these two steps to do so:

1. In the News project, open the `AppShell.xaml` file.

2. Alter the file so that it looks like the following code:

```xml
<?xml version="1.0" encoding="UTF-8"?>
<Shell xmlns="http://schemas.microsoft.com/dotnet/2021/maui"
       xmlns:x="http://schemas.microsoft.com/winfx/2009/xaml"
    xmlns:local="clr-namespace:News"
    x:Class="News.AppShell">
    <FlyoutItem Title="Home" Icon="{StaticResource HomeIcon}">
      <ShellContent Title="Headlines" Icon="{StaticResource
HeadlinesIcon}" >
        <ContentPage Title="Headlines" />
      </ShellContent>
      <Tab Title="News" Icon="{StaticResource NewsIcon}">
         <ContentPage Title="Local" />
         <ContentPage Title="Global" />
      </Tab>
    </FlyoutItem>


    <FlyoutItem Title="Settings" Icon="{StaticResource
SettingsIcon}">
        <ContentPage Title="Settings" />
    </FlyoutItem>


    <ShellContent Title="About" Icon="{StaticResource
AboutIcon}">
        <ContentPage Title="About"/>
    </ShellContent>
</Shell>
```

Let's break this down. First, by default, the .NET MAUI Shell template disables the flyout menu. Since we want to use it in this app, you must remove the line that disables it. The direct children of Shell itself are two `FlyoutItem` objects and one `ShellContent` object. All three of these have `Title` and `Icon` properties defined, as shown in the following screenshot. The icons are referenced to the Font Awesome resources we created earlier. This will render a flyout, as shown in the following screenshot:

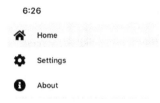

Figure 4.7 – The app flyout

The flyout is accessible by swiping in from the left. Flyout objects can have multiple children, while the ShellContent element can only have a single child.

The **Home** flyout item is the most complex example. It has two children; ShellContent with a page with the title **Headlines** and a tab that defines two child pages on its own. The first level of children will render the tab bar at the bottom of the app, as shown in the following screenshot. The second level of children, under the **Tab** element with the **News** title, will be rendered as a tab bar directly under the title of the navigation bar at the top:

Figure 4.8 – The Shell tabs and pages

The **Settings** and **About** flyouts will simply render the pages that they define.

Making the app run

It's time to try the app out and see whether it looks like the screenshots shown throughout this chapter. The app should now run. If not, stay calm and simply go through the code again. Once you are done navigating around the app, we can create a news service to fetch news from, and extend all those views that we've created.

Creating the news service

To find cool content, we'll consume an already existing News API provided by `newsapi.org`. To do so, we must register for an API key that we can use to request the news. If you aren't comfortable with doing so, you can mock the news service if you would like, instead of using the API.

The first thing we must do is obtain an API key.

Obtaining an API key

The process of registering is pretty straightforward. Be aware, however, that the UI of `newsapi.org` might have changed by the time you read this.

OK – let's get that key:

1. Browse to `https://newsapi.org/`.

2. Click **Get API key**.

3. Fill out the form, as illustrated in the following screenshot:

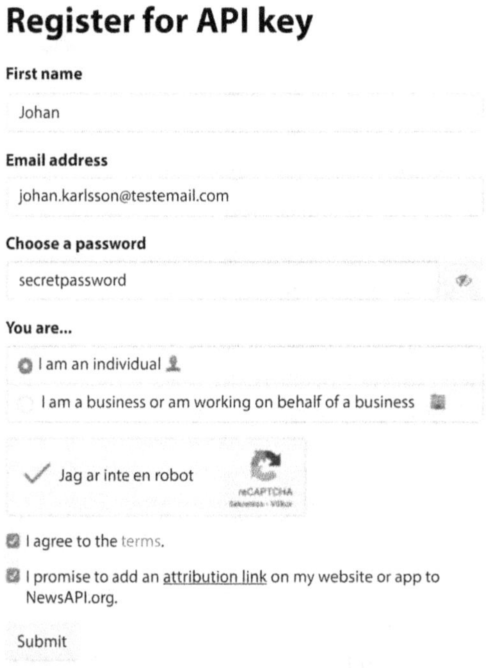

Register for API key

First name

Johan

Email address

johan.karlsson@testemail.com

Choose a password

secretpassword

You are...

I am an individual 👤

I am a business or am working on behalf of a business 🏢

✓ Jag ar inte en robot

reCAPTCHA

☑ I agree to the terms.

☑ I promise to add an attribution link on my website or app to NewsAPI.org.

Submit

Figure 4.9 – Register for API key

4. Copy the API key provided on the next page, as illustrated in the following screenshot:

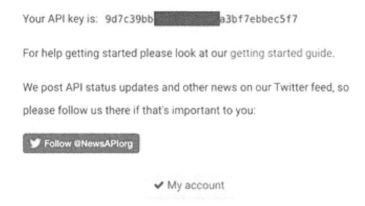

Registration complete

Your API key is: 9d7c39bb ▮▮▮▮▮▮▮ a3bf7ebbec5f7

For help getting started please look at our getting started guide.

We post API status updates and other news on our Twitter feed, so please follow us there if that's important to you:

Follow @NewsAPIorg

✔ My account

Figure 4.10 – Registration complete

Now, we need a place to store the key for easy access. We'll create a static class that will hold the key for us. Proceed as follows:

1. In the News project, create a new class called Settings in the root folder.

2. Add the code shown in the following snippet, replacing the placeholder text with the API key you obtained in the preceding steps:

```
namespace News;
internal static class Settings
{
    public static string NewsApiKey => "<Your APIKEY Here>";
}
```

The important thing here is that you copy and paste the key into the file. Now, we need models.

> **A note on tokens and other secrets**
>
> This is not a recommended way to store an API key or any other token that should be secured in your app. To securely store tokens and other data, you should use **secure storage** (see https://learn.microsoft.com/en-us/dotnet/maui/platform-integration/storage/secure-storage) and obtain the data from a secure server, preferably gated by some form of user authentication. You could also require the user to provide the API key through a settings page – hint, hint.

Creating the models

The data returned from the API needs to go somewhere, and the most convenient way of accessing it would be to deserialize the data into **Plain Old CLR Objects** (**POCO**, also known as regular C# classes). We usually call these POCO objects models, and they usually like to live in a folder called Models. Let's create our models:

1. In the News project, create a new folder called Models.

2. In the Models folder, add a new class called NewsApiModels.

3. Add the following code to the class:

```csharp
namespace News.Models;
using System;
using System.Collections.Generic;
using System.Text.Json.Serialization;

public class Source
{
    [JsonPropertyName("id")]
    public string Id { get; set; }
    [JsonPropertyName("name")]
    public string Name { get; set; }
}
public class Article
{
    [JsonPropertyName("source")]
    public Source Source { get; set; }
    [JsonPropertyName("author")]
    public string Author { get; set; }
    [JsonPropertyName("title")]
    public string Title { get; set; }
    [JsonPropertyName("description")]
```

```
        public string Description { get; set; }
        [JsonPropertyName("url")]
        public string Url { get; set; }

        [JsonPropertyName("urlToImage")]
        public string UrlToImage { get; set; }
        [JsonPropertyName("publishedAt")]
        public DateTime PublishedAt { get; set; }
        [JsonPropertyName("content")]
        public string Content { get; set; }
    }

    public class NewsResult
    {
        [JsonPropertyName("status")]
        public string Status { get; set; }
        [JsonPropertyName("totalResults")]
        public int TotalResults { get; set; }
        [JsonPropertyName("articles")]
        public List<Article> Articles { get; set; }
    }
```

The JsonPropertyName attributes on each property allow the System.Text.Json deserializer to map the name from the JSON received from the web API into the C# object. When we call the API, the API will return a NewsResult object that will, in turn, contain a list of articles. The next step is to create a service that will wrap the API and allow us to access the latest news.

Tip for creating POCO classes

If you ever need to create a class model out of a bunch of **JavaScript Object Notation (JSON)**, you can use the **Paste JSON as Classes** tool in Visual Studio for Windows (**Edit | Paste Special | Paste JSON as Classes**).

Creating a service class

The service class will wrap the API so that we can access it in a nice .NET-ish kind of way.

But we'll start by defining an enum that will define the scope of the news that we request.

Creating the NewsScope enum

The NewsScope enum defines the different kinds of news that our service supports. Let's add it by following these few steps:

1. In the News project, create a new folder called Services.

2. In the Services folder, add a new file called NewsScope.cs.

3. Add the following code to that file:

```
namespace News.Services;

public enum NewsScope
{
    Headlines,
    Local,
    Global
}
```

The next step is to create the NewsService class that will wrap the call to the News API.

Creating the NewsService class

The purpose of the NewsService class is to wrap the HTTP calls to the news REST API and make them easily accessible to our code in the form of regular .NET method calls. To make it easier to replace the source of the news – for example, to use a mock in tests – we will use an interface.

To create the INewsService interface, follow these steps:

1. In the Services folder, create a new interface called INewsService.

2. Edit the interface so that it looks like this:

```
namespace News.Services;
using News.Models;
public interface INewsService
{
    public Task<NewsResult> GetNews(NewsScope scope);
}
```

Creating the NewsService class is now quite straightforward. Follow these steps:

1. In the Services folder, create a new class called NewsService.

2. Edit the class so that it looks like this:

```
namespace News.Services;
using News.Models;
```

```csharp
using System.Net.Http.Json;
public class NewsService : INewsService, IDisposable
{
    private bool disposedValue;
    const string UriBase = "https://newsapi.org/v2";
    readonly HttpClient httpClient = new() {
        BaseAddress = new(UriBase),
        DefaultRequestHeaders = { { "user-agent", "maui-
projects-news/1.0" } }
    };

    public async Task<NewsResult> GetNews(NewsScope scope)
    {
        NewsResult result;
        string url = GetUrl(scope);
        try
        {
            result = await httpClient.
GetFromJsonAsync<NewsResult>(url);
        }
        catch (Exception ex) {
            result = new() { Articles = new() { new() { Title =
$"HTTP Get failed: {ex.Message}", PublishedAt = DateTime.Now} }
};
        }
        return result;
    }

    private string GetUrl(NewsScope scope) => scope switch
    {
        NewsScope.Headlines => Headlines,
        NewsScope.Global => Global,
        NewsScope.Local => Local,
        _ => throw new Exception("Undefined scope")
    };

    private static string Headlines => $"{UriBase}/
top-headlines?country=us&apiKey={Settings.NewsApiKey}";

    private static string Local => $"{UriBase}/
everything?q=local&apiKey={Settings.NewsApiKey}";

    private static string Global => $"{UriBase}/
everything?q=global&apiKey={Settings.NewsApiKey}";

    protected virtual void Dispose(bool disposing)
```

```
        {
            if (!disposedValue)
            {
                if (disposing)
                {
                    httpClient.Dispose();
                }
                disposedValue = true;
            }
        }

        public void Dispose()
        {
            // Do not change this code. Put cleanup code in
    'Dispose(bool disposing)' method
            Dispose(disposing: true);
            GC.SuppressFinalize(this);
        }
    }
```

The NewsService class is composed of five methods; I know there are technically eight, but we will get to that in a moment.

The first method, GetNews, is the method that we will eventually call from our app. It takes one parameter, scope, which is the enum that we created earlier. Depending on the value of this parameter, we will get different types of news. The first thing this method does is resolve the URL to call, and it does so by calling the GetUrl method with the scope.

The GetUrl method uses a switch expression to resolve the URL and, depending on the value of the scope parameter passed, returns one of three URLs. The URL points to the REST API of the News API with some predefined query parameters and the API key that we registered for.

When we've resolved the correct URL, we are ready to make the HTTP request and download the news in the form of JSON. The built-in HttpClient class in .NET does a fine job of fetching the JSON for us. All that is left after acquiring the data is to deserialize it into the news models that we defined earlier.

Let's say a word or two about the remainder of the methods and the HttpClient class. HttpClient is now the recommended class to use when requesting data from the web. It is a much safer implementation than was previously available. It has been shipping with .NET 5+ and is available as a separate NuGet package for previous versions. With that, there are a few peculiarities when using HttpClient.

First, HttpClient can hold onto native resources, so it must be properly disposed of. To properly dispose of HttpClient, we need to derive from and implement IDisposable. This is the reason for the extra methods that are in the class, Dispose(bool) and Dispose(). All they do is make sure that the instance of HttpClient is disposed of properly.

Second, `HttpClient` will pool those native resources, so it is recommended that you reuse an instance of `HttpClient` as much as possible. This is the reason for creating the `HttpClient` instance in the `NewsService` constructor.

Final word – since the `GetFromJsonAsync` call can throw an exception and it is called from an `async` method, you must handle the exception; otherwise, it will be lost on the executing thread and your only indication that something is wrong will be the fact that you have no items. For this app, we are just going to create a `NewsResult` object that contains one `Article` that has the exception in it so that something is displayed. There are much better ways of handling errors, but this will do for this app.

The next step is to wire up the `NewsService` class.

Wiring up the NewsService class

We are now ready to wire up the `NewsService` class in our app and integrate it with a real news source. We will extend all the existing `ViewModels` that we have and define the UI elements to be able to render the news in the `Views`.

Extending the HeadlinesViewModel class

In MVVM, `ViewModel` is the place to handle the logic of the app. The model is the news data that we will get from our `NewsService` class. We will now extend the `HeadlinesViewModel` class so that it uses `NewsService` to fetch news:

1. In the News project, expand the `ViewModels` folder and open the `HeadlinesViewModel.cs` file.

2. Add the following code marked in bold and resolve the references:

```
namespace News.ViewModels;
using System.Threading.Tasks;
using System.Web;
using CommunityToolkit.Mvvm.ComponentModel;
using CommunityToolkit.Mvvm.Input;
using News.Models;
using News.Services;

public partial class HeadlinesViewModel : ViewModel
{
    private readonly INewsService newsService;
    [ObservableProperty]
    private NewsResult currentNews;
    public HeadlinesViewModel(INewsService newsService)
    {
        this.newsService = newsService;
```

```
        }
        public async Task Initialize(string scope) =>
            await Initialize(scope.ToLower() switch
            {
                "local" => NewsScope.Local,
                "global" => NewsScope.Global,
                "headlines" => NewsScope.Headlines,
                _ => NewsScope.Headlines
            });
        public async Task Initialize(NewsScope scope)
        {
            CurrentNews = await newsService.GetNews(scope);
        }


        [RelayCommand]
        public void ItemSelected(object selectedItem)
        {
            var selectedArticle = selectedItem as Article;
            var url = HttpUtility.UrlEncode(selectedArticle.Url);
            // Placeholder for more code later on
        }
    }
```

Since we are using (constructor) dependency injection, we need to inject our dependencies into the constructor. The only dependency this `ViewModel` has is `NewsService`, and we store it internally in the class as a field.

The `CurrentNews` property is defined to get something to bind the UI to.

Then, we have two `Initialize` methods – one that takes `scope` as an enum and one that takes `scope` as a string. The string overload of the `Initialize` method will be used in XAML. It simply translates the string into the enum representation of `scope` and then calls the other `Initialize` method, which, in turn, calls the `GetNews(...)` method on the news service.

The property at the end, `ItemSelected`, returns a .NET MAUI command that we will wire up to respond to when the user of the app selects an item. Half of the method is implemented from the start. The selected item will be passed into the method. Then, we encode the URL to the article since we will be passing it as a query parameter when we navigate within the app. We will get back to the navigation part in a while.

If you are curious about the `ObservableProperty` and `RelayCommand` attributes, refresh your memory by reviewing *Chapter 2, Building Our First .NET MAUI App*.

Now that we have the code to fetch the data, we'll move on to defining the UI to display it.

Extending *HeadlinesView*

`HeadlinesView` is a shared view that will be used in several places in the app. The purpose of this view is to display a list of articles and to allow for navigation from one article into a web browser that will display the entire article.

To extend `HeadlinesView` we must do two things – first, must we edit the XAML and define the UI; then, we need to add some code to initialize it. Proceed as follows:

1. In the `News` project, expand the `Views` folder and open the `HeadlinesView.xaml` file.
2. Edit the XAML, as shown in the following code block:

```xml
<?xml version="1.0" encoding="UTF-8"?>
<ContentPage xmlns="http://schemas.microsoft.com/dotnet/2021/
maui"
             xmlns:x="http://schemas.microsoft.com/winfx/2009/
xaml"
    xmlns:views="clr-namespace:News.Views"
    xmlns:viewModels="clr-namespace:News.ViewModels"
    xmlns:models="clr-namespace:News.Models"
    x:Name="headlinesview"
    x:Class="News.Views.HeadlinesView"
             x:DataType="viewModels:HeadlinesViewModel"
    Title="Home" Padding="14">
  <CollectionView ItemsSource="{Binding CurrentNews.Articles}">
    <CollectionView.EmptyView>
      <Label Text="Loading" />
    </CollectionView.EmptyView>
    <CollectionView.ItemTemplate>
      <DataTemplate x:DataType="models:Article">
        <ContentView>
          <ContentView.GestureRecognizers>
            <TapGestureRecognizer Command="{Binding
BindingContext.ItemSelectedCommand, Source={x:Reference
headlinesview}}" CommandParameter="{Binding .}" />
          </ContentView.GestureRecognizers>
          <views:ArticleItem />
        </ContentView>
      </DataTemplate>
    </CollectionView.ItemTemplate>
  </CollectionView>
</ContentPage>
```

`HeadlinesView` uses `CollectionView` to display the list of articles. The `ItemsSource` property is set to the `CurrentNews.Articles` property of `ViewModel`, which, after loading news, should contain a list of news. While the list is empty or loading, we display a loading label, defined within the `CollectionView.EmptyView` element. You could, of course, create any valid UI inside of that tag to create a cooler loading screen.

Each article in the `CurrentNews.Articles` list will be rendered using whatever is inside the `CollectionView.ItemTemplate` element, and it's what's inside of the `ContentView` element that will represent the actual item. The article will be rendered using an `ArticleItem` view, which is a custom control that we defined earlier. We will define this view after we are done with this view.

To enable navigation from the view, we need to detect when a user clicks on a specific article. We can do so by adding `TapGestureRecognizer` and binding it to the `ItemSelectedCommand` property of the root `ViewModel`. The `Source={x:Reference headlinesview}}` snippet is what references the current context back to the root of the page, not the current article in the list that we are iterating. If we didn't specify the source, the binding engine would try to bind the `ItemSelectedCommand` property to a property defined in the current article of the `CurrentNews.Articles` property.

That's all for the GUI part. Now, we need to alter the code-behind to enable initialization based on data that we will pass from the XAML itself. Follow these steps to make it happen:

1. In the News project, open the `HeadlinesView.xaml.cs` code-behind file.

2. Add the following code marked in bold to the file:

```
namespace News.Views

using System.Threading.Tasks;
using News.Services;
using News.ViewModels;

public partial class HeadlinesView : ContentPage
{
    readonly HeadlinesViewModel viewModel;
    public HeadlinesView(HeadlinesViewModel viewModel)
    {
        this.viewModel = viewModel;
        InitializeComponent();
        Task.Run(async () => await
Initialize(GetScopeFromRoute()));
    }

    private async Task Initialize(string scope)
    {
        BindingContext = viewModel;
        await viewModel.Initialize(scope);
    }
```

```
        private string GetScopeFromRoute()
        {
            var route = Shell.Current.CurrentState.Location
            .OriginalString.Split("/").LastOrDefault();
            return route;
        }
    }
```

Usually, we don't want to add code to the code-behind of a view directly, but we need to make an exception so that arguments can be passed from XAML to our `ViewModel`.

Based on the routing information that was used to create the view, we will initialize `ViewModel` differently. The `GetScopeFromRoute` method will parse the location information from `Shell` to determine which scope is used to query the news service. Then, we can call a private method that creates an instance of `HeadlinesViewModel` for us, sets it as the binding context of the view, and calls the `Initialize()` method on `ViewModel`, which makes the REST call to the News API. We will define the routes when we edit the shell file.

But first, we need to extend `ContentView` of `ArticleItem` so that it displays a single-row item in the news lists.

Extending ContentView of ArticleItem

`ContentView` of `ArticleItem` represents one item in a list of news, as shown in the following screenshot:

NFL Week 17 bold predictions: DROY candidates Sauce
Gardner, Tariq Woolen each record INT - NFL.com
December 30. 2022

Figure 4.11 – A sample news item

To create the layout shown in *Figure 4.11*, we will use a `Grid` control. Follow these steps to create the layout:

1. In the News project, expand the `Views` folder and open the `ArticleItem.xaml` file.

2. Edit the XAML code, as shown in the following code block:

```
<?xml version="1.0" encoding="UTF-8"?>
<ContentView xmlns="http://schemas.microsoft.com/dotnet/2021/
maui"
    xmlns:models="clr-namespace:News.Models"
            xmlns:x="http://schemas.microsoft.com/winfx/2009/
xaml"
    x:Class="News.Views.ArticleItem"
    x:DataType="models:Article">
    <Grid Margin="0">
```

```
        <Grid.RowDefinitions>
          <RowDefinition Height="10" />
          <RowDefinition Height="40" />
          <RowDefinition Height="15" />
          <RowDefinition Height="10" />
          <RowDefinition Height="1" />
        </Grid.RowDefinitions>
        <Grid.ColumnDefinitions>
          <ColumnDefinition Width="65" />
          <ColumnDefinition Width="*" />
        </Grid.ColumnDefinitions>

        <Label Grid.Row="1" Grid.Column="1" Text="{Binding Title}"
Padding="10,0" FontSize="Small" FontAttributes="Bold" />

        <Label Grid.Row="2" Grid.Column="1" Text="{Binding
PublishedAt, StringFormat='{0:MMMM d, yyyy}'}"
Padding="10,0,0,0" FontSize="Micro" />

        <Border Grid.Row="1" Grid.RowSpan="2"
StrokeShape="RoundRectangle 15,15,15,15" Padding="0"
Margin="0,0,0,0" BackgroundColor="#667788" >
          <Image Source="{Binding UrlToImage}" Aspect="AspectFill"
HeightRequest="55" HorizontalOptions="Center"
VerticalOptions="Center" />
        </Border>

        <BoxView Grid.Row="4" Grid.ColumnSpan="2"
BackgroundColor="LightGray" />
      </Grid>
    </ContentView>
```

The preceding XAML code defines a grid layout with two columns and five rows. The `Grid.Row` and `Grid.Column` attributes position child elements into the grid, while the `Grid.ColumnSpan` attribute allows for a control to span multiple columns.

A rounded image can be achieved using a `Border` element with `StrokeShape="RoundRectangle 15,15,15,15"` specified in conjunction with the `Image Aspect` attribute set to `AspectFill`.

Strings in labels can be formatted directly in the binding statement. Check out the `Text="{Binding PublishedAt, StringFormat='{0:MMMM d, yyyy}'}"` line of code, which formats a date into a specific string format.

Lastly, the gray divider line is `BoxView` at the end of the XAML code.

Now that we have created `NewsService` and fixed all its related views, it's time to make use of them.

Adding to dependency injection

Since `HeadlinesViewModel` depends on `INewsService`, we need to register it in our dependency injection container (please see the *Wiring up dependency injection* section in *Chapter 2* for more details on dependency injection in .NET MAUI). Follow these steps to do so:

1. In the News project, open the `MauiProgram.cs` file.

2. Locate the `RegisterAppTypes()` method and add the following line marked in bold:

    ```
    public static MauiAppBuilder RegisterAppTypes(this
    MauiAppBuilder mauiAppBuilder)
    {
        // Services
         mauiAppBuilder.Services.AddSingleton<Services.
    INewsService>((serviceProvider) => new Services.NewsService());

        // ViewModels
        mauiAppBuilder.Services.AddTransient<ViewModels.
    HeadlinesViewModel>();

        //Views
        mauiAppBuilder.Services.AddTransient<Views.AboutView>();
        mauiAppBuilder.Services.AddTransient<Views.ArticleView>();
        mauiAppBuilder.Services.AddTransient<Views.HeadlinesView>();

        return mauiAppBuilder;
    }
    ```

This will allow for the dependency injection of `NewsService`.

Adding a ContentTemplate attribute

So far, we only have placeholder code in our `AppShell` file. Let's replace this with actual content, as follows:

1. In the News project, open `AppShell.xaml`.

2. Locate the `FlyoutItem` element with the title set to Home.

3. Edit the XAML so that the `ShellContent` element becomes self-closing, add the following `ContentTemplate` attribute marked in bold, and replace the contents of the `Tab` element:

    ```
    <Shell
        x:Class="News.AppShell"
        xmlns="http://schemas.microsoft.com/dotnet/2021/maui"
        xmlns:x="http://schemas.microsoft.com/winfx/2009/xaml"
        xmlns:views="clr-namespace:News.Views"
    ```

```xml
            xmlns:local="clr-namespace:News">

        <FlyoutItem Title="Home" Icon="{StaticResource HomeIcon}">
            <ShellContent Title="Headlines"
Route="headlines" Icon="{StaticResource HeadlinesIcon}"
ContentTemplate="{DataTemplate views:HeadlinesView}" />
            <Tab Title="News" Route="news" Icon="{StaticResource
NewsIcon}">
                <ShellContent Title="Local" Route="local"
ContentTemplate="{DataTemplate views:HeadlinesView}" />
                <ShellContent Title="Global" Route="global"
ContentTemplate="{DataTemplate views:HeadlinesView}" />
            </Tab>
        </FlyoutItem>

        <FlyoutItem Title="Settings" Icon="{StaticResource
SettingsIcon}">
            <ContentPage Title="Settings" />
        </FlyoutItem>

        <ShellContent Title="About" Icon="{StaticResource
AboutIcon}">
            <ContentPage Title="About"/>
        </ShellContent>
    </Shell>
```

Two things are going on here. The first is that we specify the content of `ShellContent` using the `ContentTemplate` attribute. This means that we point to the type of view that we want to have created when the shell becomes visible. Usually, you want to defer the creation of a view until right before it is going to be displayed, and the `ContentTemplate` attribute gives you that. Notice that the `Route` attribute for that `FlyoutItem` is set to `headlines`.

The second thing is that we have the same thing going on below for the `Local` and `Global` news but instead using the `local` and `global` routes.

If you run the app at this point, you should end up with something that looks like the following screenshot:

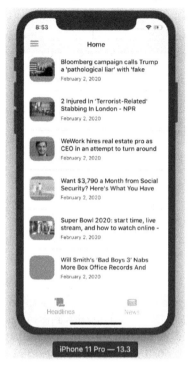

Figure 4.12 – The main list view

The last thing we need to implement is how we can view the articles when we tap on the item in the list.

Handling navigation

We are on the home stretch now for this app. The only thing we need to do is implement navigation to the article view, which will display the entire article in a web view. Since we are using Shell, we will be navigating using routes. Routes can be registered directly in the Shell markup – for example, in the `AppShell.xaml` file. We can do this by using `Route` attributes on the `ShellContent` elements, as we did in the previous section.

In the following code, we will add a route programmatically and register a view to handle it for us. We will also create a navigation service to abstract the concept of navigation a bit.

So, buckle up, and let's finish this app!

Creating the navigation service

The first step is to define an interface that will wrap .NET MAUI navigation. Why do we do this? Well, it's a good practice to separate the interfaces from the implementation; it makes unit testing easier, and so on.

Creating the INavigation interface

The `INavigation` interface is simple, and we will overshoot the target a little bit. We are only interested in the `NavigateTo` method, but we will add the `PushModal()` and `PopModal()` methods since it is likely that you will use them if you continue extending the app.

Adding the navigation interface is simple, as the following steps illustrate:

1. In the News project, expand the `ViewModels` folder and add a new file called `INavigate.cs`.

2. Add the following code to the file:

```
namespace News.ViewModels;

public interface INavigate
{
    Task NavigateTo(string route);
    Task PushModal(Page page);
    Task PopModal();
}
```

The `NavigateTo()` method declaration takes the route we want to navigate to. This is the method that we will be calling. The `PushModal()` method adds a new page on top of the navigation stack as a modal page, forcing the user only to interact with this specific page. The `PopModal()` method removes it from the navigation stack. So, if you use the `PushModal()` method, make sure that you give the user a way to pop it off the stack.

Otherwise, you will be stuck viewing the modal page forever.

That's all for the interface. Let's create an implementation using .NET MAUI Shell.

Implementing the INavigate interface using .NET MAUI Shell

The implementation is very straightforward since each method is only calling .NET MAUI static methods provided by the Shell API.

Create the `Navigator` class, as shown in the following steps:

1. In the News project, add a new class called `Navigator`.

2. Add the following code to the class:

```
namespace News;

using News.ViewModels;

public class Navigator : INavigate
{
    public async Task NavigateTo(string route) => await Shell.
Current.GoToAsync(route);
```

```
    public async Task PushModal(Page page) => await Shell.
Current.Navigation.PushModalAsync(page);

    public async Task PopModal() => await Shell.Current.
Navigation.PopModalAsync();
}
```

This is simply pass-through code that calls already existing methods. Now, we need to register the type with our dependency injection container so that it can be consumed by the ViewModel class.

Registering the Navigator class with dependency injection

For the ViewModel class and its descendants to have access to the Navigator instance, we will have to register it with the container, just like we did with NewService; just follow these steps:

1. In the News project, open the MauiProgram.cs file.

2. Find the RegisterAppTypes method and add the following highlighted code:

```
public static MauiAppBuilder RegisterAppTypes(this
MauiAppBuilder mauiAppBuilder)
{
    // Services
    mauiAppBuilder.Services.AddSingleton<Services.
INewsService>((serviceProvider) => new Services.NewsService());
    mauiAppBuilder.Services.AddSingleton<ViewModels.
INavigate>((serviceProvider) => new Navigator());

    // ViewModels
    ...
}
```

Now, we can add the INavigate interface to the ViewModel class and its descendants.

Adding the INavigate interface to the ViewModel class

To be able to access Navigator, we must extend the ViewModel base class, making it available to all ViewModels. Proceed as follows:

1. In the News project, open the ViewModels folder, and then open the ViewModel.cs file.

2. Add the following highlighted code to the class:

```
public abstract class ViewModel
{
    public INavigate Navigation { get; init; }

    internal ViewModel(INavigate navigation) => Navigation =
navigation;
}
```

3.   Open the `HeadlinesViewModel.cs` file and make the highlighted changes to the constructor:

```
public HeadlinesViewModel(INewsService newsService, INavigate
navigation) : base (navigation)
```

The base `ViewModel` now exposes the `Navigator` property through the `INavigate` interface. At this point, we are ready to wire up the navigation to our `Article` view.

Navigating using routes

Routes are a very handy way to navigate since they abstract the creation of the page away. All we need to know is the route to the view we want to navigate to – .NET MAUI Shell takes care of the rest for us. If you are familiar with how web navigation works, you might recognize the way that we pass arguments in routes. They are passed as query parameters.

Finalizing the ItemSelected command

Previously, we defined the `ItemSelected` method in the `HeadlinesViewModel` class. Now, it's time to add code that will perform the navigation to `ArticleView`:

1.   In the News project, expand the `ViewModels` folder and open `HeadlinesViewModel.cs`.

2.   Locate the `ItemSelected` method and add the following line marked in bold:

```
[RelayCommand]
public async Task ItemSelected(object selectedItem)
{
    var selectedArticle = selectedItem as Article;
    var url = HttpUtility.UrlEncode(selectedArticle.Url);
    await Navigation.NavigateTo($"articleview?url={url}");
}
```

Here, we define a route called `articleview` that takes a query-line parameter called `url` that points to the URL of the article itself. It will look something like this: `articleview?url=www.mypage.com`. Only the data passed after the `url=` argument must be **URL encoded**, meaning that we need to encode it to replace characters that would interfere with the route itself. The URL encoding is done through the `HttpUtility.UrlEncode()` method, which is defined in `System.Web` for us.

The preceding `NavigateTo()` method call uses this encoded data in the query parameter. On the recipient side of the navigation call, we need to handle the incoming `url` parameter.

Extending ArticleView so that it receives query data

`ArticleView` is responsible for rendering the article for us. To keep things simple (and also illustrate that you don't always need `ViewModel`), we will not define `ViewModel` for this class; instead, we will define `BindingContext` as an instance of the `UrlWebViewSource` class.

Add the following code to the `ArticleView.xaml.cs` file:

1. In the News project, expand the `Views` folder and open the `ArticleView.xaml.cs` file.
2. Add the following code marked in bold to the file:

```
namespace News.Views;

using System.Web;

[QueryProperty("Url", "url")]
public partial class ArticleView : ContentPage
{
    public string Url
    {
        set
        {
            BindingContext = new UrlWebViewSource
            {
                Url = HttpUtility.UrlDecode(value)
            };
        }
    }

    public ArticleView()
    {
        InitializeComponent();
    }
}
```

`ArticleView` is dependent on a URL being set, and we do so by defining a set-only property called `Url`. When this property is set, it creates a new instance of `UrlWebViewSource` with the value of the property and assigns it to the page's `BindingContext`. This setter is called by the Shell framework since we added an attribute called `QueryProperty` to the class itself. It takes two arguments – the first is which property to set, while the second is the name of the `url` query parameter.

Since the data comes URL encoded, we need to decode it using the `HttpUtility.UrlDecode()` method.

With that, we have a binding context that points to the web page we want to display. Now, we just need to define `WebView` in the XAML.

Extending ArticleView with WebView

This page only has one purpose, and that is to display the web page from the URL that we passed to it. Let's add a `WebView` control to the page, as follows:

1. In the `News` project, expand the `Views` folder and open `ArticleView.xaml`.

2. Add the following highlighted XAML:

```xml
<?xml version="1.0" encoding="utf-8" ?>
<ContentPage xmlns="http://schemas.microsoft.com/dotnet/2021/
maui"
             xmlns:x="http://schemas.microsoft.com/winfx/2009/
xaml"
    x:Class="News.Views.ArticleView"
    Title="ArticleView">
    <WebView Source="{Binding .}" />
</ContentPage>
```

The `WebView` control will take up all available space in the view. The source will be set to `.`, meaning that it will be the same as the `BindingContext` property of `ViewModel`. In this case, the `BindingContext` property is a `UrlWebViewSource` instance, which is exactly what `WebView` needs to navigate and display the content.

We've only got one step left – our app needs to know about the `ArticleView` route and what to do with it.

Registering the route

As mentioned previously, routes can be added declaratively in the XAML (`Route="MyDucks"`) or code, as shown here:

1. In the `News` project, open the `AppShell.xaml.cs` file.

2. Add the following lines of code marked in bold:

```csharp
namespace News;

public partial class AppShell : Shell
{
    public AppShell()
    {
        InitializeComponent();
        Routing.RegisterRoute("articleview", typeof(Views.
ArticleView));
    }
}
```

The `RegisterRoute()` method takes two arguments – the first is the route we want to use, and it's the one that we specify in the `NavigateTo()` calls. The second is the type of page (view) that we want to create – in our case, we want to create `ArticleView`.

Cool! That wraps it up. The app should now run, and you should be able to navigate to the article from any of your `CollectionViews`. Good work!

Summary

In this chapter, we learned how to define a navigation structure using .NET MAUI Shell, how to navigate to views using routes, and how to pass arguments between views in the form of a query string. There is a lot more to Shell, but this should get you started and confident enough to start exploring the Shell APIs. Also, keep in mind that the Shell APIs are constantly evolving, so make sure you check out the latest features available.

We also learned how to create an API client for an arbitrary REST API, which always comes in handy since most of the apps you will write need to communicate with a server at some point. There is a very good chance that the server will expose its data and functionality through a REST API.

If you are interested in extending the app even further, try designing your own **About** page, or allow the News API key to be set through the settings.

The next project will be about creating a match-making app, and how to create a swiping-enabled yes/no image selector app using nothing but .NET MAUI to render and animate cross-platform UI controls.

5

A Matchmaking App with a Rich UX Using Animations

In this chapter, we will create the base functionality for a matchmaking app. We won't be rating people, however, because of privacy issues. Instead, we will download images from a random source on the internet. This project is for anyone who wants an introduction to how to write reusable controls. We will also look at using animations to make our application feel nicer to use. This app will not be a **Model-View-ViewModel (MVVM)** application since we want to isolate the creation and usage of a control from the slight overhead of MVVM.

The following topics will be covered in this chapter:

- Creating a custom control
- Styling the app to look like a photo, with descriptive text beneath it
- Creating animations using .NET MAUI
- Subscribing to custom events
- Reusing the custom control over and over again
- Handling pan gestures

Technical requirements

To be able to complete this chapter's project, you will need to have Visual Studio for Mac or Windows installed, as well as the necessary .NET MAUI workloads. See *Chapter 1, Introduction to .NET MAUI*, for more details on how to set up your environment.

You can find the full source for the code in this chapter at `https://github.com/PackPublishing/MAUI-Projects-3rd-Edition`.

Project overview

Many of us have been there, faced with the conundrum of whether to swipe left or right. All of a sudden, you may find yourself wondering: *How does this work? How does the swipe magic happen?* Well, in this project, we're going to learn all about it. We will start by defining a `MainPage` file in which the images of our application will reside. After that, we will implement the image control, and gradually add the **graphical user interface** (**GUI**) and functionality to it until we have nailed the perfect swiping experience.

The build time for this project is about 90 minutes.

Creating the matchmaking app

In this project, we will learn more about creating reusable controls that can be added to an **Extensible Application Markup Language** (**XAML**) page. To keep things simple, we will not be using MVVM, but bare-metal .NET MAUI without any data binding. What we aim to create is an app that allows the user to swipe images, either to the right or the left, just as most popular matchmaking applications do.

Well, let's get started by creating the project!

Setting up the project

This project, like all the rest, is a **File** | **New** | **Project…**-style project. This means that we will not be importing any code at all. So, this first section is all about creating the project and setting up the basic project structure.

Let's get started!

Creating the new project

So, let's begin.

The first step is to create a new .NET MAUI project:

1. Open Visual Studio 2022 and select **Create a new project**:

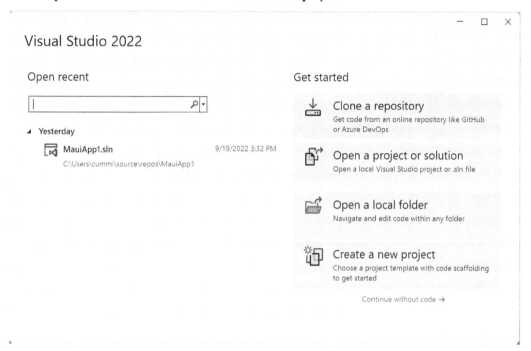

Figure 5.1 – Visual Studio 2022

This will open the **Create a new project** wizard.

2.   In the search field, type maui and select the **.NET MAUI App** item from the list:

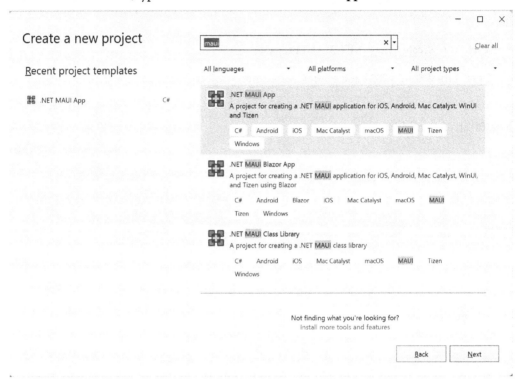

Figure 5.2 – Create a new project

3.   Click **Next**.

4.   Complete the next step of the wizard by naming your project. We will be calling our application Swiper in this case. Move on to the next dialog box by clicking **Create**, as illustrated in the following screenshot:

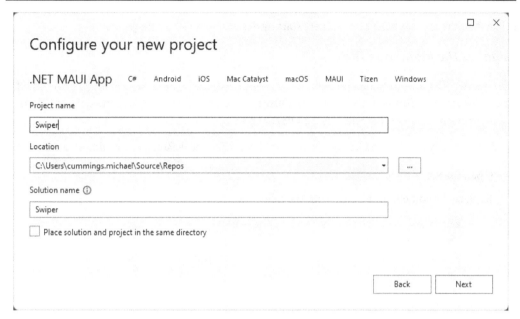

Figure 5.3 – Configure your new project

5. Click **Next**.

6. The last step will prompt you for the version of .NET Core to support. At the time of writing, .NET 6 is available as **Long-Term Support (LTS)**, and .NET 7 is available as **Standard Term Support**. For this book, we will assume that you will be using .NET 7:

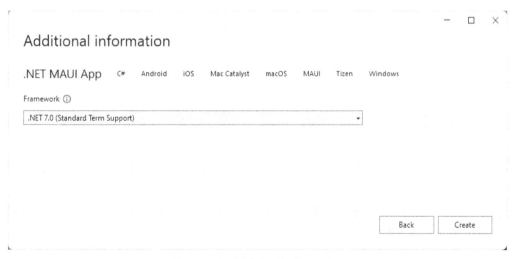

Figure 5.4 – Additional information

7. Finalize the setup by clicking **Create** and wait for Visual Studio to create the project.

Just like that, the app has been created. Let's start by designing the `MainPage` file.

Designing the MainPage file

A brand new .NET MAUI Shell app named `Swiper` has been created, with a single page called `MainPage.xaml`. This is in the root of the project. We will need to replace the default XAML template with a new layout that will contain our `Swiper` control.

Let's edit the already existing `MainPage.xaml` file by replacing the default content with what we need:

1. Open the `MainPage.xaml` file.

2. Replace the content of the page with the following highlighted XAML code:

```xml
<?xml version="1.0" encoding="utf-8"?>
<ContentPage
    xmlns=http://schemas.microsoft.com/dotnet/2021/maui
    xmlns:x="http://schemas.microsoft.com/winfx/2009/xaml"
    xmlns:local="clr-namespace:Swiper"
    x:Class="Swiper.MainPage">
    <Grid Padding="0,40" x:Name="MainGrid">
      <Grid.RowDefinitions>
        <RowDefinition Height="400" />
        <RowDefinition Height="*" />
      </Grid.RowDefinitions>
      <Grid Grid.Row="1" Padding="30">
          <!-- Placeholder for later -->
      </Grid>
    </Grid>
</ContentPage>
```

The XAML code within the `ContentPage` node defines two grids in the application. A grid is simply a container for other controls. It positions those controls based on rows and columns. The outer grid, in this case, defines two rows that will cover the entire available area of the screen. The first row is 400 units high and the second row, with `Height="*"`, uses the rest of the available space.

The inner grid, which is defined within the first grid, is assigned to the second row with the `Grid.Row="1"` attribute. The row and column indexes are zero-based, so `"1"` actually refers to the second row. We will add some content to this grid later in this chapter, but we'll leave it empty for now.

Both grids define their padding. You could enter a single number, meaning that all sides will have the same padding, or – as in this case – enter two numbers. We have entered `0,40`, which means that the left- and right-hand sides should have 0 units of padding and the top and bottom should have 40 units of padding. There is also a third option, with four digits, which sets the padding of the *left-hand* side, the *top*, the *right-hand* side, and the *bottom*, in that specific order.

The last thing to notice is that we give the outer grid a name, `x:Name="MainGrid"`. This will make it directly accessible from the code-behind defined in the `MainPage.xaml.cs` file. Since we are not using MVVM in this example, we will need a way to access the grid without data binding.

Creating the Swiper control

The main part of this project involves creating the `Swiper` control. A control, in a general sense, is a self-contained **user interface** (**UI**) with a code-behind to go with it. For .NET MAUI, a control is implemented using `ContentView`, as opposed to `ContentPage`, which is what XAML pages are. It can be added to any XAML page as an element, or in code in the code-behind file. We will be adding the control from code in this project.

Creating the control

Creating the `Swiper` control is a straightforward process. We just need to make sure that we select the correct item template, which is **Content View**, by doing the following:

1. In the `Swiper` project, create a folder called `Controls`.

2. Right-click on the `Controls` folder, select **Add**, and then click **New item...**.

3. Select **C# Items** and then **.NET MAUI** from the left pane of the **Add New Item** dialog box.

4. Select the **.NET MAUI ContentView (XAML)** item. Make sure you don't select the **.NET MAUI ContentView (C#)** option; this only creates a C# file and not an XAML file.

5. Name the control `SwiperControl.xaml`.

6. Click **Add**.

Refer to the following screenshot to view the preceding information:

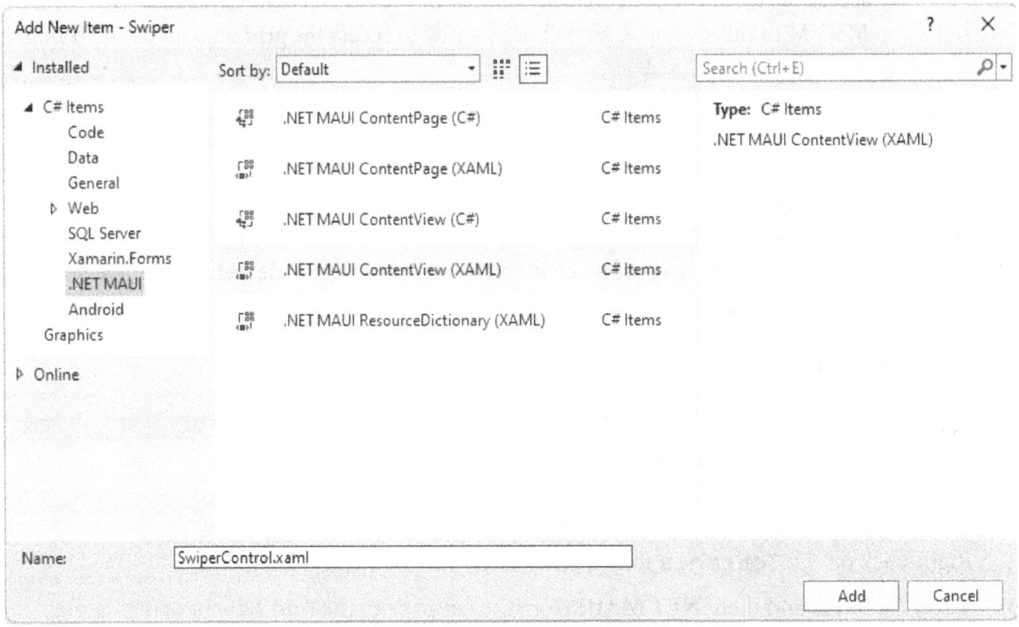

Figure 5.5 – Add New Item

This adds an XAML file for the UI and a C# code-behind file. It should look as follows:

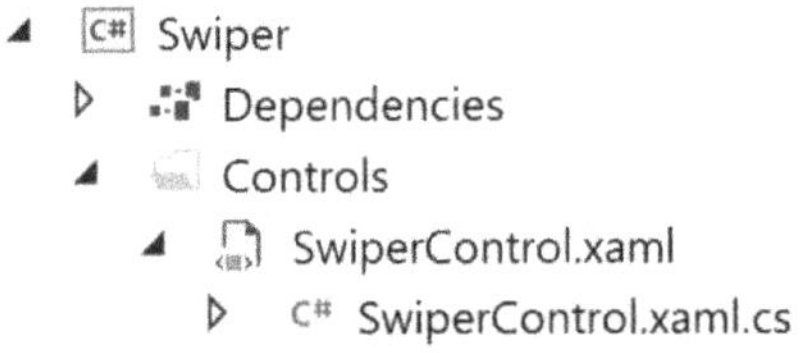

Figure 5.6 – Solution layout

Defining the main grid

Let's set the basic structure of the `Swiper` control:

1. Open the `SwiperControl.xaml` file.

2. Replace the content with the highlighted code in the following code block:

    ```
    <?xml version="1.0" encoding="UTF-8"?>
    <ContentView xmlns=" http://schemas.microsoft.com/dotnet/2021/
    maui"
    ```

```
                    xmlns:x="http://schemas.microsoft.com/winfx/2009/
xaml"
                    x:Class="Swiper.Controls.SwiperControl">
<ContentView.Content>
  <Grid>
    <Grid.ColumnDefinitions>
      <ColumnDefinition Width="100" />
      <ColumnDefinition Width="*" />
      <ColumnDefinition Width="100" />
    </Grid.ColumnDefinitions>

    <!-- ContentView for photo here -->

    <!-- StackLayout for like here -->

    <!-- StackLayout for deny here -->
  </Grid>
</ContentView.Content>
</ContentView>
```

This defines a grid with three columns. The leftmost and rightmost columns will take up 100 units of space, and the center will occupy the rest of the available space. The spaces on the sides will be areas in which we will add labels to highlight the choice that the user has made. We've also added three comments that act as placeholders for the XAML code to come in.

We will continue by adding additional XAML to create the photo layout.

Adding a content view for the photo

Now, we will extend the `SwiperControl.xaml` file by adding a definition of what we want the photo to look like. Our final result will look like *Figure 5.7*. Since we are going to pull images off the internet, we'll display a loading text to make sure that the user gets feedback on what's going on. To make it look like an instantly printed photo, we added some handwritten text under the photo, as can be seen in the following figure:

A horrible picture of grandpa

Figure 5.7 – The photo UI design

The preceding figure shows what we would like the photo to look like. To make this a reality, we need to add some XAML code to the `SwiperControl` file by doing the following:

1. Open `SwiperControl.xaml`.

2. Add the highlighted XAML code following the `<!-- ContentView for photo here -->` comment. Make sure that you do not replace the entire `ContentView` control for the page; just add this under the comment, as illustrated in the following code block. The rest of the page should be untouched:

```xaml
<!-- ContentView for photo here -->
    <ContentView x:Name="photo" Padding="40" Grid.ColumnSpan="3" >
      <Grid x:Name="photoGrid" BackgroundColor="Black" Padding="1"
>
        <Grid.RowDefinitions>
          <RowDefinition Height="*" />
          <RowDefinition Height="40" />
        </Grid.RowDefinitions>

        <BoxView Grid.RowSpan="2" BackgroundColor="White" />
        <Image x:Name="image" Margin="10"
BackgroundColor="#AAAAAA" Aspect="AspectFill" />

        <Label x:Name="loadingLabel" Text="Loading..."
TextColor="White" FontSize="Large" FontAttributes="Bold"
HorizontalOptions="Center" VerticalOptions="Center" />

        <Label Grid.Row="1" x:Name="descriptionLabel"
Margin="10,0" Text="A picture of grandpa" FontFamily="Bradley
Hand" />
```

```
    </Grid>
  </ContentView>
```

A `ContentView` control defines a new area where we can add other controls. One very important feature of a `ContentView` control is that it only takes one child control. Most of the time, we would add one of the available layout controls. In this case, we'll use a `Grid` control to lay out the control, as shown in the preceding code.

The grid defines two rows:

- A row for the photo itself, which takes up all the available space when the other rows have been allocated space
- A row for the comment, which will be exactly 40 units in height

The `Grid` control itself is set to use a black background and a padding of 1. This, in combination with a `BoxView` control, which has a white background, creates the frame that we see around the control. The `BoxView` control is also set to span both rows of the grid (`Grid.RowSpan="2"`), taking up the entire area of the grid, minus the padding.

The `Image` control comes next. It has a background color set to a nice gray tone (#AAAAAA) and a margin of 40, which will separate it a bit from the frame around it. It also has a hardcoded name (`x:Name="image"`), which will allow us to interact with it from the code-behind. The last attribute, called `Aspect`, determines what we should do if the image control isn't of the same ratio as the source image. In this case, we want to fill the entire image area, but not show any blank areas. This effectively crops the image either in terms of height or width.

We finish off by adding two labels, which also have hardcoded names for later reference.

That's a wrap on the XAML for now; let's move on to creating a description for the photo.

Creating the DescriptionGenerator class

At the bottom of the image, we can see a description. Since we don't have any general descriptions of the images from our upcoming image source, we need to create a generator that makes up descriptions. Here's a simple, yet fun, way to do it:

1. Create a folder called `Utils` in the `Swiper` project.
2. Create a new class called `DescriptionGenerator` in that folder.
3. Add the following code to this class:

```
internal class DescriptionGenerator
{
  private string[] _adjectives = { "nice", "horrible", "great",
"terribly old", "brand new" };
  private string[] _other = { "picture of grandpa", "car",
"photo of a forest", "duck" };
```

```
private static Random random = new();

public string Generate()
{
    var a = _adjectives[random.Next(_adjectives.Count())];
    var b = _other[random.Next(_other.Count())];
    return $"A {a} {b}";
}
}
```

This class only has one purpose: it takes one random word from the _adjectives array and combines it with a random word from the _other array. By calling the Generate() method, we get a fresh new combination. Feel free to enter your own words in the arrays. Note that the Random instance is a static field. This is because if we create new instances of the Random class that are too close to each other in time, they get seeded with the same value and return the same sequence of random numbers.

Now that we can create a fun description for the photo, we need a way to capture the image and description for the photo.

Creating a Picture class

To abstract all the information about the image we want to display, we'll create a class that encapsulates this information. There isn't much information in our Picture class, but it is good coding practice to do this. Proceed as follows:

1. Create a new class called Picture in the Utils folder.

2. Add the following code to the class:

```
public class Picture
{
    public Uri Uri { get; init; }
    public string Description { get; init; }

    public Picture()
    {
        Uri = new Uri($"https://picsum.
photos/400/400/?random&ts={DateTime.Now.Ticks}");

        var generator = new DescriptionGenerator();
        Description = generator.Generate();
    }
}
```

The `Picture` class has the following two public properties:

- The **Uniform Resource Identifier** (**URI**) of an image exposed as the `Uri` property, which points to its location on the internet

- A description of that image, exposed as the `Description` property

In the constructor, we create a new URI, which points to a public source of test photos that we can use. The width and height are specified in the query string part of the URI. We also append a random timestamp to avoid the images being cached by .NET MAUI. This generates a unique URI each time we request an image.

We then use the `DescriptionGenerator` class that we created previously to generate a random description for the image.

Note that the properties don't define a `set` method, but instead use `init`. Since we never need to change the values of URL or `Description` after the object is created, the properties can be read-only. `init` only allows the value to be set before the constructor completes. If you try to set the value after the constructor has run, the compiler will generate an error.

Now that we have all the pieces we need to start displaying images, let's start pulling it all together.

Binding the picture to the control

Let's begin to wire up the `Swiper` control so that it starts displaying images. We need to set the source of an image, and then control the visibility of the loading label based on the status of the image. Since we are using an image fetched from the internet, it might take a couple of seconds to download. A good UI will provide the user with proper feedback to help them avoid confusion regarding what is going on.

We will begin by setting the source for the image.

Setting the source

The `Image` control (referred to as `image` in the code) has a `source` property. This property is of the `ImageSource` abstract type. There are a few different types of image sources that you can use. The one we are interested in is the `UriImageSource` type, which takes a URI, downloads the image, and allows the image control to display it.

Let's extend the `Swiper` control so that we can set the source and description:

1. Open the `Controls/Swiper.Xaml.cs` file (the code-behind for the `Swiper` control).
2. Add a `using` statement for `Swiper.Utils` (`using Swiper.Utils;`) since we will be using the `Picture` class from that namespace.

3. Add the following highlighted code to the constructor:

```
public SwiperControl()
{
    InitializeComponent();
    var picture = new Picture();
    descriptionLabel.Text = picture.Description;
    image.Source = new UriImageSource() { Uri = picture.Uri };
}
```

Here, we create a new instance of a `Picture` class and assign the description to the `descriptionLabel` control in the GUI by setting the text property of that control. Then, we set the source of the image to a new instance of the `UriImageSource` class, and assign the URI from the `picture` instance. This will cause the image to be downloaded from the internet, and display it as soon as it is downloaded.

Next, we will change the visibility of the loading label for positive user feedback.

Controlling the loading label

While the image is downloading, we want to show a loading text centered over the image. This is already in the XAML file that we created earlier, so what we need to do is hide it once the image has been downloaded. We will do this by controlling the `IsVisibleProperty` property (yes, the property is actually named `IsVisibleProperty`) of the `loadingLabel` control by setting a binding to the `IsLoading` property of the image. Any time the `IsLoading` property is changed on the image, the binding changes the `IsVisible` property on the label. This is a nice fire-and-forget approach.

You might have noticed that we are using a binding when we said that we wouldn't be using bindings at the beginning of this chapter. This is used as a shortcut, to avoid us having to write the code that would do essentially the same thing as this binding does. And to be fair, while we did say no MVVM and data binding, we are binding to ourselves, not between classes, so all the code is self-contained inside the `Swiper` control.

Let's add the code needed to control the `loadingLabel` control, as follows:

1. Open the `Swiper.xaml.cs` code-behind file.
2. Add the following code marked in bold to the constructor:

```
public SwiperControl()
{
    InitializeComponent();
    var picture = new Picture();
    descriptionLabel.Text = picture.Description;

    image.Source = new UriImageSource() { Uri = picture.Uri };
```

```
loadingLabel.SetBinding(IsVisibleProperty, "IsLoading");
loadingLabel.BindingContext = image;
}
```

In the preceding code, the `loadingLabel` control sets a binding to the `IsVisibleProperty` property, which belongs to the `VisualElement` class that all controls inherit from. It tells `loadingLabel` to listen to changes in the `IsLoading` property of whichever object is assigned to the binding context. In this case, this is the image control.

Next, we will allow the user to "swipe right" or "swipe left."

Handling pan gestures

A core feature of this app is the pan gesture. A pan gesture is when a user presses on the control and moves it around the screen. We will also add a random rotation to the `Swiper` control to make it look like there are photos in a stack when we add multiple images.

We will start by adding some fields to the `SwiperControl` class, as follows:

1. Open the `SwiperControl.xaml.cs` file.

2. Add the following fields in the code to the class:

    ```
    private readonly double _initialRotation;
    private static readonly Random _random = new Random();
    ```

The first field, `_initialRotation`, stores the initial rotation of the image. We will set this in the constructor. The second field is a `static` field containing a `Random` object. As you might remember, it's better to create one static random object to make sure multiple random objects don't get created with the same seed. The seed is based on time, so if we create objects too close in time to each other, they get the same random sequence generated, so it wouldn't be that random at all.

The next thing we have to do is create an event handler for the `PanUpdated` event that we will bind to at the end of this section, as follows:

1. Open the `SwiperControl.xaml.cs` code-behind file.

2. Add the `OnPanUpdated` method to the class, like this:

    ```
    private void OnPanUpdated(object sender, PanUpdatedEventArgs e)
    {
      switch (e.StatusType)
      {
        case GestureStatus.Started: PanStarted();
        break;

        case GestureStatus.Running: PanRunning(e);
    ```

```
        break;

        case GestureStatus.Completed: PanCompleted();
        break;
    }
}
```

This code is straightforward. We handle an event that takes a `PanUpdatedEventArgs` object as the second argument. This is a standard method of handling events. We then have a `switch` clause that checks which status the event refers to.

A pan gesture can have the following three states:

- `GestureStatus.Started`: The event is raised once with this state when the panning begins
- `GestureStatus.Running`: The event is then raised multiple times, once for each time you move your finger
- `GestureStatus.Completed`: The event is raised one last time when you let go

For each of these states, we call specific methods that handle the different states. We'll continue with adding those methods now:

1. Open the `SwiperControl.xaml.cs` code-behind file.
2. Add the following three methods to the class, like this:

```
private void PanStarted()
{
    photo.ScaleTo(1.1, 100);
}

private void PanRunning(PanUpdatedEventArgs e)
{
    photo.TranslationX = e.TotalX;
    photo.TranslationY = e.TotalY;
    photo.Rotation = _initialRotation + (photo.TranslationX / 25);
}

private void PanCompleted()
{
    photo.TranslateTo(0, 0, 250, Easing.SpringOut);
    photo.RotateTo(_initialRotation, 250, Easing.SpringOut);
    photo.ScaleTo(1, 250);
}
```

Let's start by looking at `PanStarted()`. When the user starts dragging the image, we want to add the effect of it raising a little bit over the surface. This is done by scaling the image by 10%. .NET MAUI has a set of excellent functions to do this. In this case, we call the `ScaleTo()` method on the image control (named `Photo`) and tell it to scale to `1.1`, which corresponds to 10% of its original size. We also tell it to do this within a duration of `100` **milliseconds (ms)**. This call is also awaitable, which means we can wait for the control to finish animating before executing the next call. In this case, we are going to use a fire-and-forget approach.

Next, we have `PanRunning()`, which is called multiple times during the pan operation. This takes an argument, called `PanUpdatedEventArgs`, from the event handler that `PanRunning()` is called from. We could also just pass in *X* and *Y* values as arguments to reduce the coupling of the code. This is something that you can experiment with. The method extracts the *X* and *Y* components from the `TotalX/TotalY` properties of the event and assigns them to the `TranslationX/TranslationY` properties of the image control. We also adjust the rotation slightly, based on how far the image has been moved.

The last thing we need to do is restore everything to its initial state when the image is released. This can be done in `PanCompleted()`. First, we translate (or move) the image back to its original local coordinates (`0,0`) in `250` ms. We also added an easing function to make it overshoot the target a bit and then animate back. We can play around with the different predefined easing functions; these are useful for creating nice animations. We do the same to move the image back to its initial rotation. Finally, we scale it back to its original size in `250` ms.

Now, it's time to add the code in the constructor that will wire up the pan gesture and set some initial rotation values. Proceed as follows:

1. Open the `SwiperControl.xaml.cs` code-behind file.

2. Add the highlighted code to the constructor. Note that there is more code in the constructor, so don't overwrite the whole method; just add the bold text shown in the following code block:

```
public SwiperControl()
{
    InitializeComponent();

    var panGesture = new PanGestureRecognizer();
    panGesture.PanUpdated += OnPanUpdated;
    this.GestureRecognizers.Add(panGesture);

    _initialRotation = _random.Next(-10, 10);
    photo.RotateTo(_initialRotation, 100, Easing.SinOut);

    var picture = new Picture();
    descriptionLabel.Text = picture.Description;
    image.Source = new UriImageSource() { Uri = picture.Uri };
```

```
        loadingLabel.SetBinding(IsVisibleProperty, "IsLoading");
        loadingLabel.BindingContext = image;
}
```

All .NET MAUI controls have a property called GestureRecognizers. There are different types of gesture recognizers, such as TapGestureRecognizer or SwipeGestureRecognizer. In our case, we are interested in the PanGestureRecognizer type. We create a new PanGestureRecognizer instance and subscribe to the PanUpdated event by hooking it up to the OnPanUpdated() method we created earlier. Then, we add it to the Swiper controls GestureRecognizers collection.

After this, we set an initial rotation of the image and make sure we store the current rotation value so that we can modify the rotation, and then rotate it back to the original state.

Next, we will wire up the control temporarily so that we can test it out.

Testing the control

We now have all the code written to take the control for a test run. Proceed as follows:

1. Open MainPage.xaml.cs.

2. Add a using statement for the Swiper.Controls (using Swiper.Controls;).

3. Add the following code marked in bold to the constructor:

    ```
    public MainPage()
    {
        InitializeComponent();
        MainGrid.Children.Add(new SwiperControl());
    }
    ```

If all goes well with the build, we should end up with a photo like the one shown in the following figure:

Figure 5.8 – Testing the app

We can also drag the photo around (pan it). Notice the slight lift effect when you begin dragging, and the rotation of the photo based on the amount of translation, which is the total movement. If you let go of the photo, it animates back into place.

Now that we have the control displaying the photo and can swipe it left or right, we need to act on those swipes.

Creating decision zones

A matchmaking app is nothing without those special drop zones on each side of the screen. We want to do a few things here:

- When a user drags an image to either side, text should appear that says LIKE or DENY (the decision zones)
- When a user drops an image on a decision zone, the app should remove the image from the page

We will create these zones by adding some XAML code to the `SwiperControl.xaml` file and then add the necessary code to make this happen. It is worth noting that the zones are not hotspots for dropping the image, but rather for displaying labels on top of the control surface. The actual drop zones are calculated and determined based on how far you drag the image.

The first step is to add the UI for the left and right swipe actions.

Extending the grid

The `Swiper` control has three columns (left, right, and center) defined. We want to add some kind of visual feedback to the user if the image is dragged to either side of the page. We will do this by adding a `StackLayout` control with a `Label` control on each side.

We will add the right-hand side first.

Adding the StackLayout for liking photos

The first thing we need to do is add the `StackLayout` control for liking photos on the right-hand side of the control:

1. Open `Controls/SwiperControl.xaml`.

2. Add the following code under the `<!-- StackLayout for like here -->` comment:

    ```
    <StackLayout Grid.Column="2" x:Name="likeStackLayout"
    Opacity="0" Padding="0, 100">
        <Label Text="LIKE" TextColor="Lime" FontSize="30"
    Rotation="30" FontAttributes="Bold" />
    </StackLayout>
    ```

The `StackLayout` control is the container of child elements that we want to display. It has a name and is assigned to be rendered in the third column (it says `Grid.Column="2"` in the code due to the zero indexing). The `Opacity` property is set to `0`, making it completely invisible, and the `Padding` property is adjusted to make it move down a bit from the top.

Inside the `StackLayout` control, we'll add the `Label` control.

Now that we have the right-hand side, let's add the left.

Adding the StackLayout for denying photos

The next step is to add the `StackLayout` control for denying photos on the left-hand side of the control:

1. Open `Controls/SwiperControl.xaml`.

2. Add the following code under the `<!-- StackLayout for deny here -->` comment:

    ```
    <StackLayout x:Name="denyStackLayout" Opacity="0" Padding="0,
    100" HorizontalOptions="Start">
        <Label Text="DENY" TextColor="Red" FontSize="30"
    Rotation="-20" FontAttributes="Bold" />
    </StackLayout>
    ```

The setup for the left-hand side `StackLayout` is the same, except that it should be in the first column, which is the default, so there is no need to add a `Grid.Column` attribute. We have also specified `HorizontalOptions="End"`, which means that the content should be right-justified.

With the UI all set, we can now work on the logic for providing the user visual feedback by adjusting the opacity of the LIKE or DENIED text controls as the photo is panned.

Determining the screen size

To be able to calculate the percentage of how far the user has dragged the image, we need to know the size of the control. This is not determined until the control is laid out by .NET MAUI.

We will override the OnSizeAllocated() method and add a _screenWidth field in the class to keep track of the current width of the window:

1. Open SwiperControl.xaml.cs.

2. Add the following code to the file, putting the field at the beginning of the class and the OnSizeAllocated() method below the constructor:

```
private double _screenWidth = -1;
protected override void OnSizeAllocated(double width, double height)
{
  base.OnSizeAllocated(width, height);
  if (Application.Current.MainPage == null)
  {
    return;
  }
  _screenWidth = Application.Current.MainPage.Width;
}
```

The _screenWidth field is used to store the width as soon as we have resolved it. We do this by overriding the OnSizeAllocated() method that is called by .NET MAUI when the size of the control is allocated. This is called multiple times. The first time it's called is actually before the width and height have been set and before the MainPage property of the current app is set. At this time, the width and height are set to -1, and the Application.Current.MainPage property is null. We look for this state by null-checking Application.Current.MainPage and returning if it is null. We could also have checked for -1 values on the width. Either method would work. If it does have a value, however, we want to store it in our _screenWidth field for later use.

.NET MAUI will call the OnSizeAllocated() method any time the frame of the app changes. This is most relevant for **WinUI** apps since they are in a window that a user can easily change. Android and iOS apps are less likely to get a call to this method a second time since the app will take up the entire screen's real estate.

Adding code to calculate the state

To calculate the state of the image, we need to define what our zones are, and then create a function that takes the current amount of movement and updates the opacity of the GUI decision zones based on how far we panned the image.

Defining a method for calculating the state

Let's add the `CalculatePanState()` method to calculate how far we have panned the image, and if it should start to affect the GUI, by following these few steps:

1. Open `Controls/SwiperControl.xaml.cs`.

2. Add the properties at the top and the `CalculatePanState()` method anywhere in the class, as shown in the following code block:

```csharp
private const double DeadZone = 0.4d;
private const double DecisionThreshold = 0.4d;

private void CalculatePanState(double panX)
{
    var halfScreenWidth = _screenWidth / 2;
    var deadZoneEnd = DeadZone * halfScreenWidth;
    if (Math.Abs(panX) < deadZoneEnd)
    {
        return;
    }
    var passedDeadzone = panX < 0 ? panX + deadZoneEnd : panX - deadZoneEnd;

    var decisionZoneEnd = DecisionThreshold * halfScreenWidth;
    var opacity = passedDeadzone / decisionZoneEnd;

    opacity = double.Clamp(opacity, -1, 1);

    likeStackLayout.Opacity = opacity;
    denyStackLayout.Opacity = -opacity;
}
```

We define the following two values as constants:

*   DeadZone, which defines that 40% (0.4) of the available space on either side of the center point is a dead zone when panning an image. If we release the image in this zone, it simply returns to the center of the screen without any action being taken.

- The next constant is `DecisionThreshold`, which defines another 40% (`0.4`) of the available space. This is used for interpolating the opacity of `StackLayout` on either side of the layout.

We then use these values to check the state of the panning action whenever the panning changes. If the absolute panning value of X (panX) is less than the dead zone, we return without any action being taken. If not, we calculate how far over the dead zone we have passed and how far into the decision zone we are. We calculate the opacity values based on this interpolation and clamp the value between `-1` and `1`.

Finally, we set the opacity to this value for both `likeStackLayout` and `denyStackLayout`.

Wiring up the pan state check

While the image is being panned, we want to update the state, as follows:

1. Open `Controls/SwiperControl.xaml.cs`.

2. Add the following code in bold to the `PanRunning()` method:

```
private void PanRunning(PanUpdatedEventArgs e)
{
    photo.TranslationX = e.TotalX; photo.TranslationY = e.TotalY;
    photo.Rotation = _initialRotation + (photo.TranslationX / 25);

    CalculatePanState(e.TotalX);
}
```

This addition to the `PanRunning()` method passes the total amount of movement on the *x axis* to the `CalculatePanState()` method, to determine if we need to adjust the opacity of either `StackLayout` on the right or the left of the control.

Adding exit logic

So far, all is well, except for the fact that if we drag an image to the edge and let go, the text stays. We need to determine when the user stops dragging the image, and, if so, whether or not the image is in a decision zone.

Let's add the code needed to animate the photo back to its original position.

Checking if the image should exit

We want a simple function that determines if an image has panned far enough for it to count as an exit of that image. To create such a function, proceed as follows:

1. Open `Controls/SwiperControl.xaml.cs`.

2. Add the `CheckForExitCriteria()` method to the class, as shown in the following code snippet:

```
private bool CheckForExitCriteria()
{
  var halfScreenWidth = _screenWidth / 2;
  var decisionBreakpoint = DeadZone * halfScreenWidth;
  return (Math.Abs(photo.TranslationX) > decisionBreakpoint);
}
```

This function calculates whether we have passed over the dead zone and into the decision zone. We need to use the `Math.Abs()` method to get the total absolute value to compare it against. We could have used < and > operators as well, but we are using this approach as it is more readable. This is a matter of code style and taste – feel free to do it your way.

Removing the image

If we determine that an image has panned far enough for it to exit, we want to animate it off the screen and then remove the image from the page. To do this, proceed as follows:

1. Open `Controls/SwiperControl.xaml.cs`.

2. Add the `Exit()` method to the class, as shown in the following code block:

```
private void Exit()
{
  MainThread.BeginInvokeOnMainThread(async () =>
  {
    var direction = photo.TranslationX < 0 ? -1 : 1;
    await photo.TranslateTo(photo.TranslationX + (_screenWidth *
direction), photo.TranslationY, 200, Easing.CubicIn);
    var parent = Parent as Layout;
    parent?.Children.Remove(this);
  });
}
```

Let's break down the preceding code block to understand what the `Exit()` method does:

1. We begin by making sure that this call is done on the UI thread, which is also known as the `MainThread` thread. This is because only the UI thread can do animations.

2. We also need to run this thread asynchronously so that we can kill two birds with one stone. Since this method is all about animating the image to either side of the screen, we need to determine in which direction to animate it. We do this by determining if the total translation of the image is positive or negative.

3. Then, we use this value to await a translation through the `photo.TranslateTo()` call.

4. We `await` this call since we don't want the code execution to continue until it's done. Once it has finished, we remove the control from the parent's collection of children, causing it to disappear from existence forever.

Updating PanCompleted

The decision regarding whether the image should disappear or simply return to its original state is triggered in the `PanCompleted()` method. Here, we will wire up the two methods that we created in the previous two sections. Proceed as follows:

1. Open `Controls/SwiperControl.xaml.cs`.

2. Add the following code in bold to the `PanCompleted()` method:

```
private void PanCompleted()
{
    if (CheckForExitCriteria())
    {
        Exit();
    }

    likeStackLayout.Opacity = 0;
    denyStackLayout.Opacity = 0;

    photo.TranslateTo(0, 0, 250, Easing.SpringOut);
    photo.RotateTo(_initialRotation, 250, Easing.SpringOut);
    photo.ScaleTo(1, 250);
}
```

The last step in this section is to use the `CheckForExitCriteria()` method, and the `Exit()` method if those criteria are met. If the exit criteria are not met, we need to reset the state and the opacity of `StackLayout` to make everything go back to normal.

Now that we can swipe left or swipe right, let's add some events to raise when the user has swiped.

Adding events to the control

The last thing we have left to do in the control itself is add some events that indicate whether the image has been *liked* or *denied*. We are going to use a clean interface, allowing for simple use of the control while hiding all the implementation details.

Declaring two events

To make the control easier to interact with from the application itself, we'll need to add events for `Like` and `Deny`, as follows:

1. Open `Controls/SwiperControl.xaml.cs`.

2. Add two event declarations at the beginning of the class, as shown in the following code snippet:

```
public event EventHandler OnLike;
public event EventHandler OnDeny;
```

These are two standard event declarations with out-of-the-box event handlers.

Raising the events

We need to add code in the `Exit()` method to raise the events we created earlier, as follows:

1. Open `Controls/SwiperControl.xaml.cs`.

2. Add the following code in bold to the `Exit()` method:

```
private void Exit()
{
  MainThread.BeginInvokeOnMainThread(async () =>
  {
    var direction = photo.TranslationX < 0 ? -1 : 1;

    if (direction > 0)
    {
      OnLike?.Invoke(this, new EventArgs());
    }

    if (direction < 0)
    {
      OnDeny?.Invoke(this, new EventArgs());
    }

    await photo.TranslateTo(photo.TranslationX + (_screenWidth *
  direction), photo.TranslationY, 200, Easing.CubicIn);
    var parent = Parent as Layout;
    parent?.Children.Remove(this);
  });
}
```

Here, we inject the code to check whether we are liking or denying an image. We then raise the correct event based on this information.

We are now ready to finalize this app; the `Swiper` control is complete, so now, we need to add the right initialization code to finish it.

Wiring up the Swiper control

We have now reached the final part of this chapter. In this section, we are going to wire up the images and make our app a closed-loop app that can be used forever. We will add 10 images that we will download from the internet when the app starts up. Each time an image is removed, we'll simply add another one.

Adding images

Let's start by creating some code that will add the images to the `MainView` class. First, we will add the initial images; then, we will create a logic model for adding a new image to the bottom of the stack each time an image is liked or denied.

Adding initial photos

To make the photos look like they are stacked, we need at least 10 of them. Proceed as follows:

1. Open `MainPage.xaml.cs`.

2. Add the `AddInitalPhotos()` method and `InsertPhotoMethod()` to the class, as illustrated in the following code block:

    ```
    private void AddInitialPhotos()
    {
      for (int i = 0; i < 10; i++)
      {
        InsertPhoto();
      }
    }

    private void InsertPhoto()
    {
      var photo = new SwiperControl();
      this.MainGrid.Children.Insert(0, photo);
    }
    ```

First, we create a method called `AddInitialPhotos()` that will be called upon startup. This method simply calls the `InsertPhoto()` method 10 times and adds a new `SwiperControl` to the `MainGrid` each time. It inserts the control at the first position in the stack, effectively putting it at the bottom of the pile since the collection of controls is rendered from the beginning to the end.

Making the call from the constructor

We need to call this method for the magic to happen, so follow these steps to do so:

1. Open `MainPage.xaml.cs`.

2. Add the following code in bold to the constructor:

```
public MainPage()
{
    InitializeComponent();
    AddInitialPhotos();
}
```

There isn't much to say here. Once the `MainPage` object has been initialized, we call the method to add 10 random photos that we will download from the internet.

Adding count labels

We want to add some values to the app as well. We can do this by adding two labels below the collection of `Swiper` controls. Each time a user rates an image, we will increment one of two counters and display the result.

So, let's add the XAML code needed to display the labels:

1. Open `MainPage.xaml`.

2. Replace the `<!-- Placeholder for later -->` comment with the following code marked in bold:

```
<Grid Grid.Row="1" Padding="30">
  <Grid.RowDefinitions>
    <RowDefinition Height="auto" />
    <RowDefinition Height="auto" />
    <RowDefinition Height="auto" />
    <RowDefinition Height="auto" />
  </Grid.RowDefinitions>
  <Label Text="LIKES" />
  <Label x:Name="likeLabel" Grid.Row="1" Text="0"
FontSize="Large" FontAttributes="Bold" />
  <Label Grid.Row="2" Text="DENIED" />
  <Label x:Name="denyLabel" Grid.Row="3" Text="0"
FontSize="Large" FontAttributes="Bold" />
</Grid>
```

This code adds a new `Grid` control with four auto-height rows. This means that we calculate the height of the content of each row and use this for the layout. It is the same thing as `StackLayout`, but we wanted to demonstrate a better way of doing this.

We add a `Label` control in each row and name two of them `likeLabel` and `denyLabel`. These two named labels will hold information about how many images have been liked and how many have been denied.

Subscribing to events

The last step is to wire up the `OnLike` and `OnDeny` events and display the total count to the user.

Adding methods to update the GUI and respond to events

We need some code to update the GUI and keep track of the count. Proceed as follows:

1. Open `MainPage.xaml.cs`.
2. Add the following code to the class:

```
private int _likeCount;
private int _denyCount;

private void UpdateGui()
{
  likeLabel.Text = _likeCount.ToString();
  denyLabel.Text = _denyCount.ToString();
}

private void Handle_OnLike(object sender, EventArgs e)
{
  _likeCount++;
  InsertPhoto();
  UpdateGui();
}

private void Handle_OnDeny(object sender, EventArgs e)
{
  _denyCount++;
  InsertPhoto();
  UpdateGui();
}
```

The two fields at the top of the preceding code block keep track of the number of likes and denies. Since they are value-type variables, they default to zero.

To make the changes of these labels show up in the UI, we've created a method called `UpdateGui()`. This takes the value of the two aforementioned fields and assigns it to the `Text` properties of both labels.

The two methods that follow are the event handlers that will be handling the `OnLike` and `OnDeny` events. They increase the appropriate field, add a new photo, and then update the GUI to reflect the change.

Wiring up events

Each time a new `SwiperControl` instance is created, we need to wire up the events, as follows:

1. Open `MainPage.xaml.cs`.

2. Add the following code in bold to the `InsertPhoto()` method:

```
private void InsertPhoto()
{
    var photo = new SwiperControl();
    photo.OnDeny += Handle_OnDeny;
    photo.OnLike += Handle_OnLike;

    this.MainGrid.Children.Insert(0, photo);
}
```

The added code wires up the event handlers that we defined earlier. The events make it easy to interact with our new control. Try it for yourself and have a play around with the app that you have created.

Summary

Good job! In this chapter, we learned how to create a reusable, good-looking control that can be used in any .NET MAUI app. To enhance the **user experience** (**UX**) of the app, we used some animations that give the user more visual feedback. We also got creative with the use of XAML to define a GUI of the control that looks like a photo, with a hand-written description.

After that, we used events to expose the behavior of the control back to the `MainPage` page to limit the contact surface between your app and the control. Most importantly of all, we touched on the subject of `GestureRecognizers`, which can make our life much easier when dealing with common gestures.

Looking for ideas on how to make this app even better? Try this out: keep a history of the likes and dislikes and add a view to display each collection.

In the next chapter, we will create a photo gallery app using the `CollectionView` and `CarouselView` controls. The app will also allow you to favorite photos you like by using storage to keep the favorites list between app runs.

Building a Photo Gallery App Using CollectionView and CarouselView

In this chapter, we will build an app that shows photos from the camera roll (photo gallery) of a user's device. The user will also be able to select photos as favorites. We will then look at the different ways to display photos—in carousels and in multi-column grid control. By using the .NET MAUI `CarouselView` control to display a group of images, the user can swipe through them to view each image. To display a large group of images, we will use the .NET MAUI `CollectionView` control and vertical scrolling to allow the user to view all the images. By learning how to use these controls, we will be able to use them in a lot of other cases when we build real-world apps.

The following topics will be covered in this chapter:

- Requesting permissions from the user to access data
- How to import photos from the iOS and Mac Catalyst photo gallery
- How to import photos from the Android photo gallery
- How to import photos from the Windows photo gallery
- How to use `CarouselView` in .NET MAUI
- How to use `CollectionView` in .NET MAUI

Technical requirements

To be able to complete this project, you will need to have Visual Studio for Mac or Windows installed, as well as the necessary .NET MAUI workloads. See *Chapter 1, Introduction to .NET MAUI*, for more details on how to set up your environment.

To build an iOS app using Visual Studio for your PC, you have to have a **Macintosh** (**Mac**) device connected. If you don't have access to a Mac at all, you can just follow the Android and Windows part of this project.

You can find the full source for the code in this chapter at `https://github.com/PacktPublishing/MAUI-Projects-3rd-Edition/`.

Project overview

Almost all apps visualize collections of data, and in this chapter, we will focus on two of the .NET MAUI controls that can be used to display data collections—`CollectionView` and `CarouselView`. Our app will show the photos that users have on their devices; to do that, we need to create a photo importer for each platform—one for iOS and Mac Catalyst, one for Windows, and one for Android.

The build time for this project is about 60 minutes.

Building the photo gallery app

This project, like all the rest, is a **File | New | Project...**-style project, which means that we will not be importing any code at all. So, this first section is all about creating the project and setting up the basic project structure.

It's time to start building the app using the following steps. Let's begin!

Creating the new project

The first step is to create a new .NET MAUI project:

1. Open Visual Studio 2022 and select **Create a new project**:

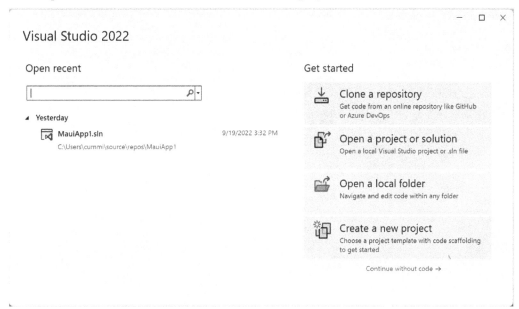

Figure 6.1 – Visual Studio 2022

This will open the **Create a new project** wizard.

2.   In the search field, type in maui and select the **.NET MAUI App** item from the list:

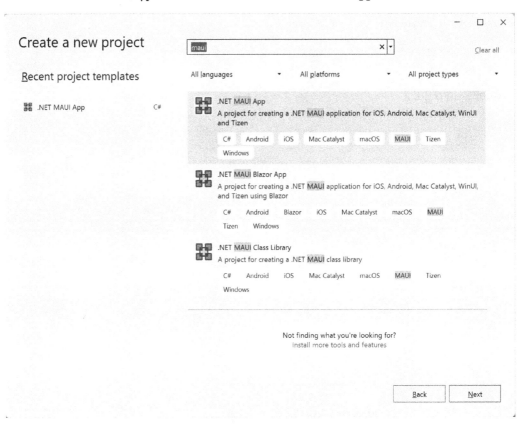

Figure 6.2 – Create a new project

3.   Click **Next**.

4.   Complete the next step of the wizard by naming your project. We will be calling our application GalleryApp in this case. Move on to the next dialog box by clicking **Next**, as shown in the following screenshot:

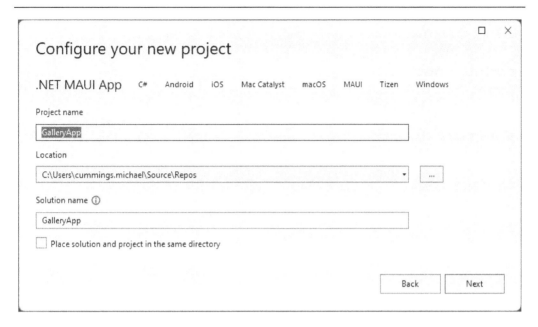

Figure 6.3 – Configure your new project

5. Click **Next**.

6. The last step will prompt you for the version of .NET Core to support. At the time of writing, .NET 6 is available as **Long-Term Support** (**LTS**), and .NET 7 is available as **Standard-Term Support**. For the purposes of this book, we will assume that you will be using .NET 7.

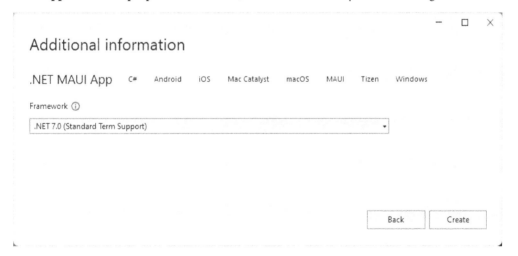

Figure 6.4 – Additional information

7. Finalize the setup by clicking **Create** and wait for Visual Studio to create the project.

Just like that, the app is created. Let's start by getting some photos to display.

Importing photos

The importing of photos is something that is carried out for all the platforms, so we will create a photo importer interface. The interface will have two `Get` methods—one that supports paging and one that gets photos with specified filenames. Both methods will also take a quality argument, but we will only use that argument in the iOS photo importer. The quality argument will be an `enum` type with two options—`High` and `Low`. However, before we create the interface, we will create a model class that will represent an imported photo using the following steps:

1. Create a new folder named `Models` in the `GalleryApp` project.

2. Create a new class named `Photo` in the recently created folder:

   ```
   namespace GalleryApp.Models;

   public class Photo
   {
     public string Filename { get; set; }
     public byte[] Bytes { get; set; }
   }
   ```

Now that we have created the model class, we can continue to create the interface:

1. Create a new folder named `Services` in the project.

2. Create a new interface named `IPhotoImporter` in the `Services` folder:

   ```
   namespace GalleryApp.Services;

   using System.Collections.ObjectModel;
   using GalleryApp.Models;

   public interface IPhotoImporter
   {
     Task<ObservableCollection<Photo>> Get(int start, int  count,
   Quality quality = Quality.Low);
     Task<ObservableCollection<Photo>> Get(List<string> filenames,
   Quality quality = Quality.Low);
   }
   ```

3. In the `Services` folder, add a new file and create an `enum` type named `Quality` with two members—Low and High:

   ```
   namespace GalleryApp.Services;
   ```

```
public enum Quality
{
  Low,
  High
}
```

4. In the `Services` folder, create a new class named `PhotoImporter`:

```
namespace GalleryApp.Services;

using GalleryApp.Models;
using System.Collections.ObjectModel;

internal partial class PhotoImporter : IPhotoImporter
{
  private partial Task<string[]> Import();

  public partial Task<ObservableCollection<Photo>> Get(int
start, int count, Quality quality);

  public partial Task<ObservableCollection<Photo>>
Get(List<string> filenames, Quality quality);

}
```

This class provides us with the base for the platform-specific implementations. By marking it `partial`, we are telling the compiler that there is more to this class in other files. We will be putting the implementation in the platform-specific folders later.

Now that we have the interface, we can add app permissions.

Requesting app permissions

If your app doesn't require any of the device's extra features, such as location, camera, or internet, then you will need to use permissions to request access to those resources. While each platform implements permissions slightly differently, .NET MAUI maps the platform-specific permissions into a common set of permissions to make things simpler. The permission system in .NET MAUI is also extensible so that you can create your own permissions to best suit your app.

Let's look at a specific example to see how requesting permissions works. `GalleryApp` displays images from a device's photo library. In the case of iOS and Android, the app must declare and request access to the photo library before it can use it. While these permissions are configured and named differently, .NET MAUI defines a `Photo` permission that hides those implementation details.

Follow these steps to add a permission check to GalleryApp:

1. Create a new class named AppPermissions in the GalleryApp project.

2. Modify the class definition to add a partial modifier, and remove the default constructor:

    ```
    namespace GalleryApp;

    internal partial class AppPermissions
    {
    }
    ```

3. Add the following class definition to the AppPermissions class:

    ```
    internal partial class AppPermissions
    {
      internal partial class AppPermission : Permissions.Photos
      {
      }
    }
    ```

    This creates a type named AppPermission that derives from the default .NET MAUI Photos permission class. It is also marked partial to allow for platform-specific implementation details to be added. Spoiler alert: we will need to have some platform-specific permissions.

4. Add the following method to the AppPermissions class:

    ```
    public static async Task<PermissionStatus>
    CheckRequiredPermission() => await Permissions.
    CheckStatusAsync<AppPermission>();
    ```

    The CheckRequiredPermission method is used to ensure our app has the right permissions before we attempt any operations that might fail if we don't. Its implementation is to call the .NET MAUI CheckSyncStatus with our AppPermission type. It returns a PermissionStatus, which is an enum type. We are mostly interested in the Denied and Granted values.

5. Add the CheckAndRequestRequiredPermission method to the AppPermissions class:

    ```
    public static async Task<PermissionStatus>
    CheckAndRequestRequiredPermission()
    {
      PermissionStatus status = await Permissions.
    CheckStatusAsync<AppPermission>();

      if (status == PermissionStatus.Granted)
            return status;
    ```

```
   if (status == PermissionStatus.Denied && DeviceInfo.Platform
== DevicePlatform.iOS)
   {
      // Prompt the user to turn on in settings
      // On iOS once a permission has been denied it may not be
requested again from the application
      await App.Current.MainPage.DisplayAlert("Required App
Permissions", "Please enable all permissions in Settings for
this App, it is useless without them.", "Ok");
   }

   if
   (Permissions.ShouldShowRationale<AppPermission>())
   {
      // Prompt the user with additional information as to why the
permission is needed
      await App.Current.MainPage.DisplayAlert("Required App
Permissions", "This is a Photo gallery app, without these
permissions it is useless.", "Ok");
   }

   status = await MainThread.InvokeOnMainThreadAsync(Permissions.
RequestAsync<AppPermission>);
   return status;
   }
}
```

The `CheckAndRequestRequiredPermission` method handles the intricacies of requesting access from the user. The first step is to simply check whether the permission has already been granted, and if it has, return the status. Next, if you are on iOS and the permission has been denied, it cannot be requested again, so you must instruct the user on how to grant permission to the app by using the settings panel. In the request behavior, Android includes the ability to nag the user if they have denied access. This behavior is exposed through .NET MAUI with the `ShouldShowRationale` method. It will return `false` for any platform that does not support this behavior; on Android, it will return `true` the first time the user denies access and `false` if the user denies it a second time. Finally, we request access for `AppPermission` from the user. Again, .NET MAUI is hiding all the platform implementation details from us, making checking and requesting access to certain resources very straightforward.

Look familiar?

If the preceding code looks familiar, it might be. This is the same implementation that is described in the .NET MAUI documentation. You can find it at `https://learn.microsoft.com/en-us/dotnet/maui/platform-integration/appmodel/permissions`.

Now that we have the shared `AppPermissions` in place, we can start with the platform implementations.

Importing photos from the iOS photo gallery

First, we will write the iOS code. To access photos, we need permission from the user, and we need to explain why we are asking for permission. To do that, we add text that explains why we need permission to the `info.plist` file. This text will be displayed when we ask the users for permission. To open the `info.plist` file, right-click on the file in the `Platforms/iOS` folder and click **Open With…**, then select **XML (Text) Editor**; this bypasses the default graphical `Info.plist` editor. Add the following text to the end of the `<dict>` element:

```
<key> NSPhotoLibraryUsageDescription </key>
<string> We want to show your photos in this app </string>
```

The first thing we will do is implement the `Import` method that reads what photos can be loaded:

1. In the `GalleryApp` project in the `Platforms/iOS` folder, create a new class called `PhotoImporter`.

2. Change the namespace declaration from `GalleryApp.Platforms.iOS` to `GalleryApp.Services`.

3. The partial class definitions must be in the same namespace, even though they are in different folders.

4. Add the `partial` modifier.

5. Resolve all the references.

6. Create a `private` field with a `PHAsset` dictionary named `assets`. This will be used to store photo information:

    ```
    private Dictionary<string,PHAsset> assets;
    ```

7. Create a new `private partial` method named `Import`:

    ```
    private partial async Task<string[]> Import()
    {
    }
    ```

8. In the `Import` method, request authorization using the `AppPermissions.Check AndRequestRequiredPermission` method:

    ```
    var status = await AppPermissions.
    CheckAndRequestRequiredPermission();
    ```

9. If the user has granted access, fetch all the image assets by using `PHAsset.FetchAssets`:

    ```
    internal partial class PhotoImporter
    {
    ```

```
private Dictionary<string,PHAsset> assets;

private partial async Task<string[]> Import()
{
   var status = await AppPermissions.
CheckAndRequestRequiredPermission();
   if (status == PermissionStatus.Granted)
   {

      assets = PHAsset.FetchAssets(PHAssetMediaType.Image, null)
      .Select(x => (PHAsset)x)
      .ToDictionary(asset => asset.
ValueForKey((NSString)"filename").ToString(), asset => asset);
   }
   return assets?.Keys.ToList().ToArray();
}
```

Now, we have fetched `PHAssets` for all the photos, but to show the photo, we need to get the actual photo. On iOS, to do that, we need to request the image for the asset. This is something that is carried out asynchronously, so we will use `ObservableCollection`:

```
private void AddImage(ObservableCollection<Photo> photos, string path,
PHAsset asset, Quality quality)
{
  var options = new PHImageRequestOptions()
  {
    NetworkAccessAllowed = true,
    DeliveryMode = quality == Quality.Low ?
    PHImageRequestOptionsDeliveryMode.FastFormat :

    PHImageRequestOptionsDeliveryMode.HighQualityFormat
  };

  PHImageManager.DefaultManager.RequestImageForAsset(asset,
PHImageManager.MaximumSize, PHImageContentMode.AspectFill, options,
(image, info) =>
  {
    using NSData imageData = image.AsPNG();
    var bytes = new byte[imageData.Length];
    System.Runtime.InteropServices.Marshal.Copy(imageData.
Bytes, bytes, 0, Convert.ToInt32(imageData.Length));
    photos.Add(new Photo()
    {
```

```
      Bytes = bytes,
      Filename = Path.GetFileName(path)
    });
  });
}
```

Now, we have what we need to start implementing the two `Get` methods from the interface. We will start with the partial `Task<ObservableCollection<Photo>> Get(int start, int count, Quality quality = Quality.Low)` method, which will be used to get photos from the `CollectionView` view that loads photos incrementally:

```csharp
public partial async Task<ObservableCollection<Photo>> Get(int start,
int count, Quality quality)
{
  var photos = new ObservableCollection<Photo>();

  var status = await AppPermissions.
CheckAndRequestRequiredPermission();
  if (status == PermissionStatus.Granted)
  {
    var result = await Import();
    if (result.Length == 0)
    {
      return photos;
    }

    Index startIndex = start;
    Index endIndex = start + count;

    if (endIndex.Value >= result.Length)
    {
      endIndex = result.Length;
    }
    if (startIndex.Value > endIndex.Value)
    {
      return photos;
    }

    foreach (var path in result[startIndex..endIndex])
    {
      AddImage(photos, path, assets[path], quality);
    }
  }
```

```
        return photos;
    }
```

The other method from the IPhotoImporter interface, Task<ObservableCollection<Photo>> Get(List<string> filenames, Quality quality = Quality.Low), is very similar to the Task<ObservableCollection<Photo>> Get(int start, int count, Quality quality = Quality.Low) method. The only difference is that there is no code to handle indexes, and the foreach loop that loops through the result's array contains an if statement that checks whether the filename is the same as the current PHAsset object, and if it is, it calls the AddImage method:

```
public partial async Task<ObservableCollection<Photo>>
Get(List<string> filenames, Quality quality)
{
    var photos = new ObservableCollection<Photo>();

    var result = await Import();
    if (result?.Length == 0)
    {
        return photos;
    }

    foreach (var path in result)
    {
        if (filenames.Contains(path))
        {
            AddImage(photos, path, assets[path], quality);
        }
    }

    return photos;
}
```

In the preceding code, we set NetworkAccessAllowed = true. We do this to make it possible to download photos from **iCloud**.

Now, one of the four photo importers of our project is complete. The Mac Catalyst importer will be the next one we implement.

Importing photos from the Mac Catalyst photo gallery

The Mac Catalyst importer is exactly the same as what we just did for iOS. However, there isn't a convenient way to say, "*I need this class for just iOS and Mac Catalyst, and nothing else.*" So we will take the path of least resistance and just copy the class into the Mac Catalyst platform folder:

1. Right-click the `PhotoImporter.cs` file in the `Platforms/iOS` folder in the project and select **Copy**.

2. Right-click the `Platforms/MacCatalyst` folder and select **Paste**.

3. Right-click on the `Info.plist` file in the `Platforms/MacCatalyst` folder and click **Open With…**, then select **XML (Text) Editor**, and add the following text to the end of the `<dict>` element:

```
<key> NSPhotoLibraryUsageDescription </key>
<string> We want to show your photos in this app </string>
```

That concludes the Mac Catalyst implementation of the `PhotoImporter` class. Next, we will work on the Android platform.

Importing photos from the Android photo gallery

Now that we have created an implementation for iOS, we will do the same for Android. Before we jump right into the importer, we need to address the permissions on Android.

In Android API version 33, three new permissions were added to enable read access to media files: `ReadMediaImages`, `ReadMediaVideos`, and `ReadMediaAudio`. Prior to API version 33, all that was required was the `ReadExternalStorage` permission. To properly request the correct permission for the API version of the device, create a new file named `AppPermissions` in the `Platform/Android` folder and modify it to look like the following:

```
using Android.OS;

[assembly: Android.App.UsesPermission(Android.Manifest.Permission.
ReadMediaImages)]
[assembly: Android.App.UsesPermission(Android.Manifest.Permission.
ReadExternalStorage, MaxSdkVersion = 32)]

namespace GalleryApp;

internal partial class AppPermissions
{
  internal partial class AppPermission : Permissions.Photos
  {
    public override (string androidPermission, bool isRuntime)[]
RequiredPermissions
```

```
    {
       get
       {
          List<(string androidPermission, bool isRuntime)> perms =
new();

          if (Build.VERSION.SdkInt >= BuildVersionCodes.Tiramisu)
                    perms.Add((global::Android.Manifest.Permission.
ReadMediaImages, true));
          else
                    perms.Add((global::Android.Manifest.Permission.
ReadExternalStorage, true));

          return perms.ToArray();
       }
    }
  }
}
```

The first two lines add the required permissions to the AndroidManifet.xml file, similar to what we did manually with the iOS info.plist file. However, we only need ReadMediaImages for API 33+ and ReadExternalStorage for API versions below 33, so we set MaxSdkVersion for the ReadExternalStorage property. Then, we extend the AppPermission class with an implementation of the RequirePermissions property. In RequirePermissions, we return an array containing the ReadMediaImages permissions if the API version is 33 or higher, or the ReadExternalStorage permission if the API version is lower than 33. The Boolean value that is part of the perms array indicates whether the permission requires requesting access at runtime from the user. Now, when the app launches, it will request access for the correct permission based on the API level of the device.

Now that we have the Android-specific permissions sorted, we can import images using the following steps:

1. Create a new class named PhotoImporter in the project in the Platforms/ Android folder.
2. Change the namespace declaration from GalleryApp.Platforms.Android to GalleryApp.Services.
3. The partial class definitions must be in the same namespace, even though they are in different folders.
4. Add the partial modifier.
5. Add a using statement for GalleryApp.Models to use the Photo class later.

6. Like the iOS implementation, we will start by implementing the `Import` method. Add a new method named `Import` as follows:

```csharp
private partial async Task<string[]> Import()
{
    var paths = new List<string>();
    return paths.ToArray();
}
```

7. Request permissions from the user to get the photos (highlighted in the following code block):

```csharp
private partial async Task<string[]> Import()
{
    var paths = new List<string>();

    var status = await AppPermissions.
CheckAndRequestRequiredPermission();
    if (status == PermissionStatus.Granted)
    {
    }
    return paths.ToArray();
}
```

8. Now, use `ContentResolver` to query for the files and add them to the result:

```csharp
private partial async Task<string[]> Import()
{
    var paths = new List<string>();

    var status = await AppPermissions.
CheckAndRequestRequiredPermission();
    if (status == PermissionStatus.Granted)
    {
        var imageUri = MediaStore.Images.Media.ExternalContentUri;
        var projection = new string[] { MediaStore.IMediaColumns.
Data };
        var orderBy = MediaStore.Images.IImageColumns.DateTaken;
        var cursor = Platform.CurrentActivity.ContentResolver.
Query(imageUri, projection, null, null, orderBy);
        while (cursor.MoveToNext())
        {
            string path = cursor.GetString(cursor.
GetColumnIndex(MediaStore.IMediaColumns.Data));
            paths.Add(path);
        }
```

```
        }
        return paths.ToArray();
    }
```

We will then start editing the `Task<ObservableCollection<Photo>> Get(int start, int count, Quality quality = Quality.Low)` method. If the import succeeds, we will continue to write the code that handles which photos should be imported in this loading of images. Conditions are specified with the `start` and `count` parameters. Use the following code listing to implement the first `Get` method:

```
public partial async Task<ObservableCollection<Photo>> Get(int start,
int count, Quality quality)
{
    var photos = new ObservableCollection<Photo>();

    var result = await Import();

    if (result.Length == 0)
    {
        return photos;
    }

    Index startIndex = start;
    Index endIndex = start + count;

    if (endIndex.Value >= result.Length)
    {
        endIndex = result.Length;
    }
    if (startIndex.Value > endIndex.Value)
    {
        return photos;
    }

    foreach (var path in result[startIndex..endIndex])
    {
        photos.Add(new()
        {
            Bytes = File.ReadAllBytes(path),
            Filename = Path.GetFileName(path)
        });
    }
}
```

```
    return photos;
}
```

Let's review the preceding code. The first step is to call the `Import` method and verify that there are photos to import. If there are none, we simply return an empty list. If there are photos to import, then we need to know `startIndex` and `endIndex` in the `photos` array to import. The code defaults to `endIndex` being `startIndex` plus the count of photos to import. If the count of photos to import is greater than the number of photos returned from the `Import` method, then `endindex` is adjusted to the length of the photos returned from the `Import` method. If `startIndex` is greater than `endIndex`, then we return the list of photos. Finally, we can read the images from `startIndex` to `endIndex` from the array of photos and return the bytes from the file and just the name of the file for each entry.

Now, we will continue with the other `Task<ObservableCollection<Photo>> Get (List<string> filenames, Quality quality = Quality.Low)` method.

Create a `foreach` loop to loop through all the photos and to check whether each photo is specified in the `filenames` parameter. If a photo is specified in the `filenames` parameter, read the photo from the path, as in the first `Get` method:

```
public partial async Task<ObservableCollection<Photo>>
Get(List<string> filenames, Quality quality)
{
    var photos = new ObservableCollection<Photo>();

    var result = await Import();

    if (result.Length == 0)
    {
        return photos;
    }

    foreach (var path in result)
    {
        var filename = Path.GetFileName(path);

        if (!filenames.Contains(filename))
        {
            continue;
        }

        photos.Add(new Photo()
        {
            Bytes = File.ReadAllBytes(path),
```

```
        Filename = filename
    });
  }

  return photos;
}
```

With the Android importer finished, we can move to the final importer for Windows.

Importing photos from the Windows photo gallery

The final importer that we need is for the Windows platform. The code will follow the same pattern as for the other platforms; however, for Windows, we will use the **Windows Search** service to get the list of photos. Let's see how this platform is implemented by following these steps:

1. Import the `tlbimp-Windows.Search.Interop` and `System.Data.OleDB` NuGet packages. These are needed to search the filesystem for images.

2. Open the `GalleryApp` project by double-clicking it in **Solution Explorer**; edit the new imports to add a condition:

    ```
    <PackageReference Include="System.Data.OleDb" Version="7.0.0"
    Condition="$([MSBuild]::GetTargetPlatformIdentifier('$(Target
    Framework)'))  == 'windows'" />
    <PackageReference Include="tlbimp-Microsoft.Search.Interop"
    Version="1.0.0" Condition="$([MSBuild]::GetTargetPlatform
    Identifier('$(TargetFramework)'))  == 'windows'" />
    ```

    We will use these two NuGet packages to search for image files on the local computer. But since we only need these assemblies on Windows, we limit `PackageReference` so that it is only used when `TargetPlatformIdentifier` is `'windows'`.

3. Create a new class called `PhotoImporter` in the `Windows` platform folder, and mark it as `partial`.

4. Change the namespace declaration from `GalleryApp.Platforms.Windows` to `GalleryApp.Services`.

5. The `partial` class definitions must be in the same namespace, even though they are in different folders.

6. Add `using` directives, so that we can use the classes in those namespaces:

    ```
    using GalleryApp.Models;
    using Microsoft.Search.Interop;
    using System.Data.OleDb;
    ```

7. Add a `private` field to hold the `QueryHelper` reference:

    ```
    ISearchQueryHelper queryHelper;
    ```

8. Like the previous implementations, we will start by implementing the `Import` method, so add a new method named `Import` as follows:

    ```
    private partial async Task<string[]> Import()
    {
        var paths = new List<string>();
        return paths.ToArray();
    }
    ```

9. Request permissions from the user to get the photos (highlighted in the following code block):

    ```
    private partial async Task<string[]> Import()
    {
        var paths = new List<string>();

        var status = await AppPermissions.
    CheckAndRequestRequiredPermission();
        if (status == PermissionStatus.Granted)
        {
        }
        return paths.ToArray();
    }
    ```

10. Now, using `QueryHelper`, get all the image paths:

    ```
    private partial async Task<string[]> Import()
    {
        var paths = new List<string>();

        var status = await AppPermissions.
    CheckAndRequestRequiredPermission();
        if (status == PermissionStatus.Granted)
        {
            string sqlQuery = queryHelper.GenerateSQLFromUserQuery(" ");

            using OleDbConnection conn = new(queryHelper.
    ConnectionString);
            conn.Open();

            using OleDbCommand command = new(sqlQuery, conn);
            using OleDbDataReader WDSResults = command.ExecuteReader();
    ```

```
        while (WDSResults.Read())
        {
            var itemUrl = WDSResults.GetString(0);
            paths.Add(itemUrl);
        }
    }
    return paths.ToArray();
}
```

Here, QueryHelper is used to create a SQL query, and we use OleDbConnection to query the search index for all the matching files.

We can now start editing the Task<ObservableCollection<Photo>> Get(int start, int count, Quality quality = Quality.Low) method. Add the following declaration to the PhotoImporter class:

```
public partial async Task<ObservableCollection<Photo>> Get(int start,
int count, Quality quality)
{
}
```

Now, we will start the implementation of the method by setting up file patterns and the locations we will search:

```
string[] patterns = { ".png", ".jpeg", ".jpg" };

string[] locations = {
Environment.GetFolderPath(Environment.SpecialFolder.MyPictures),
      Environment.GetFolderPath(Environment.SpecialFolder.
CommonPictures),
      Path.Combine(Environment.GetFolderPath(Environment.SpecialFolder.
UserProfile),"OneDrive","Camera Roll")
    };
```

These arrays define the file extensions and folders that we will search for photos. We will then create QueryHelper from the tlbimp-Windows.Search.Interop NuGet package and, using the arrays, configure the query parameters:

```
queryHelper = new CSearchManager().GetCatalog("SystemIndex").
GetQueryHelper();

queryHelper.QueryMaxResults = start + count;

queryHelper.QuerySelectColumns = "System.ItemPathDisplay";
```

```
queryHelper.QueryWhereRestrictions = "AND (";
foreach (var pattern in patterns)
     queryHelper.QueryWhereRestrictions += " Contains(System.
FileExtension, '" + pattern + "') OR";
queryHelper.QueryWhereRestrictions = queryHelper.
QueryWhereRestrictions[..^2];
queryHelper.QueryWhereRestrictions += ")";

queryHelper.QueryWhereRestrictions += " AND (";
foreach (var location in locations)
     queryHelper.QueryWhereRestrictions += " scope='" + location + "'
OR";
queryHelper.QueryWhereRestrictions = queryHelper.
QueryWhereRestrictions[..^2];
queryHelper.QueryWhereRestrictions += ")";

queryHelper.QuerySorting = "System.DateModified DESC";
```

QueryMaxResults is set so that we only retrieve the results we are looking for. Then, we specify that the only data column to return is "System.ItemPathDisplay". Next, we set QueryWhereRestictions from our list of extensions. Note the use of the range operator to remove the trailing "OR" in the query string. We use the same technique to add the locations to QueryWhereRestrictions. Finally, we set the sort order.

The remainder of the method is going to be very similar to those of the previous platforms. If the import succeeds, we will continue to handle what photos should be imported in this loading of images. Conditions are specified with the start and count parameters. Use the following code listing to complete the implementation of the first Get method:

```
var photos = new ObservableCollection<Photo>();

var result = await Import();
if (result?.Length == 0)
{
  return photos;
}

Index startIndex = start;
Index endIndex = start + count;

if (endIndex.Value >= result.Length)
{
  endIndex = result.Length;
}
```

```
if (startIndex.Value > endIndex.Value)
{
  return photos;
}

foreach (var uri in result[startIndex..endIndex])
{
  var path = new System.Uri(uri).AbsolutePath;
  photos.Add(new()
  {
    Bytes = File.ReadAllBytes(path),
    Filename = Path.GetFileName(path)
  });
}
return photos;
```

Let's quickly review the preceding code. The first step is to call the `Import` method and verify that there are photos to import. If there are none, we simply return an empty list. If there are photos to import, then we need to know `startIndex` and `endIndex` in the `photos` array to import. `startIndex` and `endIndex` are adjusted to make sure they are valid for the photos to import. Then, we can read the images from `startIndex` to `endIndex` from the array of photos and return the bytes from the file and just the name of the file for each entry.

Now, we will continue with the other `Task<ObservableCollection<Photo>>` `Get(List<string> filenames, Quality quality = Quality.Low)` method. Add the following declaration to the `PhotoImporter` class:

```
public partial async Task<ObservableCollection<Photo>>
Get(List<string> filenames, Quality quality)
{
}
```

Now, we will start the implementation of the method by setting the search parameters:

```
queryHelper = new CSearchManager().GetCatalog("SystemIndex").
GetQueryHelper();

queryHelper.QuerySelectColumns = "System.ItemPathDisplay";

queryHelper.QueryWhereRestrictions = "AND (";
foreach (var filename in filenames)
    queryHelper.QueryWhereRestrictions += " Contains(System.Filename,
'" + filename + "') OR";
```

```
queryHelper.QueryWhereRestrictions = queryHelper.
QueryWhereRestrictions[..^2];
queryHelper.QueryWhereRestrictions += ")";
```

For this method, we only need to add all the filenames to `QueryWhereRestrictions`. Following that, call the `Import` method, and if it returns results, then use a `foreach` loop to loop through all the photos and to check whether each photo is specified in the `filenames` parameter. If a photo is specified in the `filenames` parameter, read the photo from the path, as in the first `Get` method:

```
var photos = new ObservableCollection<Photo>();

var result = await Import();
if (result?.Length == 0)
{
  return photos;
}

foreach (var uri in result)
{
  var path = new System.Uri(uri).AbsolutePath;
  var filename = Path.GetFileName(path);
  if (filenames.Contains(filename))
  {
    photos.Add(new()
    {
      Bytes = File.ReadAllBytes(path),
      Filename = filename
    });
  }
}
return photos;
```

The photo importers are now finished, and we are ready to write the rest of the app, which will mostly involve adding code that is shared between the platforms.

Writing the app-initializing code

We have now written the code that we will use to get data to the app. Let's continue to build the app, starting with initializing the core parts of the app.

Wiring up dependency injection

By using dependency injection as a pattern, we can keep our code cleaner and more testable. This app will use constructor injection, which means that all the dependencies that a class has must be passed through its constructor. The container then constructs objects for you, so you don't have to care too much about the dependency chain. Since .NET MAUI already includes a dependency injection framework, **Microsoft.Extensions.DependencyInjection**, there is nothing extra to install.

> **Confused about dependency injection?**
>
> Check out the *Wiring up a dependency injection* section in *Chapter 2, Building Our First .NET MAUI App*, for more details on dependency injection.

While it is recommended to use extension methods to group the types together, we have very few types in this app to register so we will use a different method in the next section.

Registering PhotoImporter with dependency injection

Let's add the required code to register the types we have created so far, as shown in the following steps:

1. In the `GalleryApp` project, open `MauiProgram.cs`.

2. Make the following changes to the `MauiProgram` class (the changes are highlighted):

```
using GalleryApp.Services;
using Microsoft.Extensions.Logging;

public static class MauiProgram
{
  public static MauiApp CreateMauiApp()
  {
    var builder = MauiApp.CreateBuilder();
    builder
      .UseMauiApp<App>()
      .ConfigureFonts(fonts =>
      {
        fonts.AddFont("OpenSans-Regular.ttf",
"OpenSansRegular");
        fonts.AddFont("OpenSans-Semibold.ttf",
"OpenSansSemibold");
      });

#if DEBUG
    builder.Logging.AddDebug();
#endif
  builder.Services.AddSingleton<IPhotoImporter>(serviceProvider
=> new PhotoImporter());
    return builder.Build();
  }
```

```
    }
```

The .NET MAUI `MauiAppBuilder` class exposes the `Services` property, which is the dependency injection container. We simply need to add the types we want dependency injection to know about and the container will do the rest for us. Think of a builder as something that collects a lot of information on what needs to be done, and then builds the object we need. It's a very useful pattern on its own, by the way.

We only use the builder for one thing at the moment. Later on, we will use it to register any class in the assembly that inherits from our abstract `ViewModel` class and our views. The container is now prepared for us to ask for these types.

Creating a shell

The main navigation for this app will be tabs at the bottom of the screen. The app will have a fly-out menu with two options—**Home** and **Gallery**:

1. Create a new folder named `Views` in the project.
2. In the `Views` folder, create two new files using the **.NET MAUI ContentPage (XAML)** template—one named `MainView` and one named `GalleryView`.
3. Delete the `MainPage.Xaml` and `MainPage.Xaml.cs` files from the root of the project since we won't be needing those.
4. Open the `AppShell.xaml` file in the root of the project.

    Add both views to the `Shell` object using the `ContentTemplate` property of `ShellContent`. Use the `DataTemplate` markup extension to load the view from the dependency injection container:

    ```xml
    <?xml version="1.0" encoding="UTF-8" ?>
    <Shell
        x:Class="GalleryApp.AppShell"
          xmlns="http://schemas.microsoft.com/dotnet/2021/maui"
         xmlns:x="http://schemas.microsoft.com/winfx/2009/xaml"
        xmlns:local="clr-namespace:GalleryApp"
        xmlns:views="clr-namespace:GalleryApp.Views">

        <ShellContent Title="Home" ContentTemplate="{DataTemplate
    views:MainView}" />
        <ShellContent Title="Gallery" ContentTemplate="{DataTemplate
    views:GalleryView}" />
    </Shell>
    ```

5. Since the views are loaded via `DataTemplates`, they must be registered with dependency injection. Add the highlighted code to `MauiProgram.cs`, after the `IPhotoInmporter` line:

```
        builder.Services.
AddSingleton<IPhotoImporter>(serviceProvider => new
PhotoImporter());

        builder.Services.AddTransient<Views.MainView>();
        builder.Services.AddTransient<Views.GalleryView>();

return builder.Build();
```

Now that we have created a shell, let's continue with some other base code before we start to create the views.

Creating a base view model

Before we create an actual view model, we will create an abstract base view model that all view models can inherit from. The idea behind this base view model is that we can write common code in it. In this case, we will implement the `INotifyPropertyChanged` interface by going through the following steps:

1. In the `GalleryApp` project, create a folder named `ViewModels`.

2. Add a NuGet reference to `CommunityToolkit.Mvvm`; we use `CommunityToolkit.Mvvm` for implementing the `INotifyPropertyChanged` interface, as we have in other chapters.

3. Create a new abstract class named `ViewModel`:

```
namespace GalleryApp.ViewModels;

using CommunityToolkit.Mvvm.ComponentModel;

public abstract partial class ViewModel: ObservableObject
{
    [ObservableProperty]
    [NotifyPropertyChangedFor(nameof(IsNotBusy))]
    private bool isBusy;

    public bool IsNotBusy => !IsBusy;

    abstract protected internal Task Initialize();
}
```

In this app's `ViewModel` class, we have added an abstract method for `Initialize`. Each `ViewModel` implementation will override this method and load the images asynchronously for display. The `IsBusy` and `NotIsBusy` properties are used as flags to indicate when the data has completed loading.

Now, we have a `ViewModel` base that we can use for all `ViewModel` instances that we will create later in this project.

Creating the gallery view

Now, we will start to build the views. We will start with the gallery view, which will show the photos as a grid. We will start with `GalleryViewModel`, and then create `GalleryView`. Creating the view model first allows Visual Studio to use the `GalleryViewModel` definition to check the syntax of the data bindings in the XAML file.

Creating GalleryViewModel

`GalleryViewModel` is the class that will be responsible for fetching the data and handling the view logic. Because photos will be added asynchronously to the photo collection, we don't want to set `IsBusy` to `false` immediately after we call the `Get` method of `PhotoImporter`. We will instead wait 3 seconds first. However, we will also add an event listener to the collection so that we can listen for changes. If the collection changes and there are items in it, we will set `IsBusy` to `false`. Create a class named `GalleryViewModel` in the `ViewModels` folder and add the following code to implement this:

```
namespace GalleryApp.ViewModels;

using CommunityToolkit.Mvvm.ComponentModel;
using GalleryApp.Models;
using GalleryApp.Services;
using System.Collections.ObjectModel;
using System.Threading.Tasks;

public partial class GalleryViewModel : ViewModel
{
    private readonly IPhotoImporter photoImporter;

    [ObservableProperty]
    public ObservableCollection<Photo> photos;

    public GalleryViewModel(IPhotoImporter photoImporter) : base()
    {
        this.photoImporter = photoImporter;
    }
```

```
    override protected internal async Task Initialize()
    {
        IsBusy = true;
        Photos = await photoImporter.Get(0, 20);

        Photos.CollectionChanged += Photos_CollectionChanged;
        await Task.Delay(3000);
        IsBusy = false;
    }

    private void Photos_CollectionChanged(object sender, System.
Collections.Specialized.NotifyCollectionChangedEventArgs e)
    {
        if (e.NewItems != null && e.NewItems.Count > 0)
        {
            IsBusy = false;
            Photos.CollectionChanged -= Photos_CollectionChanged;
        }
    }
}
```

Finally, register `GalleryViewModel` with dependency injection in `MauiProgram`:

```
    builder.Services.AddSingleton<IphotoImporter>(serviceProvider
=> new PhotoImporter());
    builder.Services.AddTransient<ViewModels.GalleryViewModel>();

builder.Services.AddTransient<Views.MainView>();
builder.Services.AddTransient<Views.GalleryView>();

return builder.Build();
```

Now, `GalleryViewModel` is ready, so we can start to create `GalleryView`.

Creating GalleryView

First, we will create a converter that will convert `byte[]` to `Microsft.Maui.Controls.ImageSource`. In the `GalleryApp` project, create a new folder named `Converters`, and inside the folder, create a new class named `BytesToImageConverter`:

```
namespace GalleryApp.Converters;

using System.Globalization;

internal class BytesToImageConverter : IValueConverter
```

```
{
    public object Convert(object value, Type targetType, object
parameter, CultureInfo culture)
    {
        if (value != null)
        {
            var bytes = (byte[])value;
            var stream = new MemoryStream(bytes);
            return ImageSource.FromStream(() => stream);
        }
        return null;
    }

    public object ConvertBack(object value, Type targetType, object
parameter, CultureInfo culture)
    {
        throw new NotImplementedException();
    }
}
```

To use the converter, we need to add it as a resource. We will do this by adding it to a `Resource Dictionary` object in the `Resources` property of `GalleryView`.

Open `GalleryView.xaml`, and add the following highlighted code to the view:

```
<ContentPage  xmlns="http://schemas.microsoft.com/dotnet/2021/maui"
              xmlns:x="http://schemas.microsoft.com/winfx/2009/xaml"
    xmlns:converters="clr-namespace:GalleryApp.Converters"
    x:Class="GalleryApp.Views.GalleryView"
    Title="GalleryView">
    <ContentPage.Resources>
      <ResourceDictionary>
        <converters:BytesToImageConverter x:Key="ToImage" />
      </ResourceDictionary>
    </ContentPage.Resources>
</ContentPage>
```

To be able to bind to `ViewModel`, we will set `BindingContext` to `GalleryViewModel`. Use the constructor dependency injection in `GalleryView.xaml.cs` to create an instance of `GalleryViewModel`.

Open `GalleryView.xaml.cs`, and add the following highlighted code to the class:

```
public GalleryView(GalleryViewModel viewModel)
{
```

```
    InitializeComponent();
    BindingContext = viewModel;
    MainThread.InvokeOnMainThreadAsync(viewModel.Initialize);
}
```

First, we modify the constructor so that dependency injection will provide an instance of GalleryViewModel. That instance is set as BindingContext for the page. This object will be used in the XAML bindings of the view. Finally, we initialize the view model asynchronously.

What we will show in this view is a grid with three columns. To build this with .NET MAUI, we will use the CollectionView control. To specify the layout that CollectionView should have, add a GridItemsLayout element to the ItemsLayout property of CollectionView. Follow these steps to build this view:

1. Navigate to GalleryView.xaml.

    Import the namespaces for GalleryApp.ViewModels and GalleryApp.Models as viewModels and models, respectively:

    ```
    xmlns:converters="clr-namespace:GalleryApp.Converters"
    xmlns:models="clr-namespace: GalleryApp.Models"
    xmlns:viewModels="clr-namespace: GalleryApp.ViewModels"
    x:Class=" GalleryApp.Views.GalleryView"
    ```

2. On ContentPage, set x:DataType to viewModels:GalleryViewModel. This makes the bindings compile, which will make our view faster to render:

    ```
    <CollectionView x:Name="Photos" ItemsSource="{Binding Photos}">
      <CollectionView.ItemsLayout>
        <GridItemsLayout Orientation="Vertical" Span="3"
    HorizontalItemSpacing="0" />
      </CollectionView.ItemsLayout>
      <CollectionView.ItemTemplate>
        <DataTemplate x:DataType="models:Photo">
          <Grid>
            <Image Aspect="AspectFill" Source="{Binding Bytes,
    Converter={StaticResource ToImage}}" HeightRequest="120" />
          </Grid>
        </DataTemplate>
      </CollectionView.ItemTemplate>
    </CollectionView>
    ```

Now, we can see the photos in the view. However, we will also need to create the content that will be shown when we don't have any photos to show as they have not been loaded yet, or if there are no photos available. Add the following highlighted code to create a `DataTemplate` object to show when `CollectionView` doesn't have any data:

```
<CollectionView :Name="Photos" ItemsSource="{Binding Photos}"
EmptyView="{Binding}">
...
    <CollectionView.EmptyViewTemplate>
      <DataTemplate x:DataType="viewModels:GalleryViewModel">
        <Grid>
          <ActivityIndicator IsVisible="{Binding IsBusy}" />
          <Label Text="No photos to import could be found"
IsVisible="{Binding IsNotBusy}" HorizontalOptions="Center"
VerticalOptions="Center" HorizontalTextAlignment="Center" />
        </Grid>
      </DataTemplate>
    </CollectionView.EmptyViewTemplate>
</CollectionView>
```

Now, we can run the app. The next step is to load more photos when a user reaches the end of the view.

Loading photos incrementally

To load more than the first 20 items, we will load photos incrementally so that when users scroll to the end of `CollectionView`, it will start to load more items. `CollectionView` has built-in support for loading data incrementally. Because we get an `ObservableCollection` object back from the photo importer and data is added asynchronously to it, we need to create an event listener to handle when items are added to the photo importer so that we can add it to the `ObservableCollection` instance that we bound to `CollectionView`. Create the event listener by navigating to `GalleryViewModel.cs` and adding the following code at the end of the class:

```
private int itemsAdded;
private void Collection_CollectionChanged(object sender, System.
Collections.Specialized.NotifyCollectionChangedEventArgs args)
{
  foreach (Photo photo in args.NewItems)
  {
    itemsAdded++;
    Photos.Add(photo);
  }
  if (itemsAdded == 20)
  {
    var collection = (ObservableCollection<Photo>)sender;
    collection.CollectionChanged -= Collection_CollectionChanged;
```

```
    }
}

private int currentStartIndex = 0;

[RelayCommand]
public async Task LoadMore()
{
    currentStartIndex += 20;
    itemsAdded = 0;
    var collection = await photoImporter.Get(currentStartIndex, 20);
    collection.CollectionChanged += Collection_CollectionChanged;
}
```

The only thing we have left to do to get the incremental load to work is to bind `CollectionView` to the code we created in `ViewModel`. The following code will trigger the loading of more photos when the user has just five items left:

```
<CollectionView x:Name="Photos" EmptyView="{Binding}"
ItemsSource="{Binding Photos}" RemainingItemsThreshold="5"
RemainingItemsThresholdReachedCommand="{Binding LoadMoreCommand}">
```

Now that we have a view that shows photos and loads them incrementally, we can make it possible to add photos as favorites.

Saving favorites

In `GalleryView`, we want to be able to select favorites that we can show in `MainView`. To do that, we need to store the photos that we have selected so that it remembers our selection. Create a new interface in the `GalleryApp` project named `ILocalStorage` in the `Services` folder:

```
public interface ILocalStorage
{
    void  Store(string filename);
    List<string> Get();
}
```

The easiest way to store/persist data in .NET MAUI is to use the built-in property store. `Preferences` is a static class in the `Microsoft.Maui.Storage` namespace. Follow these steps to use it:

1. Create a new class named `MauiLocalStorage` in the `Services` folder.
2. Implement the `ILocalStorage` interface:

    ```
    namespace GalleryApp.Services;
    ```

```
using System.Text.Json;

public class MauiLocalStorage : ILocalStorage
{
  public const string FavoritePhotosKey = "FavoritePhotos";

  public List<string> Get()
  {
    if (Preferences.ContainsKey(FavoritePhotosKey))
    {
      var filenames = Preferences.Get(FavoritePhotosKey,string.
Empty);
      return JsonSerializer.
Deserialize<List<string>>(filenames);
    }
    return new List<string>();
  }

  public void Store(string filename)
  {
    var filenames = Get();
    filenames.Add(filename);

    var json = JsonSerializer.Serialize(filenames);

    Preferences.Set(FavoritePhotosKey, json);
  }
}
```

To be able to use ILocalStorage with constructor injection, we need to register it with the container. Navigate to the MauiProgram class and add the following highlighted code:

```
builder.Services.AddSingleton<IPhotoImporter>(serviceProvider => new
PhotoImporter());
builder.Services.AddTransient<ILocalStorage>(ServiceProvider => new
MauiLocalStorage());

builder.Services.AddTransient<ViewModels.MainViewModel>();
```

Now, we are ready to use the local storage.

Navigate to the `GalleryViewModel` class, add the `ILocalStorage` interface to the constructor, and assign it to a field:

```
private readonly IPhotoImporter photoImporter;
private readonly ILocalStorage localStorage;

public GalleryViewModel(IPhotoImporter photoImporter, ILocalStorage
localStorage)
{
  this.photoImporter = photoImporter;
  this.localStorage = localStorage;
}
```

The next step is to create a command that we can bind to from the view when we select photos. The command will monitor which photos we have selected and notify other views that we have added favorite photos. We will use `WeakReferenceManager` from `CommunityToolkit` to send messages from `GalleryViewModel` to `MainViewModel`.

Follow these steps to implement the `GalleryViewModel` side:

1. Create a new class in the `Services` folder named `Messages`:

    ```
    namespace GalleryApp.Services;

    internal static class Messages
    {
      public const string FavoritesAddedMessage =
    nameof(FavoritesAddedMessage);
    }
    ```

    This is used to define the message type we are sending to `MainViewModel`.

2. Navigate to `GalleryViewModel`.

3. Create a new method named `AddFavorites` that is attributed to the `RelayCommand` type.

4. Add the following code:

    ```
    [RelayCommand]
    public void AddFavorites(List<Photo> photos)
    {
      foreach (var photo in photos)
      {
        localStorage.Store(photo.Filename);
      }
            WeakReferenceMessenger.Default.Send<string>(Messages.
    ```

```
FavoritesAddedMessage);
}
```

Now, we are ready to start working with the view. The first thing we will do is make it possible to select photos. Navigate to `GalleryView.xaml` and set the `SelectionMode` mode of `CollectionView` to `Multiple` to make it possible to select multiple items:

```
<CollectionView x:Name="Photos"
EmptyView="{Binding}" ItemsSource="{Binding Photos}"
SelectionMode="Multiple" RemainingItemsThreshold="5"
RemainingItemsThresholdReachedCommand="{Binding LoadMore}">
```

When a user selects a photo, we want it to be clear which photos have been selected. To achieve this, we will use `VisualStateManager`. We will do this by creating a style for `Grid` and setting `Opacity` to `0.5`, as in the following code. Add the code to `Resources` of the page:

```
<ContentPage.Resources>
  <ResourceDictionary>
    <converters:BytesToImageConverter x:Key="ToImage" />
      <Style TargetType="Grid">
        <Setter Property="VisualStateManager.VisualStateGroups">
          <VisualStateGroupList>
            <VisualStateGroup x:Name="CommonStates">
              <VisualState x:Name="Normal" />
                <VisualState x:Name="Selected">
                  <VisualState.Setters>
                    <Setter Property="Opacity" Value="0.5" />
                  </VisualState.Setters>
                </VisualState>
              </VisualStateGroup>
            </VisualStateGroupList>
          </Setter>
        </Style>
  </ResourceDictionary>
</ContentPage.Resources>
```

To save the selected photos, we will create a toolbar item that the user can tap:

1. Add `ToolbarItem` with the `Text` property set to `Select`.

2. Add an event handler named `SelectToolBarItem_Clicked`:

    ```
    <ContentPage.ToolbarItems>
        <ToolbarItem Text="Select" Clicked="SelectToolBarItem_
    Clicked" />
    </ContentPage.ToolbarItems>
    ```

3. Navigate to the code behind the `GalleryView.xaml.cs` file.

4. Add the following `using` statements:

```
using GalleryApp.Models;
using GalleryApp.ViewModels;
```

5. Create an event handler named `SelectToolBarItem_Clicked`:

```
private void SelectToolBarItem_Clicked(object sender, EventArgs
e)
{
  if (!Photos.SelectedItems.Any())
  {
    DisplayAlert("No photos", "No photos selected", "OK");
    return;
  }
  var viewModel = (GalleryViewModel)BindingContext;
  viewModel.AddFavoritesCommand.Execute(Photos.
SelectedItems.Select(x =>(Photo)x).ToList());

  DisplayAlert("Added", "Selected photos have been added to
favorites", "OK");
}
```

Now that we are done with `GalleryView`, we will continue with the main view, which will show the latest photos and the favorite photos in two carousels.

Creating the carousels for MainView

The last view in this app is `MainView`, which is the view that is visible when users start the app. This view will show two carousel views—one with recent photos and one with favorite photos.

Creating the view model for MainView

We will start by creating `ViewModel` that we will use for the view. In the `ViewModel` folder, create a new class named `MainViewModel`:

```
namespace GalleryApp.ViewModels;

using CommunityToolkit.Mvvm.ComponentModel;
using CommunityToolkit.Mvvm.Messaging;
using GalleryApp.Models;
using GalleryApp.Services;
using System.Collections.ObjectModel;
```

```csharp
public partial class MainViewModel : ViewModel
{
  private readonly IPhotoImporter photoImporter;
  private readonly ILocalStorage localStorage;

  [ObservableProperty]
  private ObservableCollection<Photo> recent;

  [ObservableProperty]
  private ObservableCollection<Photo> favorites;

  public MainViewModel(IPhotoImporter photoImporter, ILocalStorage
localStorage)
  {
    this.photoImporter = photoImporter;
    this.localStorage = localStorage;
  }

  override protected internal async Task Initialize()
  {
    var photos = await photoImporter.Get(0, 20, Quality.Low);

    Recent = photos;
    await LoadFavorites();
        WeakReferenceMessenger.Default.Register<string>(this, async
(sender, message) => {
      if( message == Messages.FavoritesAddedMessage )
      {
        await MainThread.InvokeOnMainThreadAsync(LoadFavorites);
      }
    });
  }

  private async Task LoadFavorites()
  {
    var filenames = localStorage.Get();
    var favorites = await photoImporter.Get(filenames, Quality.Low);

    Favorites = favorites;
  }
}
```

In the preceding code, the `Initialize` method is used to register a callback with `Weak ReferenceManager`. This callback invokes the `LoadFavorites` method if the message sent was `Message.FavoritesAddedMessage`. Recall that `Messages.Favorites AddedMessage` is sent from `GalleryViewModel` after selecting new photos.

In the `LoadFavorites` method, the favorites are loaded from the storage provider instance in `localStorage`. Then, the photos from the favorites are imported using the `photoImporter` instance.

We need to add the view model to dependency injection so that we can use it in the view. Open `MauiProgram` and add the highlighted code:

```
builder.Services.AddTransient<ILocalStorage>(ServiceProvider
=> new MauiLocalStorage());

builder.Services.AddTransient<ViewModels.MainViewModel>();
builder.Services.AddTransient<ViewModels.GalleryViewModel>();
```

Now that we have created `MainViewModel`, we will continue with the latest photos.

Showing the latest photos

We are now ready to set up the carousel views. We have already created the view model, so we can use the view model to populate the view with content.

Let's look at the steps to create the view:

1. In the constructor of the code, behind the `MainView.xaml.cs` file, set `ViewModel` to `BindingContext`:

    ```
    public MainView(MainViewModel viewModel)
    {
        InitializeComponent();
        BindingContext = viewModel;
            MainThread.InvokeOnMainThreadAsync(viewModel.
    Initialize);
    }
    ```

2. Navigate to `MainView.xaml`.

3. Add the following code:

    ```
    <ContentPage xmlns="http://schemas.microsoft.com/dotnet/2021/
    maui"
                 xmlns:x="http://schemas.microsoft.com/winfx/2009/
    xaml"
        xmlns:converters="clr-namespace:GalleryApp.Converters"
        xmlns:viewModels="clr-namespace:GalleryApp.ViewModels"
    ```

```
      xmlns:models="clr-namespace:GalleryApp.Models"
      x:Class="GalleryApp.Views.MainView"
      x:DataType="viewModels:MainViewModel"
      Title="My Photos">
  <ContentPage.Resources>
    <ResourceDictionary>
      <converters:BytesToImageConverter x:Key="ToImage" />

    </ResourceDictionary>
  </ContentPage.Resources>
  <Grid>
    <Grid.RowDefinitions>
      <RowDefinition Height="*" />
      <RowDefinition Height="50" />
      <RowDefinition Height="*" />
      <RowDefinition Height="20" />
    </Grid.RowDefinitions>

    <CarouselView ItemsSource="{Binding Recent}"
PeekAreaInsets="40,0,40,0" >
      <CarouselView.ItemsLayout>
        <LinearItemsLayout
Orientation="Horizontal"  SnapPointsAlignment="Start"
SnapPointsType="Mandatory" />
      </CarouselView.ItemsLayout>
      <CarouselView.ItemTemplate>
        <DataTemplate x:DataType="models:Photo">
          <Image Source="{Binding Bytes,
Converter={StaticResource ToImage}}" Aspect="AspectFill" />
        </DataTemplate>
      </CarouselView.ItemTemplate>
    </CarouselView>
  </Grid>
</ContentPage>
```

The CarouselView control is used to present data to the user in a scrollable layout, where the user can swipe to move through the collection of items. It is very similar to CollectionView; however, the uses of the two controls are different. You would use CollectionView when you want to display a list of items with an indeterminate length, and CarouselView is used to highlight items from a list of items with a limited length. Since CarouselView shares implementations with the CollectionView control, it uses the familiar ItemTemplate property to customize how each item is displayed. It adds an ItemsLayout property to define how the collection of items is displayed. CarouselView can use either a Horizontal or Vertical layout direction, with Horizontal being the default.

In `MainView`, `CarouselView` is used to display the `Recent` photos from `MainViewModel`. `ItemsLayout` is customized to set the scrolling behavior so that items will snap into view using the start, or left edge of the image. The `SnapPointType` property set to `Mandatory` makes sure that `CarouselView` snaps the image into place after scrolling, which would ensure a single image is always in view.

`ItemsTemplate` is used to display an image that is data-bound to each photo and displays the image from the bytes in the `Photo` model. `BytesToImageConverter` converts the byte array from the `Photo` model into `ImageSource` that can be displayed by the `Image` control. The `Image` control has the `Aspect` property set to `AspectFill`, allowing the image control to resize the image, maintaining the aspect ratio of the source image to fill the available visible space.

Now that we have shown the latest photos in a carousel, the next (and the last) step is to show the favorite photos in another carousel.

Showing the favorite photos

The last thing we will do in this app is add a carousel to show favorite photos. Add the following highlighted code inside `Grid`, after the first `CarouselView`, as shown in the following code snippet:

```
<Grid>
  <!—Code omitted for brevity -->
  <CarouselView>
    <!—Code omitted for brevity -->
  </CarouselView>
  <Label Grid.Row="1" Margin="10" Text="Favorites" FontSize="Subtitle"
FontAttributes="Bold" />
  <CarouselView Grid.Row="2" ItemsSource="{Binding Favorites}"
PeekAreaInsets="0,0,40,0" IndicatorView="Indicator">
    <CarouselView.ItemsLayout>
      <LinearItemsLayout Orientation="Horizontal"
SnapPointsAlignment="Start" SnapPointsType="MandatorySingle" />
    </CarouselView.ItemsLayout>
    <CarouselView.EmptyViewTemplate>
      <DataTemplate>
        <Label Text="No favorites selected" />
      </DataTemplate>
    </CarouselView.EmptyViewTemplate>
    <CarouselView.ItemTemplate>
      <DataTemplate x:DataType="models:Photo">
        <Border Grid.RowSpan="2" StrokeShape="RoundRectangle
15,15,15,15" Padding="0" Margin="0,0,0,0" BackgroundColor="#667788" >
          <Image Source="{Binding Bytes, Converter={StaticResource
ToImage}}" Aspect="AspectFill" />
```

```
        </Border>
      </DataTemplate>
    </CarouselView.ItemTemplate>
  </CarouselView>
  <IndicatorView Grid.Row="3" x:Name="Indicator"
HorizontalOptions="Center" SelectedIndicatorColor="Red"
IndicatorColor="LightGray" />
</Grid>
```

For the Favorites photos, again, CarouselView is used with a few changes from CarouselView displaying the Recent photos. The most visible change is that the ItemsLayout property is now using MandatorySingle for the value of SnapPointsType. This forces a behavior that only allows the user to swipe one image at a time, snapping each image into view.

The ItemTemplate property has also been changed to add a rounded border around each image, with a background color.

New to this CarouselView is the EmptyViewTemplate property. This is used to display the text "No favorites selected" when the Favorites property is empty.

Finally, IndicatorView was added to provide the user with a visual cue of how many items are in CarouselView and which item is currently displayed. CarouselView is connected to IndicatorView by the IndicatorView property of CarouselView. The IndicatoryView property is set to the x:Name property of IndicatorView. The IndicatorView displays on the page as a series of horizontal light gray dots, with the dot representing the current image in red.

That is all—now, we can run the app and see both the most recent photos and the photos that have been marked as favorites.

Summary

In this chapter, we focused on photos. We learned how to import photos from the platform-specific photo galleries and how we can display them as a grid using CollectionView and in carousels using CarouselView. This makes it possible for us to build other apps and provides multiple options for presenting data to users, as we can now pick the best method for the situation.

Additionally, we learned about permissions and how to check and request permission to use protected resources in our app.

If you are interested in extending the app even further, try creating a page to view the details of the photo, or to view the photo in full screen by tapping on the photo.

In the next chapter, we will build an app using location services and look at how to visualize location data on a map.

7

Building a Location Tracking App Using GPS and Maps

In this chapter, we will create a location tracking app that saves the location of the user and displays it as a heat map. We will learn how to run tasks in the background on iOS, macOS, and Android devices. We will extend the .NET MAUI Map control to display the map with the saved locations directly in the map.

The following topics will be covered in this chapter:

- Tracking the location of a user in the background on an iOS device and a macOS device
- Tracking the location of a user in the background on an Android device
- How to show maps in a .NET MAUI app
- How to extend the functionality of .NET MAUI maps

Let's get started!

Technical requirements

To be able to complete this project, you'll need to have Visual Studio for Mac or Windows installed, as well as the .NET MAUI components. See *Chapter 1, Introduction to .NET MAUI*, for more details on how to set up your environment. To build an iOS app using Visual Studio for Windows, you must have a Mac connected. If you don't have access to a Mac at all, you can just complete the Android part of this project.

You can find the full source for the code in this chapter at `https://github.com/PacktPublishing/MAUI-Projects-3rd-Edition`.

> **Important information for Windows users**
>
> At the time of writing, there was no Map control for the Windows platform in .NET MAUI. This is due to the lack of a Map control in the underlying WinUI platform. For the latest information on Map support in Windows, visit the Map documentation at `https://learn.microsoft.com/en-us/dotnet/maui/user-interface/controls/map`.

Project overview

Many apps can be made richer by adding a map and location services. In this project, we will build a location tracking app that we will call `MeTracker`. This app will track the position of the user and save it to an SQLite database so that we can visualize the result in the form of a heat map. To build this app, we will learn how to set up processes in the background on iOS, macOS, and Android. Luckily for us, the iOS and macOS implementations are identical; however, the Android implementation is very different. For the map, we will use the .NET MAUI Maps component and extend its functionality to build a heat map.

Due to the lack of Map support on Windows, and just for some variety, this chapter will use Visual Studio for Mac screenshots and references. If you don't have a Mac, don't worry; you can still complete the project for Android on your Windows development machine. If you need help with the steps, look at some of the earlier chapters for equivalent steps.

The estimated build time for this project is 180 minutes.

Building the MeTracker app

It's time to start building the app. Use the following steps to create a project from a template:

1. Open Visual Studio for Mac and click **New**:

Figure 7.1 – Visual Studio for Mac start screen

2. In the **Choose a template for your new project** dialog, use the **.NET MAUI App** template, which is under **Multiplatform | App**; then, click **Continue**:

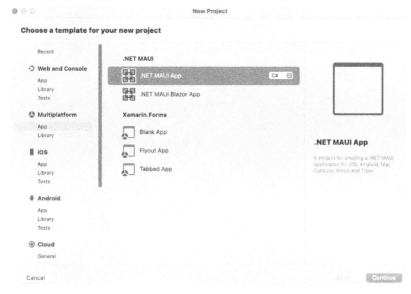

Figure 7.2 – New project

3. In the **Configure your new .NET MAUI App** dialog, ensure the **.NET 7.0** target framework is selected, then click **Continue**:

Figure 7.3 – Choosing a target framework

4.   In the **Configure your new .NET MAUI App** dialog, name the project `MeTracker`, and then click **Create**:

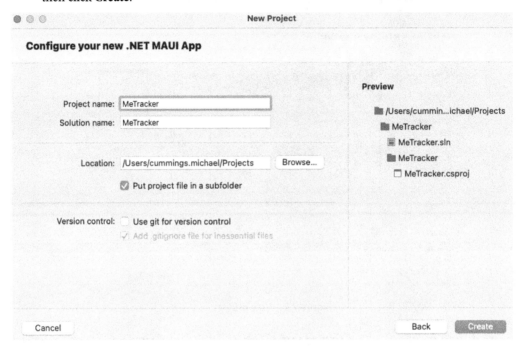

Figure 7.4 – Naming the new app

If you run the app now, you should see something like the following:

Figure 7.5 – MeTracker app on macOS

Now that we have created a project from a template, it's time to start coding!

Creating a repository to save the locations of the users

The first thing we will do is create a repository that we can use to save the locations of the users.

Creating a model for the location data

Before we create a repository, we will create a model class that will represent a user location. Follow these steps to do so:

1. Create a new `Models` folder that we can use for all our models.

2. Create a `Location` class in the `Models` folder and add properties for `Id`, `Latitude`, and `Longitude`.

3. Create two constructors – one that's empty and one that takes `latitude` and `longitude` as arguments. Use the following code to do so:

```
Using'System;

namespace MeTracker.Models;

public class Location
{
    public Location() {}

    public Location(double latitude, double longitude)
    {
        Latitude = latitude;
        Longitude = longitude;
    }

    public int Id { get; set; }
    public double Latitude { get; set; }
    public double Longitude { get; set; }
}
```

Now that we have created a model, we can start creating a repository.

Creating a repository

First, we will create an interface for the repository. Follow these steps to do so:

1. Create a new folder called `Repositories`.

2. In our new folder, create an interface called `ILocationRepository`.

3. Write the following code in the new file that we created for the interface:

```
using MeTracker.Models;
using System;
using System.Threading.Tasks;

namespace MeTracker.Repositories;

public interface ILocationRepository
{
    Task SaveAsync(Models.Location location);
}
```

Now that we have an interface, we need to create an implementation of it. Follow these steps to do so:

1. Create a new `LocationRepository` class in the `Repositories` folder.

2. Implement the `ILocationRepository` interface and add the `async` keyword to the `SaveAsync` method using the following code:

```
using System;
using System.Threading.Tasks;
using MeTracker.Models;

namespace MeTracker.Repositories;

public class LocationRepository : ILocationRepository
{
    public async Task SaveAsync(Models.Location location)
    {
    }
}
```

A word on the Async suffix

You will see in this and many other chapters in this book the use of `Async` as a suffix on methods. Appending a suffix of `Async` to all asynchronous methods is a .NET convention. How do we know whether a method is asynchronous in an interface where you can't see the `async` keyword? It will most likely return a `Task` or `ValueTask` object. There are some cases where an asynchronous method will return `void`; however, that is frowned upon, as Stephen Cleary explains in his article at https://msdn.microsoft.com/en-us/magazine/jj991977.aspx, so you won't see it used in this book.

To store the data, we will use an SQLite database and the **object-relational mapper** (**ORM**) known as SQLite-net so that we can write code against a domain model instead of using SQL to perform operations against the database. This is an open source library that was created by Frank A. Krueger. Let's set this up by going through the following steps:

1. Add a reference to `sqlite-net-pcl` by right-clicking the `Dependencies` node in **Solution Explorer**:

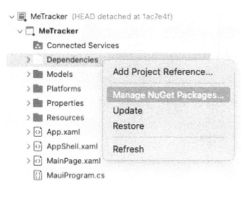

Figure 7.6 – Adding the NuGet package

I. Select **Manage NuGet Packages…** from the context menu to open the **NuGet Packages** window.

II. Check the **Include prereleases** checkbox and enter `sqlite-net-pcl` into the search box as shown next:

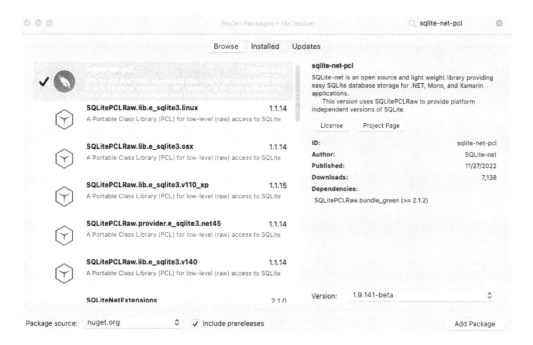

Figure 7.7 – Adding the sqlite-net-pcl package

III.   Finally, check the box next to `sqlite-net-pcl` and click **Add Package**.

2. Go to the `Location` model class and add `PrimaryKeyAttribute` and `Auto IncrementAttribute` attributes to the `Id` property. When we add these attributes, the `Id` property will be a primary key in the database, and a value for it will be automatically created. The `Location` class should now look like the following:

```
using SQLite;

namespace MeTracker.Models;

public class Location
{
    public Location() { }

    public Location(double latitude, double longitude)
    {
        Latitude = latitude;
        Longitude = longitude;
    }

    [PrimaryKey]
    [AutoIncrement]
    public int Id { get; set; }
    public double Latitude { get; set; }
    public double Longitude { get; set; }
}
```

3. Write the following code in the `LocationRepository` class to create a connection to the SQLite database. An `if` statement is used to check whether we have already created a connection. If we have, we won't create a new one; instead, we will use the connection that we've already created:

```
private SQLiteAsyncConnection connection;
private async Task CreateConnectionAsync()
{
    if (connection != null)
    {
        return;
    }

    var databasePath = Path.Combine(Environment.GetFolderPath
    (Environment.SpecialFolder .MyDocuments), "Locations.db");

    connection = new SQLiteAsyncConnection(databasePath);
```

```
        await connection.CreateTableAsync<Location>();
    }
```

Now, it's time to implement the `SaveAsync` method, which will take a `location` object as a parameter and store it in the database.

We will use the `CreateConnectionAsync` method in the `SaveAsync` method to ensure that a connection is created when we try to save data to the database. When we know that we have an active connection, we can just use the `InsertAsync` method and pass the `location` parameter of the `SaveAsync` method as an argument.

Edit the `SaveAsync` method in the `LocationRepository` class so that it looks like this:

```
public async Task SaveAsync(Models.Location location)
{
    await CreateConnectionAsync();
    await connection.InsertAsync(location);
}
```

That wraps up the repository for now, so let's move on to the location tracking service.

Creating a service for location tracking

To track a user's location, we need to write code according to the platform. .NET MAUI has methods for getting the location of a user, but it cannot be used in the background. To be able to use the code that we will write for each platform, we need to create an interface. For the `ILocationRepository` interface, there is just one implementation that will be used on both platforms (iOS and Android), whereas for the location tracking service, we will have one implementation for each platform.

Go through the following steps to create an `ILocationTrackingService` interface:

1. Create a new folder called `Services`.

2. Create a new `ILocationTrackingService` interface in the `Services` folder.

3. In the interface, add a method called `StartTracking`, as shown in the following code snippet:

    ```
    public interface ILocationTrackingService
    {
        void StartTracking();
    }
    ```

To make sure we can run and test our app while we implement the location tracking service for each platform, we will use a partial class. The main part of the class will be in the shared code section of the project and the platform-specific portions of the class will be in the platform-specific folders. We will come back to each implementation later in this chapter.

Create a class called `LocationTrackingService` in the `Services` folder, as shown:

```
public partial class LocationTrackingService :
ILocationTrackingService
{
    public void StartTracking()
    {
        StartTrackingInternal();
    }

    partial void StartTrackingInternal();
}
```

We are using an interface to abstract our implementation. We are also using a partial class to abstract each specific implementation, but providing a base implementation so that we don't have to have an implementation for every platform immediately. However, the two methods (partial classes and base class inheritance) do not play together with the same method.

An implementation of the `StartTracking` interface method requires the `public` keyword, which would look like this:

```
public void StartTracking() {}
```

Then, make it partial, like so:

```
public partial void StartTracking() {}
```

The compiler complains that there is no initial definition of the partial method – that is, one that has no implementation.

Remove the empty definition, like so:

```
public partial void StartTracking();
```

The compiler now complains because it has an accessibility modifier, `public`.

There is just no making the compiler happy in this case. Therefore, to avoid these issues, we implement the `StartTracking` interface method by calling a `StartTrackingInternal` partial method. We will visit the implementation of `StartTrackingInternal` for each platform later in this chapter; for now, the app should compile and run, even though we haven't implemented `StartTrackingInternal`.

Now that we have the interface and base implementation of the location tracking service, we can turn our attention to the app logic and user interface.

Setting up the app logic

Now that we have created the interfaces, we need to track the location of the user and save it locally on the device. It's time to write some code so that we can start tracking a user. We still don't have any code that tracks the location of the user, but it will be easier to write this if we have already written the code that starts the tracking process.

Creating a view with a map

To start with, we will create a view with a simple map that is centered on the position of the user. Let's set this up by going through the following steps:

1. Create a new folder called Views.

2. In the Views folder, create a XAML-based ContentPage template and name it MainView:

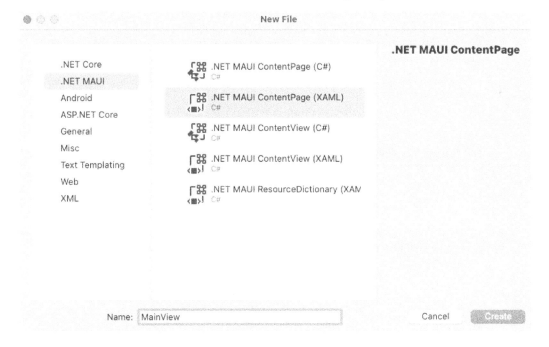

Figure 7.8 – Adding the .NET MAUI XAML ContentPage component

3. Add a reference to Microsoft.Maui.Controls.Maps by right-clicking the Dependencies node in **Solution Explorer**:

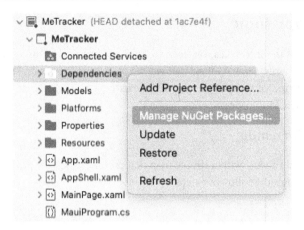

Figure 7.9 – Adding the NuGet package

I.   Select **Manage NuGet Packages…** from the context menu to open the **NuGet package manager** window.

II.   Type Microsoft.Maui.Controls.Maps into the search box as shown next:

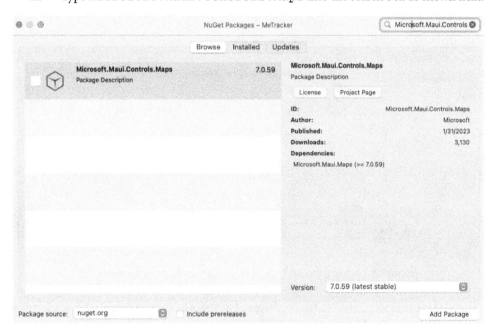

Figure 7.10 – Adding the .NET MAUI Maps package

III. Finally, check the box next to Microsoft.Maui.Controls.Maps and click **Add Package**.

4. Add the Map initialization code by opening the `MauiProgram.cs` file and making the highlighted change:

```
var builder = MauiApp.CreateBuilder();
builder
    .UseMauiApp<App>()
    .ConfigureFonts(fonts =>
    {
        fonts.AddFont("OpenSans-Regular.ttf",
"OpenSansRegular");
        fonts.AddFont("OpenSans-Semibold.ttf",
"OpenSansSemibold");
    })
    .UseMauiMaps();

    return builder.Build();
```

5. Add the namespace for `Microsoft.Maui.Controls.Maps` to `MainView` using the following highlighted code:

```
<ContentPage xmlns="http://schemas.microsoft.com/
dotnet/2021/maui"
             xmlns:x="http://schemas.microsoft.com/
winfx/2009/xaml"
    xmlns:maps="clr-namespace:Microsoft.Maui.Controls.
Maps;assembly=Microsoft.Maui.Controls.Maps"
    x:Class="MeTracker.Views.MainView"
    Title="MainView">
```

Now, we can use the map in our view. Because we want Map to cover the whole page, we can add it to the root of `ContentPage`.

6. Add Map to `ContentPage` with a name so that we can access it from the code-behind file. Name it Map, as shown in the following code snippet:

```
<ContentPage xmlns="http://schemas.microsoft.com/dotnet/2021/
maui"
             xmlns:x="http://schemas.microsoft.com/winfx/2009/
xaml"
    xmlns:maps="clr-namespace:Microsoft.Maui.Controls.
Maps;assembly=Microsoft.Maui.Controls.Maps"
    x:Class="MeTracker.Views.MainView"
    Title="MainView">
    <maps:Map x:Name="Map" />
</ContentPage>
```

Before we can start the app to see the Map control for the first time, we need to set the shell to use our new `MainView` template instead of the default `MainPage` template. But first, we will delete the `MainPage.xaml` and `MainPage.xaml.cs` files that we created when we started the project since we won't be using them here:

1. Delete the `MainPage.xaml` and `MainPage.xaml.cs` files in the project since we will be setting our `MainView` template as the first view that the user sees.

2. Edit the `AppShell.xaml` file, as shown in the following highlighted code:

```
<Shell
    x:Class="MeTracker.AppShell"
    xmlns="http://schemas.microsoft.com/dotnet/2021/maui"
    xmlns:x="http://schemas.microsoft.com/winfx/2009/xaml"
    xmlns:local="clr-namespace:MeTracker"
    xmlns:views="clr-namespace:MeTracker.Views"

    Shell.FlyoutBehavior="Disabled">

    <ShellContent
        Title="Home"
        ContentTemplate="{DataTemplate views:MainView}"
        Route="MainView" />

</Shell>
```

Could we have used the existing `MainPage` template as it was? Sure – it really doesn't make any difference to the compiler what the XAML file is named or where it is located, but for consistency and by MVVM convention in .NET MAUI, we put our *pages* in the `Views` folder and suffix page names with `Views`.

Choosing either Mac Catalyst or an iOS simulator and running the app will produce the result shown in *Figure 7.11*. Android won't work until we have completed the next section:

Figure 7.11 – Running the app after adding the Map control

Now that we have a page with the Map control on it, we will need to make sure we have permission from the user to use location information.

Declaring platform-specific location permissions

To use the Map control, we need to declare that we require permission to location information. The Map control will make the runtime request if it is required. iOS/Mac Catalyst and Android each have their own way of declaring the required permissions. We will start with iOS/Mac Catalyst, following which we will do Android.

Open the info.plist file in the Platforms/iOS folder into **Property List Editor** by double-clicking on it. Add two new entries to the file, highlighted in the next screenshot:

Figure 7.12 – Editing info.plist for iOS

Make the same changes in the `info.plist` file in the `Platforms/MacCatalyst` folder, as shown next:

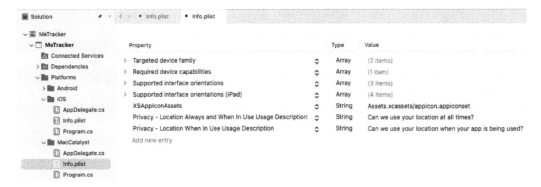

Figure 7.13 – Editing info.plist for Mac Catalyst

> **Windows users**
>
> To edit the `info.plist` files on Windows, you need to open the file in a text editor by right-clicking the file, selecting **Open With…**, and then choosing **XML Editor**. Then, add the entries highlighted in the next code snippet.

Editing the `info.plist` file using the **Property List Editor** results in the changes highlighted next:

```
<key>XSAppIconAssets</key>
<string>Assets.xcassets/appicon.appiconset</string>
<key>NSLocationAlwaysAndWhenInUseUsageDescription</key>
<string>Can we use your location at all times?</string>
<key>NSLocationWhenInUseUsageDescription</key>
<string>Can we use your location when your app is being used?</string>
</dict>
</plist>
```

To track the location of the user in the background with Android, we need to declare five permissions, as shown in the following table:

ACCESS_COARSE_LOCATION	To get an approximate location for the user
ACCESS_FINE_LOCATION	To get a precise location for the user
ACCESS_NETWORK_STATE	We need this because the location services in Android use information from a network to determine the location of the user
ACCESS_WIFI_STATE	We need this because the location services in Android use information from a Wi-Fi network to determine the location of the user
RECEIVE_BOOT_COMPLETED	So that the background job can start again after the device is rebooted

The following steps will declare the required permissions for our app:

1. Open the `MainActivity.cs` file in the `Platforms/Android` folder.

2. After the `using` declarations' block method, add the following `assembly` attributes, highlighted next:

```
using Android.App;
using Android.Runtime;

[assembly: UsesPermission(Android.Manifest.Permission.
AccessCoarseLocation)]
[assembly: UsesPermission(Android.Manifest.Permission.
AccessFineLocation)]
[assembly: UsesPermission(Android.Manifest.Permission.
AccessWifiState)]
[assembly: UsesPermission(Android.Manifest.Permission.
ReceiveBootCompleted)]

namespace MeTracker;
```

Note that we don't declare `Android.Manifest.Permission.AccessNetworkState` because it is part of the .NET MAUI template.

Now that we have declared all the permissions that we require, we can enable the map services on Android.

Android requires an **API key** for **Google Maps** to work with maps. The Microsoft documentation regarding how to obtain an API key can be found at `https://learn.microsoft.com/en-us/dotnet/maui/user-interface/controls/map?view=net-maui-7.0#get-a-google-maps-api-key`. Follow those instructions to obtain your Google Maps key, then use your key in the following steps to configure the Google Maps API key in the app:

1. Open `AndroidManifest.xml`, which is in the `Platforms/Android` folder, by right-clicking on the file and selecting **Open With…**, then selecting **XML (Text) Editor**.

2. Insert a metadata element as a child of the application element, as shown in the following highlighted code, replacing "{YourKeyHere}" with the key you obtained from Google:

```
<application android:label="MeTracker.Android">
    <meta-data android:name="com.google.android.geo.API_KEY"
android:value="{YourKeyHere}" />
</application>
```

Recent versions of Android and iOS have changed how permissions are handled. Certain permissions such as location are not granted without explicit approval from the user while the app is running. It is also possible that the user can deny permissions. Let's look at how to handle runtime permission requests in the next section.

Requesting location permission at runtime

Before we can use the location of the user, we need to request permissions from the user. .NET MAUI has cross-platform permission APIs, and we just need a tiny bit of code to make handling the request a little nicer. To implement the permission request handling, follow these steps:

1. Create a new class called AppPermissions in the root of the project.

2. Edit the new file to look like the following:

```
namespace MeTracker;

internal partial class AppPermissions
{
    internal partial class AppPermission : Permissions.
LocationWhenInUse
    {
    }

    public static async Task<PermissionStatus>
CheckRequiredPermissionAsync() => await Permissions.
CheckStatusAsync<AppPermission>();

    public static async Task<PermissionStatus>
CheckAndRequestRequiredPermissionAsync()
    {
        PermissionStatus status = await Permissions.
CheckStatusAsync<AppPermission>();

        if (status == PermissionStatus.Granted)
            return status;

        if (status == PermissionStatus.Denied && DeviceInfo.
Platform == DevicePlatform.iOS)
```

```
        {
            // Prompt the user to turn on in settings
            // On iOS once a permission has been denied it may
not be requested again from the application
            await App.Current.MainPage.DisplayAlert("Required
App Permissions", "Please enable all permissions in Settings for
this App, it is useless without them.", "Ok");
        }

        if (Permissions.ShouldShowRationale<AppPermission>())
        {
            // Prompt the user with additional information as to
why the permission is needed
            await App.Current.MainPage.DisplayAlert("Required
App Permissions", "This is a location based app, without these
permissions it is useless.", "Ok");
        }

        status = await MainThread.
InvokeOnMainThreadAsync(Permissions.
RequestAsync<AppPermission>);
        return status;
    }

}
```

This creates a type named `AppPermission` that derives from the default .NET MAUI `LocationWhenInUse` permission class.

The `CheckRequiredPermission` method is used to ensure our app has the right permissions before we attempt any operations that might fail if we don't. Its implementation is to call the .NET MAUI `CheckSyncStatus` method with our `AppPermission` type. It returns a `PermissionStatus` type, which is an enum. We are mostly interested in the `Denied` and `Granted` values.

The `CheckAndRequestRequiredPermission` method handles the intricacies of requesting access from the user. The first step is to simply check and see whether the permission has already been granted, and if it has, return the status. Next, if we are on iOS and the permission has been denied, it cannot be requested again, so you must instruct the user on how to grant permission to the app by using the settings panel. Android includes in the request behavior the ability to nag the user if they have denied access. This behavior is exposed through .NET MAUI with the `ShouldShowRationale` method. It will return `false` for any platform that does not support this behavior, and on Android, it will return `true` the first time the user denies access and `false` if the user denies it a second time. Finally, we request access to the `AppPermission` type from the user. Again, .NET MAUI is hiding all the platform implementation details from us, making checking and requesting access to certain resources very straightforward.

Now that we have the `AppPermissions` class in place, we can use it to request the current location of the user, and center that map on that location.

Centering the map on the current user location

We will center the map on the position of the user in the constructor of `MainView.xaml.cs`. Because we want to fetch the user's location asynchronously and this needs to be executed on the main thread, we will use `MainThread.BeginInvokeOnMainThread` to run an anonymous method on the main thread. Once we have the location, we can use the `MoveToRegion` method of `Map`. We can set this up by going through the following steps:

1. Open `MainView.xaml.cs`.

2. Add the highlighted code shown in the following code snippet to the constructor of the `MainView.xaml.cs` class:

```
public MainView ()
{
    InitializeComponent ();

    MainThread.BeginInvokeOnMainThread(async () =>
    {
        var status = await AppPermissions.
CheckAndRequestRequiredPermissionAsync ();
        if (status == PermissionStatus.Granted)
        {
            var location = await Geolocation.
GetLastKnownLocationAsync ();

            if (location == null)
            {
                location = await Geolocation.GetLocationAsync ();
            }

            Map.MoveToRegion(MapSpan.FromCenterAndRadius (
                location,
                Distance.FromKilometers (50)));
        }
    });
}
```

If you run the application now, it should look something like the following:

Figure 7.14 – Map centered on user location

Now that we have the map displaying our current location, let's start building the logic of the rest of the app, starting with our `ViewModel` class.

Creating a ViewModel class

Before we create an actual `ViewModel` class, we will create an abstract base view model that all view models can inherit from. The idea behind this base view model is that we can write common code in it. In this case, we will implement the `INotifyPropertyChanged` interface by using the `CommunityToolkit.Mvvm` NuGet package. To add the package, follow these steps:

1. Add a reference to `CommunityToolkit.Mvvm` by right-clicking the `Dependencies` node in **Solution Explorer**:

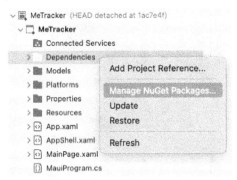

Figure 7.15 – Adding the NuGet package

2. Select **Manage NuGet Packages…** from the context menu to open the **NuGet package manager** window.

3. Type `CommunityToolkit.Mvvm` into the search box, as shown next:

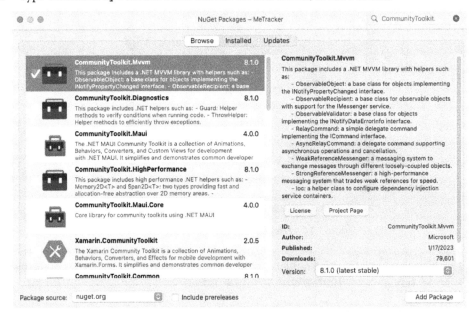

Figure 7.16 – Adding the CommunityToolkit.Mvvm package

4. Finally, check the box next to `CommunityToolkit.Mvvm` and click **Add Package**.

Now, we can create a `ViewModel` class by going through the following steps:

1. Create a folder called `ViewModels` in the project.

2. Create a new class called `ViewModel`.

3. Modify the template code to match the following:

```
using CommunityToolkit.Mvvm.ComponentModel;

namespace MeTracker.ViewModels;

public partial class ViewModel : ObservableObject
{
}
```

The next step is to create the actual view model that will use `ViewModel` as a base class. Let's set this up by going through the following steps:

1. Create a new `MainViewModel` class in the `ViewModels` folder.

2. Make the `MainViewModel` class inherit `ViewModel`.

3. Add a read-only field of the `ILocationTrackingService` type and name it `locationTrackingService`.

4. Add a read-only field of the `ILocationRepository` type and name it `locationRepository`.

5. Create a constructor with `ILocationTrackingService` and `ILocationRepository` as parameters.

6. Set the values of the fields that we created in *steps 3* and *4* with the values from the parameters, as shown in the following code snippet:

```
public class MainViewModel : ViewModel
{
    private readonly ILocationRepository locationRepository;
    private readonly ILocationTrackingService
locationTrackingService;

    public MainViewModel(ILocationTrackingService
locationTrackingService,
        ILocationRepository locationRepository)
    {
        this.locationTrackingService = locationTrackingService;
        this.locationRepository = locationRepository;
    }
}
```

To make the app start tracking the location of a user, we need to run the code that starts the tracking process on the main thread. Follow these steps:

1. In the constructor of the newly created `MainViewModel` class, add an invocation to the main thread using `MainThread.BeginInvokeOnMainThread`.

2. Call `locationService.StartTracking` in the action that we pass to the `BeginInvokeOnMainThread` method. This is shown in the following highlighted code:

```
public MainViewModel(ILocationTrackingService
locationTrackingService, ILocationRepository locationRepository)
{
    this.locationTrackingService = locationTrackingService;
    this.locationRepository = locationRepository;
    MainThread.BeginInvokeOnMainThread(() =>
    {
        locationTrackingService.StartTracking();
    });
}
```

Finally, we need to inject a `MainViewModel` class into the constructor of `MainView` and assign the `MainViewModel` instance to the binding context of the view. This will allow what data binding we've done to be processed, and the properties of `MainViewModel` will be bound to the controls in the user interface. Follow these steps:

1. Go to the constructor of the `Views/MainView.xaml.cs` file.

2. Add `MainViewModel` as a parameter of the constructor and call it `viewModel`.

3. Set `BindingContext` as the instance of `MainViewModel`, as shown in the following code snippet:

```
public MainView(MainViewModel viewModel)
{
    InitializeComponent();
    BindingContext = viewModel;

    MainThread.BeginInvokeOnMainThread(async () =>
    {
        var location = await Geolocation.
GetLastKnownLocationAsync();

        if (location == null)
        {
            location = await Geolocation.GetLocationAsync();
        }
```

```
        Map.MoveToRegion(MapSpan.FromCenterAndRadius(
            location, Distance.FromKilometers(5)));
    });
}
```

In order for .NET MAUI to locate the classes we have implemented in this section so far, we need to add them to the **dependency injection (DI)** container.

Adding classes to the DI container

Since we have added a parameter to the constructor of the view, the .NET MAUI `View` framework won't be able to construct the view automatically. So, we need to add `MainView`, `MainViewModel`, `LocationTrackingService`, and `LocationRepository` instances to the DI container. To do that, follow these steps:

1. Open the `MauiProgram.cs` file.

2. Add the following highlighted lines to the `CreateMauiApp` method:

```
public static MauiApp CreateMauiApp()
{
    var builder = MauiApp.CreateBuilder();
    builder
        .UseMauiApp<App>()
        .ConfigureFonts(fonts =>
        {
            fonts.AddFont("OpenSans-Regular.ttf",
"OpenSansRegular");
            fonts.AddFont("OpenSans-Semibold.ttf",
"OpenSansSemibold");
        })
        .UseMauiMaps();

#if DEBUG
    builder.Logging.AddDebug();
#endif
        builder.Services.AddSingleton<Services.
ILocationTrackingService, Services.LocationTrackingService>();
        builder.Services.AddSingleton<Repositories.
ILocationRepository, Repositories.LocationRepository>();

        builder.Services.AddTransient(typeof(ViewModels.
MainViewModel));
        builder.Services.AddTransient(typeof(Views.MainView));

    return builder.Build();
}
```

Now, we will be able to run the app again. We haven't changed any of the interfaces, so it should look and behave the same as before. If it doesn't, go back through the previous section carefully to make sure you have all the code correct.

Let's add some code so that we can track the user's location over time using background location tracking.

Background location tracking on iOS and Mac Catalyst

The code for location tracking is something that we need to write for each platform. For iOS and Mac Catalyst, we will use `CLLocationManager` from the `CoreLocation` namespace.

Enabling location updates in the background

When we want to perform tasks in the background in an iOS or Mac Catalyst app, we need to declare what we want to do in the `info.plist` file. The following steps show how we go about this:

1. Open `info.plist`; you will need to do this for both `Platforms/iOS/info.plist` and `Platforms/MacCatalyst/info.plist`.

2. Add the following highlighted entry using the **Property List Editor** by selecting **Required background modes** from the dropdown and **App registers for location updates**, as shown in the following screenshot:

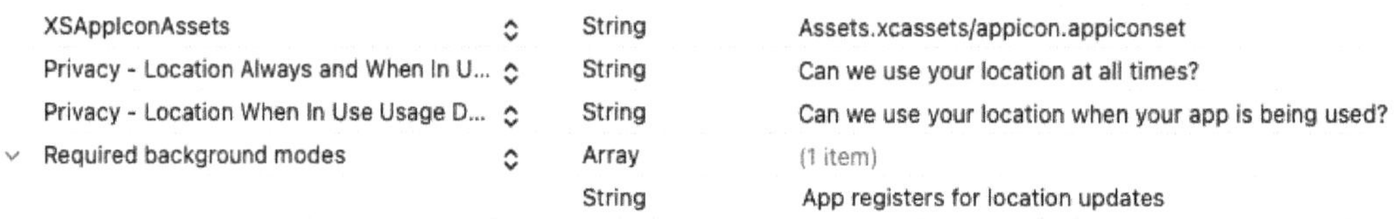

XSAppIconAssets	String	Assets.xcassets/appicon.appiconset
Privacy - Location Always and When In U...	String	Can we use your location at all times?
Privacy - Location When In Use Usage D...	String	Can we use your location when your app is being used?
∨ Required background modes	Array	(1 item)
	String	App registers for location updates

Figure 7.17 – Adding location updates

We can also enable background modes directly in the `info.plist` file if we open it with an XML editor. In this case, we will add the following XML:

```
<key>UIBackgroundModes</key>
<array>
<string>location</string>
</array>
```

Subscribing to location updates

Now that we have prepared the `info.plist` file for location tracking, it is time to write the actual code that will track the location of the user. If we don't set `CLLocationManager` to not pause location updates, location updates can be paused automatically by iOS or Mac Catalyst when the location data is unlikely to change. In this app, we don't want that to happen because we want to save the location multiple times so that we can establish whether a user visits a particular location frequently.

If you recall from earlier, we already defined the service as a partial class with a partial method; now, we will finish the service by implementing the platform-specific pieces of the service. Let's set this up:

1. Create a new folder named `Services` in the `Platforms/iOS` folder.

2. Create a new class named `LocationTrackingService` in the `Services` folder.

3. Modify the class to match the following:

    ```
    namespace MeTracker.Services;

    public partial class LocationTrackingService :
    ILocationTrackingService
    {
        partial void StartTrackingInternal()
        {
        }
    }
    ```

4. Add a private field for `CLLocationManager`.

5. Create an instance of `CLLocationManager` in the `StartTrackingInternal` method.

6. Set `PausesLocationUpdatesAutomatically` to `false`.

    Before we can start tracking the location of the user, we need to set the accuracy of the data that we want to receive from `CLLocationManager`. We will also add an event handler to handle location updates.

7. Set `DesiredAccuracy` to `CLLocation.AccuracyBestForNavigation`. One of the constraints when running the app in the background is that `DesiredAccuracy` needs to be set to either `AccuracyBest` or `AccuracyBestForNavigation`.

8. Set `AllowBackgroundLocationUpdates` to `true` (as shown in the following code snippet) so that the location updates will continue, even when the app is running in the background.

    Your changes should look like this:

    ```
    CLLocation locationManager;

    partial void StartTrackingInternal()
    {
        locationManager = new CLLocationManager
        {
            PausesLocationUpdatesAutomatically = false,
            DesiredAccuracy = CLLocation.AccuracyBestForNavigation,
            AllowsBackgroundLocationUpdates = true
        };

        // Add code here
    }
    ```

The next step is to ask the user for permission to track their location. We will request permission to track their location all the time, but the user has the option of only giving us permission to track their location when they are using the app. Because the user also has the option of denying us permission to track their location, we need to check this before we start. Let's set this up:

1. Add an event handler for LocationsUpdated just after the // Add code here comment. It should look like the code highlighted in the following snippet:

```
partial void StartTrackingInternal()
{
    locationManager = new CLLocationManager
    {
        PausesLocationUpdatesAutomatically = false,
        DesiredAccuracy = CLLocation.AccuracyBestForNavigation,
        AllowsBackgroundLocationUpdates = true
    };

    // Add code here
    locationManager.LocationsUpdated +=
async (object sender, CLLocationsUpdatedEventArgs e) =>
    {
        // Final block of code goes here
    };

};
```

2. After the event handler, call the RequestAlwaysAuthorization method of the instance that we recently created in CLLocationManager:

```
partial void StartTrackingInternal()
{
    locationManager = new CLLocationManager
    {
        PausesLocationUpdatesAutomatically = false,
        DesiredAccuracy = CLLocation.AccurracyBestForNavigation,
        AllowsBackgroundLocationUpdates = true
    };

    // Add code here
    locationManager.LocationsUpdated +=
async (object sender, CLLocationsUpdatedEventArgs e) =>
    {
        // Final block of code goes here
    };
```

```
            locationManager.RequestAlwaysAuthorization();

    };
```

3. Then, call the `StartUpdatingLocation` method of `locationManager`:

    ```
    partial void StartTrackingInternal()
    {
        locationManager = new CLLocationManager
        {
            PausesLocationUpdatesAutomatically = false,
            DesiredAccuracy = CLLocation.AccurracyBestForNavigation,
            AllowsBackgroundLocationUpdates = true
        };

        // Add code here
        locationManager.LocationsUpdated +=
    async (object sender, CLLocationsUpdatedEventArgs e) =>
        {
            // Final block of code goes here
        };

        locationManager.RequestAlwaysAuthorization();
        locationManager.StartUpdatingLocation();

    };
    ```

Tip

The higher the accuracy is, the higher the battery consumption. If we only want to track where the user has been and not how popular a place is, we could also set `AllowDeferredLocationUpdatesUntil`. This way, we can specify that the user has to move a specific distance before the location is updated. We can also specify how often we want the location to be updated using the `timeout` argument. The most power-efficient solution to track how long a user has been at a place is to use the `StartMonitoringVisits` method of `CLLocationManager`.

Now, it's time to handle the `LocationsUpdated` event. Let's go through the following steps:

1. Add a private field called `locationRepository` that is of the `ILocationRepository` type.

2. Add a constructor that has `ILocationRepository` as a parameter. Set the value of the parameter to the `locationRepository` field. Your class should resemble the following code snippet:

    ```
    using CoreLocation;
    using MeTracker.Repositories;
    ```

```
namespace MeTracker.Services;

public partial class LocationTrackingService :
ILocationTrackingService
{
    CLLocationManager locationManager;
    ILocationRepository locationRepository;

    public LocationTrackingService(ILocationRepository
locationRepository)
    {
        this.locationRepository = locationRepository;
    }

    partial void StartTrackingInternal()
    {
    // Remainder of code omitted for brevity
    }
```

3. Read the latest location of the `Locations` property of `CLLocationsUpdatedEventArgs`.

4. Create an instance of `MeTracker.Models.Location` and pass the latitude and longitude of the latest location to it.

5. Save the location using the `SaveAsync` method of `ILocationRepository`.

6. The code should be placed after the `// Final block of code goes here` comment. It should look like the code shown in bold in the following fragment:

```
locationManager.LocationsUpdated +=
async (object sender, CLLocationsUpdatedEventArgs e) =>
{
    // Final block of code goes here
    var lastLocation = e.Locations.Last();
    var newLocation = new Models.Location(lastLocation.
Coordinate.Latitude, lastLocation.Coordinate.Longitude);

    await locationRepository.SaveAsync(newLocation);
};
```

With that, we have completed the tracking part of the app for iOS. The implementation is identical for Mac Catalyst; you can either repeat the steps in this section for Mac Catalyst (but create the file as `Platforms/MacCatalyst/Services` instead of `Platforms/iOS/Services`) or copy the `Platforms/iOS/Services/LocationTrackingService.cs` file to the `Platforms/MacCatalyst/Services` folder.

Now, we will implement background tracking for Android, following which we will visualize the location tracking data.

Background location tracking with Android

The Android way to carry out background updates is very different from how we implemented it with iOS. With Android, we need to create a JobService class and schedule it.

Creating a background job

To track the location of users in the background, we need to create a background job. A background job is used by the OS to allow developers to execute code even when the app is not in the foreground or visible on the screen. Follow these steps to create a background job to capture a user's location:

1. Create a new folder named Services in the Platforms/Android folder.

2. Create a new class called LocationJobService in the Services folder.

3. Make the class inherit from Android.App.Job.JobService as a base class.

4. Add using Android.App.Job and using Android.App.Job declarations to the top of the file.

5. Implement the OnStartJob and OnStopJob abstract methods, as shown in the following code snippet:

```
using Android.App;
using Android.App.Job;

namespace MeTracker.Platforms.Android.Services;

internal class LocationJobService : JobService
{
    public override bool OnStartJob(JobParameters @params)
    {
        return true;
    }

    public override bool OnStopJob(JobParameters @params)
    {
        return true;
    }
}
```

All the Android services in the app need to be added to the AndroidManifest.xml file. We don't have to do this manually; instead, we can add an attribute to the LocationJobService class, which will then be generated in the `AndroidManifest.xml` file. We will use the `Name` and `Permission` properties to set the required information, as shown in the following code snippet:

```
[Service(Name = "MeTracker.Platforms.Android.Services.
LocationJobService", Permission = "android.permission.BIND_JOB_
SERVICE")]
internal class LocationJobService : JobService
```

Scheduling a background job

When we have created a job, we need to schedule it. We will do this from `LocationTrackingService` in the `Platforms/Android` folder. To configure the job, we will use the `JobInfo.Builder` class.

We will use the `SetPersisted` method to ensure that the job starts again after a reboot. This is why we added the `RECEIVE_BOOT_COMPLETED` permission earlier.

To schedule a job, at least one constraint is needed. In this case, we will use `SetOverrideDeadline`. This will specify that the job needs to run before the specified time (in milliseconds) has elapsed.

The `SetRequiresDeviceIdle` method can be used to make sure that a job only runs when the device is not being used by a user. We could pass `true` to the method if we want to make sure that we don't slow down the device when the user is using it.

The `SetRequiresBatteryNotLow` method can be used to specify that a job should not run when the battery level is low. We recommend that this always be set to `true` if you don't have a good reason to run the job when the battery is low. This is because we don't want our applications to drain the user's battery.

So, let's implement `LocationTrackingService`. Follow these steps to do so:

1. Create a new class named `LocationTrackingService` in the `Platforms/Android/Services` folder.

2. Modify the class to look like the following:

```
namespace MeTracker.Services;

public partial class LocationTrackingService :
ILocationTrackingService
{
    partial void StartTrackingInternal()
    {
    }
}
```

3. Create a `JobInfo.Builder` class based on an ID that we'll specify (we will use 1 here) and the component name (which we'll create from the application context and the Java class) in the `StartTrackingInternal` method. The component name is used to specify which code will run during the job.

4. Use the `SetOverrideDeadline` method and pass `1000` to it to make the job run before 1 second has elapsed from when the job was created.

5. Use the `SetPersisted` method and pass `true` to make the job persist even after the device is rebooted.

6. Use the `SetRequiresDeviceIdle` method and pass `false` so that the job will run even when a user is using the device.

7. Use the `SetRequiresBatteryLow` method and pass `true` to make sure that we don't drain the user's battery. This method was added in Android API level 26.

The code for `LocationTrackingService` should now look like this:

```
using Android.App.Job;
using Android.Content;
using MeTracker.Platforms.Android.Services;

namespace MeTracker.Services;

public partial class LocationTrackingService :
ILocationTrackingService
{
    partial void StartTrackingInternal()
    {
        var javaClass = Java.Lang.Class.
FromType(typeof(LocationJobService));
        var componentName = new ComponentName(global::Android.
App.Application.Context, javaClass);
        var jobBuilder = new JobInfo.Builder(1, componentName);

        jobBuilder.SetOverrideDeadline(1000);
        jobBuilder.SetPersisted(true);
        jobBuilder.SetRequiresDeviceIdle(false);
        jobBuilder.SetRequiresBatteryNotLow(true);

        var jobInfo = jobBuilder.Build();
    }
}
```

The last step in the `StartTrackingInternal` method is to schedule the job with the system using `JobScheduler`. The `JobScheduler` service is an Android system service. To get an instance of a system service, we will use the application context. Follow these steps to do so:

1. Use the `GetSystemService` method on `Application.Context` to get the `JobScheduler` service.

2. Cast the result to `JobScheduler`.

3. Use the `Schedule` method on the `JobScheduler` class and pass the `JobInfo` object to schedule the job, as shown in the following code snippet:

```
var jobScheduler = (JobScheduler)global::Android.
App.Application.Context.GetSystemService(Context.
JobSchedulerService);
jobScheduler.Schedule(jobInfo);
```

Now that the job is scheduled, we can start receiving location updates; let's work on that next.

Subscribing to location updates

Once we have scheduled the job, we can write the code to specify what the job should do – that is, track the location of a user. To do this, we will use `LocationManager`, which is a `SystemService` class. With `LocationManager`, we can either request a single location update or we can subscribe to location updates. In this case, we want to subscribe to location updates.

We will start by creating an instance of the `ILocationRepository` interface. We will use this to save the locations to the SQLite database. Let's set this up:

1. Create a constructor for `LocationJobService`.

2. Create a `private` read-only field for the `ILocationRepository` interface called `locationRepository`.

3. Use `Services.GetService<T>` in the constructor to create an instance of `ILocationRepository`, as shown in the following code snippet:

```
private ILocationRepository locationRepository;

public LocationJobService()
{
    locationRepository = MauiApplication.Current.Services.
GetService<ILocationRepository>();
}
```

Before we subscribe to location updates, we will add a listener. To do this, we will use the `Android.Locations.ILocationListener` interface.

Follow these steps:

1. Add the `Android.Locations.ILocationListener` interface to `LocationJobService`.

2. Add the following namespace declarations to the top of the file:

    ```
    using Android.Content;
    using Android.Locations;
    using Android.OS;
    using Android.Runtime;
    using MeTracker.Repositories;
    ```

3. Implement the interface and remove all instances of `throw new NotImplemented Exception();`. This is added to the methods when you let Visual Studio generate an implementation of the interface.

    The method implementations should look like those in the following code snippet:

    ```
    public override bool OnStartJob(JobParameters @params)
    {
        return true;
    }

    public void OnLocationChanged(global::Android.Locations.Location
    location) { }

    public override bool OnStopJob(JobParameters @params) => true;

    public void OnStatusChanged(string provider, [GeneratedEnum]
    Availability status, Bundle extras) { }

    public void OnProviderDisabled(string provider) { }

    public void OliknProviderEnabled(string provider) { }
    ```

4. In the `OnLocationChanged` method, map the `Android.Locations.Location` object to the `Model.Location` object.

5. Use the `SaveAsync` method on the `LocationRepository` class, as shown in the following code snippet:

    ```
    public void OnLocationChanged(Android.Locations.Location
    location)
    {
      var newLocation = new Models.Location(location.Latitude,
    location.Longitude);
    ```

```
locationRepository.SaveAsync(newLocation);
}
```

Now that we have created a listener, we can subscribe to location updates. Follow these steps to do so:

1. Create a `static` field of the `LocationManager` type named `locationManager`. Make sure it has the same lifetime as the app.

2. It is possible in Android that `JobService` will start before `MainView` is displayed and we request location permissions. To prevent any errors due to missing permissions, we will check for them first:

```
public override bool OnStartJob(JobParameters @params)
{
    PermissionStatus status = PermissionStatus.Unknown;
    Task.Run(async ()=> status = await AppPermissions.
CheckRequiredPermissionAsync()).Wait();
    if (status == PermissionStatus.Granted)
    {
    }
}
```

We run `CheckRequiredPermissionsAsync` inside a `Task.Run` instance because it's an async call, and we can't add `async` to the method because the return type is incompatible. The call to `Wait` turns the `async` call into a synchronous one. If the result is `Granted`, then we can continue.

3. Go to the `StartJob` method in `LocationJobService`. Get `LocationManager` by using `GetSystemService` on `ApplicationContext`.

4. To subscribe for location updates, use the `RequestLocationUpdates` method, as shown in the following code snippet:

```
public override bool OnStartJob(JobParameters @params)
{
    PermissionStatus status = PermissionStatus.Unknown;
    Task.Run(async ()=> status = await AppPermissions.
CheckRequiredPermissionAsync()).Wait();
    if (status == PermissionStatus.Granted)
    {
        locationManager = (LocationManager)ApplicationContext.
GetSystemService (Context.LocationService);
        locationManager.RequestLocationUpdates (LocationManager.
GpsProvider, 1000L, 0.1f, this);

        return true;
    }
}
```

```
        return false;
    }
```

The first argument that we pass to the `RequestLocationUpdates` method ensures that we get locations from the GPS. The second ensures that at least `1000` milliseconds will elapse between location updates. The third argument ensures that the user moves at least `0.1` meters to get a location update. The last argument specifies which listener we should use. Because the current class implements the `Android.Locations.ILocationListener` interface, we will pass `this`.

Now that we have collected location data on from the user and stored it in our SQLite database, we can now display that data on a map.

Creating a heat map

To visualize the data that we have collected, we will create a heat map. We will add lots of dots to the map and make them different colors, based on how much time a user spends in a particular place. The most popular places will have a warm color, while the least popular places will have a cold color.

Before we add the dots to the map, we need to get all locations from the repository.

Adding the GetAllAsync method to LocationRepository

In order to visualize the data, we need to write some code so that location data can be read from the database. Let's set this up:

1. Open the `ILocationRepository.cs` file.

2. Add a `GetAllAsync` method that returns a list of `Location` objects, using the following code:

    ```
    Task<List<Models.Location>> GetAllAsync();
    ```

3. Open the `LocationRepository.cs` file, which implements `ILocationRepository`.

4. Implement the new `GetAllAsync` method and return all the saved locations in the database, as shown in the following code snippet:

    ```
    public async Task<List<Location>> GetAllAsync()
    {
        await CreateConnectionAsync();
        var locations = await connection.Table<Location>
    ().ToListAsync();

        return locations;
    }
    ```

Preparing the data for visualization

Before we can visualize the data on the map, we need to prepare the data. The first thing we will do is create a new model that we can use for the prepared data. Let's set this up:

1. In the `Models` folder, create a new class called `Point`.

2. Add properties for `Location`, `Count`, and `Heat`, as shown in the following code snippet:

    ```
    namespace MeTracker.Models;

    public class Point
    {
        public Location Location { get; set; }
        public int Count { get; set; } = 1;
        public Color Heat { get; set; }
    }
    ```

    `MainViewModel` will store the locations that we will find later on. Let's add a property for storing points.

3. Open the `MainViewModel` class.

4. Add a `private` field called `points`, which is of the `List<Point>` type.

5. Add the `ObservableProperty` attribute to the `points` field, as shown in the following code snippet:

    ```
    [ObservableProperty]
    private List<Models.Point> points;
    ```

Now that we have storage for our points, we must add some code so that we can add locations. We will do this by implementing the `LoadDataAsync` method of the `MainViewModel` class and making sure that it is called on the main thread, right after location tracking has started.

The first thing we will do is group the saved locations so that all locations within 200 meters will be handled as one point. We will track how many times we have logged a position within that point so that we can decide which color the point will be on the map. Let's set this up:

1. Add an `async` method called `LoadDataAsync`. This returns a `Task` object to `Main ViewModel`.

2. Call the `LoadDataAsync` method from the constructor after the call to the `StartTracking` method on `ILocationTrackingService`, as shown in the following code snippet:

    ```
    public MainViewModel(ILocationTrackingService
    locationTrackingService, ILocationRepository locationRepository)
    {
        this.locationTrackingService = locationTrackingService;
    ```

```
                this.locationRepository = locationRepository;
                MainThread.BeginInvokeOnMainThread(async() =>
                {
                    locationTrackingService.StartTracking();
                    await LoadDataAsync();
                });
        }
```

The first step of the `LoadDataAsync` method is to read all tracked locations from the SQLite database. When we have all the locations, we will loop through them and create points.

To calculate the distance between a location and a point, we will use the `CalculateDistance` method, as shown in the following code snippet:

```
private async Task LoadDataAsync()
{
    var locations = await locationRepository.GetAll();
    var pointList = new List<Models.Point>();

    foreach (var location in locations)
    {
        //If no points exist, create a new one and continue to the
next location in the list
        if (!pointList.Any())
        {
            pointList.Add(new Models.Point() { Location = location });
            continue;
        }

        var pointFound = false;

        //try to find a point for the current location
        foreach (var point in pointList)
        {
            var distance = Location.CalculateDistance(
                new Location(point.Location.Latitude, point.Location.
Longitude),
                new Location(location.Latitude, location.Longitude),
                DistanceUnits.Kilometers);

            if (distance < 0.2)
            {
                pointFound = true;
                point.Count++;
                break;
```

```
        }
    }

    //if no point is found, add a new Point to the list of points
    if (!pointFound)
    {
        pointList.Add(new Models.Point() { Location = location });
    }

    // Next section of code goes here
    }
}
```

When we have a list of points, we can calculate the heat color for each point. We are going to use the **hue**, **saturation**, and **lightness (HSL)** representation of a color, as described here:

- **Hue**: Hue is a degree on the color wheel that goes from 0 to 360, with 0 being red and 240 being blue. Because we want our most popular places to be red (hot) and our least popular places to be blue (cold), we will calculate a value between 0 and 240 for each point, based on how many times the user has been to that point. This means that we will only use two-thirds of the scale.

- **Saturation**: Saturation is a percentage value: 0% is a shade of gray, while 100% is full color. In our app, we will always use 100% (this will be represented as 1 in the code).

- **Lightness**: Lightness is a percentage value of the amount of light: 0% is black and 100% is white. We want it to be neutral, so we will use 50% (this will be represented as 0.5 in the code).

The first thing that we need to do is find out how many times the user has been to the most popular and least popular places. Let's take a look:

1. First, check that the list of points is not empty.
2. Get the Min and Max values for the Count property in the list of points.
3. Calculate the difference between the minimum and the maximum values.
4. The code should be added after the // Next section of code goes here comment at the bottom of the LoadDataAsync method, as shown in the following code snippet:

```
private async Task LoadDataAsync()
{
    // The rest of the method has been omitted for brevity

    // Next section of code goes here
    if (pointList == null || !pointList.Any())
    {
        return;
```

```
    }

    var pointMax = pointList.Select(x => x.Count).Max();
    var pointMin = pointList.Select(x => x.Count).Min();
    var diff = (float)(pointMax - pointMin);

    // Last section of code goes here
}
```

Now, we can calculate the heat for each point, as follows:

1. Loop through all the points.

2. The following code should be added after the `// Last section of code goes here` comment, at the bottom of the `LoadDataAsync()` method (this is highlighted in the following code snippet):

```
private async Task LoadDataAsync()
{
    // The rest of the method has been omitted for brevity

    // Next section of code goes here
    if (pointList == null || !pointList.Any())
    {
        return;
    }

    var pointMax = pointList.Select(x => x.Count).Max();
    var pointMin = pointList.Select(x => x.Count).Min();
    var diff = (float)(pointMax - pointMin);

    // Last section of code goes here
    foreach (var point in pointList)
    {
        var heat = (2f / 3f) - ((float)point.Count / diff);
        point.Heat = Color.FromHsla(heat, 1, 0.5);
    }

    Points = pointList;
}
```

That's all we need to do to set up location tracking in the `MeTracker` project. Now, let's turn our attention to visualizing the data we receive.

Adding data visualizations

In .NET MAUI, the Map control can render additional information over the map. This includes pins and custom shapes, which are called MapElements. We could simply add each location that is stored in the repository as a pin; however, to get the heat map, we want to add a colored dot to the map for each location, so we will use MapElements for each location.

If the MapElements property were BindableProperty, we could use a converter to map the MainViewModel Points property to the map's MapElements property in a binding. But MapElements is not a bindable property, so it won't be that easy.

Let's start by creating a custom map control.

Creating a custom control for the map

To show the heat map on our map, we will create a new control. Since Map is a sealed class, we won't be able to subclass it directly; instead, we will encapsulate the Map control inside ContentView with BindableProperty to get access to the Points data from ViewModel.

Follow these steps to create the custom control:

1. Create a new folder called Controls.
2. Create a new class called CustomMap.
3. Add ContentView as a base class to the new class, as shown in the following code snippet:

    ```
    namespace MeTracker.Controls;

    public class CustomMap : ContentView
    {
        public CustomMap()
        {
        }
    }
    ```

Now, we need to add the Map control to the custom control. Follow these steps to add the Map control:

1. Derive the CustomMap control from the .NET MAUI Map control, as highlighted in the following code snippet:

    ```
    using Microsoft.Maui.Controls.Maps;
    using Microsoft.Maui.Maps;
    using Map = Microsoft.Maui.Controls.Maps.Map;

    namespace MeTracker.Controls;

    public class CustomMap : Map
    {
    ```

```
        public CustomMap()
        {
        }
    }
```

2. Initialize the map in the constructor, as shown with the new changes highlighted:

```
    public CustomMap()
    {
        IsScrollEnabled = true;
        IsShowingUser = true;
    }
```

If we want to have properties that we want to bind data to, we need to create a `BindableProperty` class. This should be a public static field in the class. We also need to create a regular property to hold the value. The naming of the properties is really important. The name of `BindableProperty` needs to be `{NameOfTheProperty}Property`; for example, the name of `BindableProperty` that we will create in the following steps will be `PointsProperty` because the name of the property is `Points`. A `BindableProperty` is created using the static `Create` method on the `BindableProperty` class. This requires at least four arguments, as follows:

*   `propertyName`: This is the name of the property as a string.
*   `returnType`: This is the type that will be returned from the property.
*   `declaringType`: This is the type of the class that `BindableProperty` is declared in.
*   `defaultValue`: This is the default value that will be returned if no value is set. This is an optional argument. If it is not set, .NET MAUI will use `null` as a default value.

The `set` and `get` methods for the property will call methods in the base class to set or get values from `BindableProperty`:

1. Create `BindableProperty` called `PointsProperty`, as shown in the following code snippet:

```
    public static BindableProperty PointsProperty =
    BindableProperty.Create(nameof(Points), typeof(List<Models.
    Point>), typeof(CustomMap), new List<Models.Point>());
```

2. Create a property of the `List<Models.Point>` type called `Points`. Remember to cast the result of `GetValue` so that it's the same type as the property. We need to do this because `GetValue` will return the value as a `type` object:

```
    public List<Models.Point> Points
    {
      get => GetValue(PointsProperty) as List<Models.Point>;
      set => SetValue(PointsProperty, value);
    }
```

In order to display `Points`, we need to convert them to `MapElements`. This is accomplished using a `BindingProperty` event called `PropertyChanged`. `PropertyChanged` is fired every time `BindingProperty` changes. To add the event and convert `Points` to `MapElements`, add the following highlighted code to the class:

```
public readonly static BindableProperty PointsProperty =
BindableProperty.Create(nameof(Points), typeof(List<Models.
Point>), typeof(MapView), new List<Models.Point>(), propertyChanged:
OnPointsChanged);

private static void OnPointsChanged(BindableObject bindable, object
oldValue, object newValue)
{
    var map = bindable as Map;
    if (newValue == null) return;
    if (map == null) return;
    foreach (var point in newValue as List<Models.Point>)
    {
        // Instantiate a Circle
        Circle circle = new()
        {
            Center = new Location(point.Location.Latitude, point.
Location.Longitude),
            Radius = new Distance(200),
            StrokeColor = Color.FromArgb("#88FF0000"),
            StrokeWidth = 0,
            FillColor = point.Heat
        };

        // Add the Circle to the map's MapElements collection
        map.MapElements.Add(circle);
    }
}

public List<Models.Point> Points
{
    get => GetValue(PointsProperty) as List<Models.Point>;
    set => SetValue(PointsProperty, value);
}
```

Now that we've created a custom map control, we will use it to replace the Map control in MainView. Follow these steps:

1. In the MainView.xaml file, declare the namespace for the custom control.

2. Replace the Map control with the new control that we have created.

3. Add a binding to the Points property in MainViewModel, as shown in the following code snippet:

```
<ContentPage xmlns="http://xamarin.com/schemas/2014/forms"
xmlns:x="http://schemas.microsoft.com/winfx/2009/xaml"
xmlns:map="clr-namespace:MeTracker.Controls;"
x:Class="MeTracker.Views.MainView">
<map:CustomMap x:Name="Map" Points="{Binding Points}" />
</ContentPage>
```

This concludes this section on how to extend the Maps control. The final step for our app is to refresh the map when the app resumes.

Refreshing the map when the app resumes

The last thing we will do is make sure that the map is up to date with the latest points when the app is resumed. The easiest way to do this is to set the MainPage property in the App.xaml.cs file to a new instance of AppShell, in the same way as the constructor, as shown in the following code snippet:

```
protected override void OnResume()
{
    base.OnResume();

    MainPage = new AppShell();
}
```

The `MeTracker` app is now complete – try it out. A sample screenshot is shown in *Figure 7.18*:

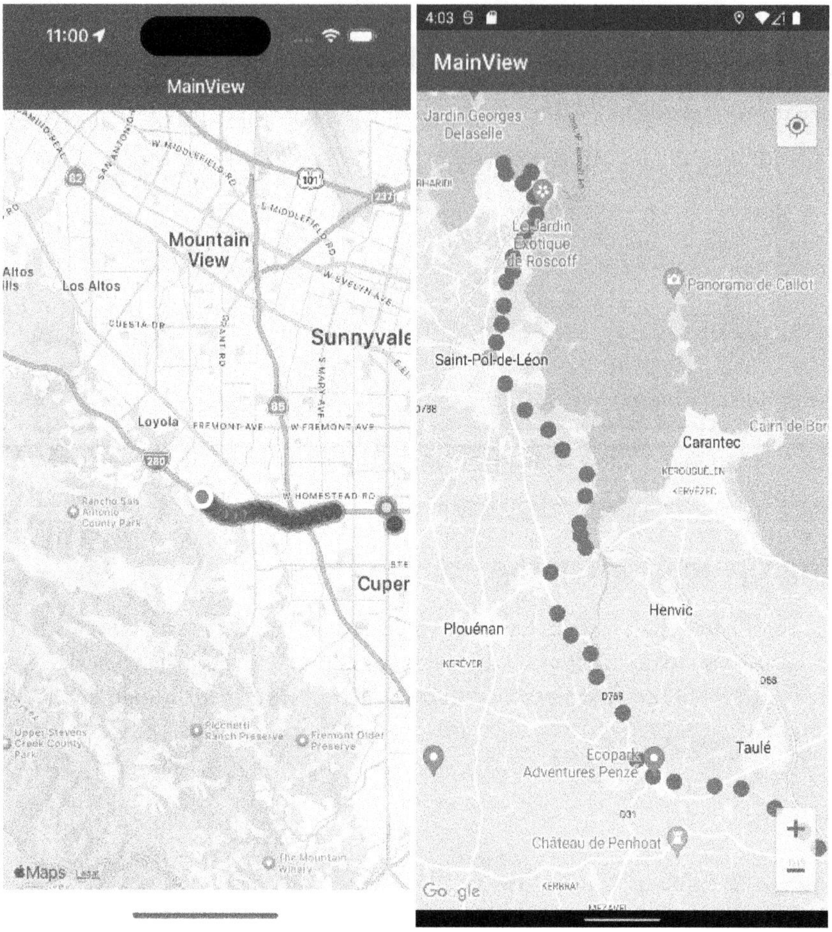

Figure 7.18 – MeTracker on iOS and Android

Summary

In this chapter, we built an app for iOS, Mac Catalyst, and Android that tracked the location of a user. When we built the app, we learned how to use maps in .NET MAUI and how to use location tracking when it's running in the background. We also learned how to extend .NET MAUI with custom controls. With this knowledge, we can create applications that perform other tasks in the background. We also learned how to extend most controls in .NET MAUI.

Here are some ways you could extend this app even further:

- Right now, the app updates the map location when the app resumes. How could you update the map when the location changes?

- Add a view that lists all locations from the database. Allow the user to remove a location from the list.

The next project will be a weather app. In the next chapter, we will use an existing weather service API to retrieve weather data and then display that data in the app.

8

Building a Weather App for Multiple Form Factors

.NET MAUI isn't just for creating apps for phones; it can also be used to create apps for tablets and desktop computers. In this chapter, we will build an app that will work on all of these platforms and optimize the user interface for each form factor. As well as using three different form factors, we are also going to be working on four different operating systems: iOS, macOS, Android, and Windows.

The following topics will be covered in this chapter:

- Using `FlexLayout` in .NET MAUI
- Using `VisualStateManager`
- Using different views for different form factors
- Using behaviors

Let's get started!

Technical requirements

To work on this project, we need to have Visual Studio for Mac or PC installed, as well as the necessary .NET MAUI components. See *Chapter 1, Introduction to .NET MAUI*, for more details on how to set up your environment. To build an iOS app using Visual Studio for PC, you need to have a Mac connected. If you don't have access to a Mac at all, you can choose to just work on the Windows and Android parts of this project. Similarly, if you only have a Mac, you can choose to work on only the iOS and Android parts of this project.

You can find the full source for the code in this chapter at `https://github.com/PacktPublishing/MAUI-Projects-3rd-Edition`.

Project overview

Applications for iOS and Android can run on both phones and tablets. Often, apps are just optimized for phones. In this chapter, we will build an app that will work on different form factors, but we aren't going to stick to just phones and tablets – we are going to target desktop computers as well. The desktop version will be for **Window UI Library (WinUI)** and macOS via Mac Catalyst.

The app that we are going to build is a weather app that displays the weather forecast based on the location of the user. For this chapter, we will be referencing Visual Studio for Mac in the instructions. If you are using Visual Studio for Windows, you should be able to follow along. Use one of the other chapters for reference if you need help.

Building the weather app

It's time to start building the app. Create a new blank .NET MAUI app using the following steps for Visual Studio for Mac:

1. Open Visual Studio for Mac and click on **New**:

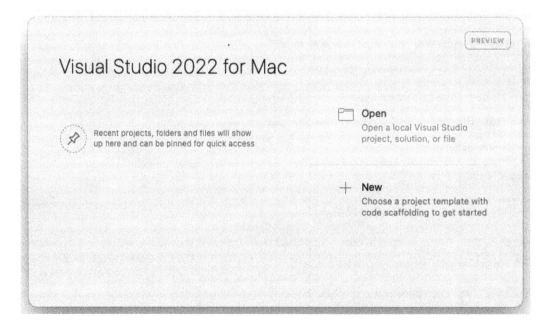

Figure 8.1 – Visual Studio 2022 for Mac start screen

2. In the **Choose a template for your new project** dialog, use the **.NET MAUI App** template, which is under **Multiplatform | App**, then click **Continue**:

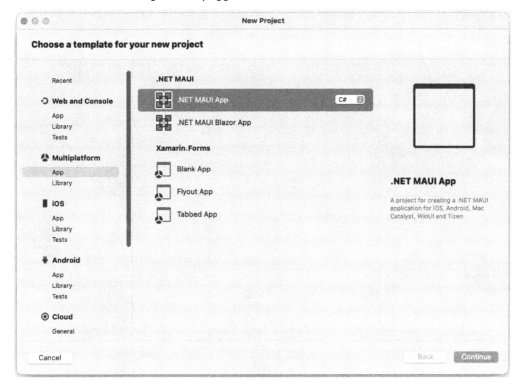

Figure 8.2 – New project

3. In the **Configure your new .NET MAUI App** dialog, ensure the **.NET 7.0** target framework is selected, then click **Continue**:

Figure 8.3 – Choosing the target framework

4. In the **Configure your new .NET MAUI App** dialog, name the project `Weather`, then click **Create**:

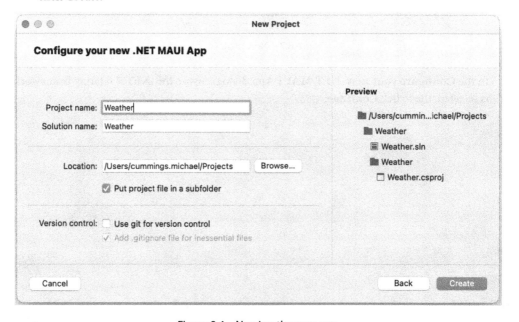

Figure 8.4 – Naming the new app

If you run the app now, you should see something like the following:

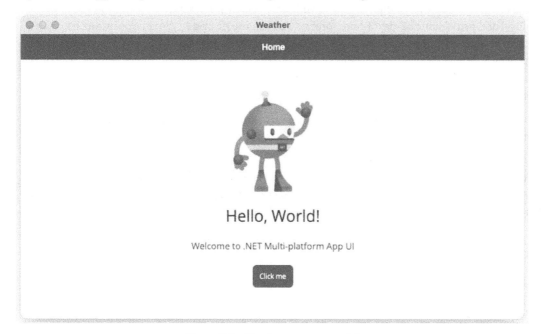

Figure 8.5 – Weather app on macOS

Now that we have created the project from a template, it's time to start coding!

Creating models for the weather data

Before we write the code to fetch data from the external weather service, we will create models to deserialize the results from the service. We will do this so that we have a common model that we can use to return data from the service.

As the data source for this app, we will use an external Weather API. This project will use **OpenWeatherMap**, a service that offers a couple of free APIs. You can find this service at `https://openweathermap.org/api`. We will use the **5 day / 3 hour forecast** service in this project, which provides a 5-day forecast in 3-hour intervals. To use the OpenWeatherMap API, we have to create an account to get an API key. If you don't want to create an API key, you can mock the data instead.

Follow the instructions at `https://home.openweathermap.org/users/sign_up` to create your account and get your API key, which you will need to call the API.

The easiest way to generate models to use when we are deserializing results from the service is to make a call to the service either in the browser or with a tool (such as **Postman**) to see the structure of the JSON. For OpenWeather's 5-day/3-hour forecast, you can use `https://api.openweathermap.org/data/2.5/forecast?lat=44.34&lon=10.99&appid={API key}` in your browser, replacing {API KEY} with your API key.

> **Note**
>
> If you got a 401 error, please wait a couple of hours before you can use your API, as mentioned at `https://openweathermap.org/faq#error401`.

We can create classes manually or use a tool that can generate C# classes from the JSON. One tool that can be used is **quicktype**, which can be found at `https://quicktype.io/`. Just paste the output from the API call into quicktype to generate your C# models.

If you generate them using a tool, make sure you set the namespace to `Weather.Models`.

As mentioned previously, you can also create these models manually. We will describe how to do this in the next section.

Adding the Weather API models manually

If you wish to add the models manually, then go through the following instructions. We will be adding a single code file called `WeatherData.cs`, which will contain multiple classes:

1. Create a folder called `Models`.
2. Add a file called `WeatherData.cs` to the newly created folder.
3. Add the following code to the `WeatherData.cs` file:

```
using System.Collections.Generic;
namespace Weather.Models
{
    public class Main
    {
        public double temp { get; set; }
        public double temp_min { get; set; }
        public double temp_max { get; set; }
        public double pressure { get; set; }
        public double sea_level { get; set; }
        public double grnd_level { get; set; }
        public int humidity { get; set; }
        public double temp_kf { get; set; }
    }
```

```csharp
public class Weather
{
    public int id { get; set; }
    public string main { get; set; }
    public string description { get; set; }
    public string icon { get; set; }
}

public class Clouds
{
    public int all { get; set; }
}

public class Wind
{
    public double speed { get; set; }
    public double deg { get; set; }
}

public class Rain
{
}

public class Sys
{
    public string pod { get; set; }
}

public class List
{
    public long dt { get; set; }
    public Main main { get; set; }
    public List<Weather> weather { get; set; }
    public Clouds clouds { get; set; }
    public Wind wind { get; set; }
    public Rain rain { get; set; }
    public Sys sys { get; set; }
    public string dt_txt { get; set; }
}
```

```
public class Coord
{
    public double lat { get; set; }
    public double lon { get; set; }
}

public class City
{
    public int id { get; set; }
    public string name { get; set; }
    public Coord coord { get; set; }
    public string country { get; set; }
}

public class WeatherData
{
    public string cod { get; set; }
    public double message { get; set; }
    public int cnt { get; set; }
    public List<List> list { get; set; }
    public City city { get; set; }
}
}
```

As you can see, there are quite a lot of classes. These classes map directly to the response we get from the service. In most cases, you only want to use these classes when communicating with the service. To represent the data in your app, you will need to use a second set of classes that exposes only the information you need in your app.

Adding the app-specific models

In this section, we will create the models that our app will translate the Weather API models into. Let's start by adding the WeatherData class (unless you created this manually in the preceding section):

1. Create a new folder called Models in the Weather project.

2. Add a new file called WeatherData.cs.

3. Paste the generated code from quicktype or write the code for the classes based on the JSON. If code other than the properties is generated, ignore it and just use the properties.

4. Rename MainClass (this is what **quicktype** names the root object) WeatherData.

Now, we will create models based on the data we are interested in. This will make the rest of the code more loosely coupled to the data source.

Adding the ForecastItem model

The first model we are going to add is `ForecastItem`, which represents a specific forecast for a point in time. We can do this as follows:

1. In the `Weather` project and the `Models` folder, create a new class called `ForecastItem`.

2. Add the following code:

    ```
    using System;
    using System.Collections.Generic;

    namespace Weather.Models
    {
        public class ForecastItem
        {
            public DateTime DateTime { get; set; }
            public string TimeAsString => DateTime.
    ToShortTimeString();
            public double Temperature { get; set; }
            public double WindSpeed { get; set; }
            public string Description { get; set; }
            public string Icon { get; set; }
        }
    }
    ```

Now that we have a model for each forecast, we need a container model that will group `ForecastItems` by `City`.

Adding the Forecast model

In this section, we'll create a model called `Forecast` that will keep track of a single forecast for a city. The `Forecast` model keeps a list of multiple `ForeCastItem` objects, each representing a forecast for a specific point in time. Let's set this up:

1. Create a new class called `Forecast` in the `Models` folder.

2. Add the following code:

    ```
    using System;
    using System.Collections.Generic;

    namespace Weather.Models;

    public class Forecast
    {
        public string City { get; set; }
    ```

```
        public List<ForecastItem> Items { get; set; }
    }
```

Now that we have our models for both the Weather API and the app, we need to fetch data from the Weather API.

Creating a service to fetch the weather data

To make it easier to change the external weather service and to make the code more testable, we will create an interface for the service. Here's how we can go about it:

1. Create a new folder called Services.

2. Create a new public interface called IWeatherService.

3. Add a method for fetching data based on the location of the user, as shown in the following code. Name the method GetForecastAsync:

    ```
    using System.Threading.Tasks;
    using Weather.Models;

    namespace Weather.Services;

    public interface IWeatherService
    {
        Task<Forecast> GetForecastAsync(double latitude, double
    longitude);
    }
    ```

Now that we have an interface, we can create an implementation for it, as follows:

1. In the Services folder, create a new class called OpenWeatherMapWeatherService.

2. Implement the interface and add the async keyword to the GetForecastAsync method:

    ```
    using System;
    using System.Globalization;
    using Weather.Models;
    using System.Text.Json;

    namespace Weather.Services;

    public class OpenWeatherMapWeatherService : IWeatherService
    {
        public async Task<Forecast> GetForecastAsync(double
    latitude, double longitude)
        {
    ```

```
        }
    }
```

Before we call the OpenWeatherMap API, we need to build a URI for the call to the Weather API. This will be a GET call, and the latitude and longitude of the position will be added as query parameters. We will also add the API key and the language that we would like the response to be in. Let's set this up:

1. Open the `OpenWeatherMapWeatherService` class.

2. Add the highlighted code in the following code snippet to the `OpenWeatherMap WeatherService` class:

```
public async Task<Forecast> GetForecastAsync(double latitude,
double longitude)
{
    var language = CultureInfo.CurrentUICulture.
TwoLetterISOLanguageName;
    var apiKey = "{AddYourApiKeyHere}";
    var uri = $"https://api.openweathermap.org/data/2.5/
forecast?lat={latitude}&lon={longitude}&units=metric&lang=
{language}&appid={apiKey}";
}
```

Replace the {AddYourApiKeyHere} with the key you obtained from the *Creating models for the weather data* section

To deserialize the JSON that we will get from the external service, we will use `System.Text.JSON`.

To make a call to the `Weather` service, we will use the `HttpClient` class and the `GetStringAsync` method, as follows:

1. Create a new instance of the `HttpClient` class.

2. Call `GetStringAsync` and pass the URL as the argument.

3. Use the `JsonSerializer` class and the `DeserializeObject` method from `System.Text.Json` to convert the JSON string into an object.

4. Map the `WeatherData` object to a `Forecast` object.

5. The code for this should look like the highlighted code shown in the following snippet:

```
public async Task<Forecast> GetForecastAsync(double latitude,
double longitude)
{
    var language = CultureInfo.CurrentUICulture.
TwoLetterISOLanguageName;
    var apiKey = "{AddYourApiKeyHere}";
```

```
        var uri = $"https://api.openweathermap.org/data/2.5/
forecast?lat={latitude}&lon={longitude}&units=metric&lang=
{language}
&appid={apiKey}";

    var httpClient = new HttpClient();
    var result = await httpClient.GetStringAsync(uri);
    var data = JsonSerializer.Deserialize<WeatherData>(result);
    var forecast = new Forecast()
    {
        City = data.city.name,
        Items = data.list.Select(x => new ForecastItem()
        {
            DateTime = ToDateTime(x.dt),
            Temperature = x.main.temp,
            WindSpeed = x.wind.speed,
            Description = x.weather.First().description,
            Icon = $"http://openweathermap.org/img/w/{x.weather.
First().icon}.png"
        }).ToList()
    };
    return forecast;
}
```

> **Performance tip**
>
> To optimize the performance of the app, we can use `HttpClient` as a singleton and reuse it for all network calls in the application. The following information is from Microsoft's documentation: "*HttpClient is intended to be instantiated once and reused throughout the life of an application. Instantiating an HttpClient class for every request will exhaust the number of sockets available under heavy loads. This will result in SocketException errors.*" This can be found at `https://learn.microsoft.com/en-gb/dotnet/api/system.net.http.httpclient?view=netstandard-2.0`.

In the preceding code, we have a call to a `ToDateTime` method, which is a method that we will need to create. This method converts the date from a Unix timestamp into a `DateTime` object, as shown in the following code:

```
private DateTime ToDateTime(double unixTimeStamp)
{
    DateTime dateTime = new DateTime(1970, 1, 1, 0, 0, 0, 0,
DateTimeKind.Utc);
    dateTime = dateTime.AddSeconds(unixTimeStamp).ToLocalTime();
    return dateTime;
}
```

Performance tip

By default, HttpClient uses the Mono implementation of HttpClient (iOS and Android). To increase performance, we can use a platform-specific implementation instead. For iOS, use NSUrlSession. This can be set in the project settings of the iOS project under the **iOS Build** tab. For Android, use **Android**. This can be set in the project settings of the Android project under **Android Options | Advanced**.

Configuring the application platforms so that they use location services

To be able to use location services, we need to carry out some configuration on each platform.

Configuring the iOS platform so that it uses location services

To use location services in an iOS app, we need to add a description to indicate why we want to use the location in the info.plist file. In this app, we only need to get the location when we are using the app, so we only need to add a description for this. Let's set this up:

1. Open info.plist in Platforms/iOS with **XML (Text) Editor**.

2. Add the NSLocationWhenInUseUsageDescription key using the following code:

    ```
    <key>NSLocationWhenInUseUsageDescription</key>
    <string>We are using your location to find a forecast for you</string>
    ```

Configuring the Android platform so that it uses location services

For Android, we need to set up the app so that it requires the following two permissions:

*   **ACCESS_COARSE_LOCATION**
*   **ACCESS_FINE_LOCATION**

We can set this in the AndroidManifest.xml file, which can be found in the Platforms\Android\ folder. However, we can also set this in the project properties on the **Android Manifest** tab, as shown in the following screenshot:

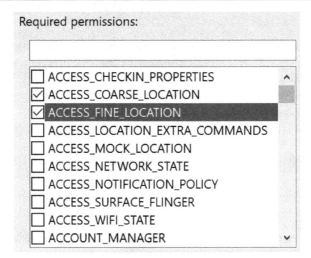

Figure 8.6 – Selecting location permissions

Configuring the WinUI platform so that it uses location services

Since we will be using location services in the WinUI platform, we need to add the `Location` capability under **Capabilities** in the `Package.appxmanifest` file of the project, which is located in the `Platforms/Windows` folder, as shown in the following screenshot:

Application	Visual Assets	Capabilities	Declarations	Content URIs	Packaging

Use this page to specify system features or devices that your app can use.

Capabilities:

- ☐ AllJoyn
- ☐ Appointments
- ☐ Background Media Playback
- ☐ Blocked Chat Messages
- ☐ Bluetooth
- ☐ Chat Message Access
- ☐ Code Generation
- ☐ Contacts
- ☐ Enterprise Authentication
- ☐ Internet (Client & Server)
- ☑ Internet (Client)
- ☑ Location
- ☐ Low Level
- ☐ Low Level Devices
- ☐ Microphone
- ☐ Music Library
- ☐ Objects 3D
- ☐ Offline Maps Management
- ☐ Phone Call

Description:

Provides access to the current location, which is obtained from dedicated hardware like a GPS sensor in the PC or derived from available network information.

More information

Figure 8.7 – Adding a location to the WinUI app

Creating the ViewModel class

Now that we have a service that's responsible for fetching weather data from the external weather source, it's time to create a `ViewModel`. First, however, we will create a base view model where we can put the code that can be shared between all the `ViewModels` of the app. Let's set this up:

1. Create a new folder called `ViewModels`.

2. Create a new class called `ViewModel`.

3. Make the new class `public` and `abstract`.

4. Add a package reference to CommunityToolkit.MVVM:

```
public abstract partial class ViewModel : ObservableObject
{
}
```

We now have a base view model. We can use this for the view model that we're about to create.

Now, it's time to create `MainViewModel`, which will be the ViewModel for our `MainView` in the app. Perform the following steps to do so:

1. In the `ViewModels` folder, create a new class called `MainViewModel`.

2. Add the abstract `ViewModel` class as a base class.

3. Because we are going to use constructor injection, we will add a constructor with the `IWeatherService` interface as a parameter.

   Create a `read-only private` field. We will use this to store the `IweatherService` instance:

```
public class MainViewModel : ViewModel
{
    private readonly IWeatherService weatherService;

    public MainViewModel(IWeatherService weatherService)
    {
        this.weatherService = weatherService;
    }
}
```

`MainViewModel` takes any object that implements `IWeatherService` and stores a reference to that service in a field. We will be adding functionality that will fetch weather data in the next section.

Getting the weather data

Now, we will create a new method for loading the data. This will be a three-step process. First, we will get the location of the user. Once we have this, we can fetch data related to that location. The final step is to prepare the data that the views can consume to create a user interface for the user.

To get the location of the user, we will use the `Geolocation` class, which exposes methods that can fetch the location of the user. Perform the following steps:

1. Create a new method called `LoadDataAsync`. Make it an asynchronous method that returns `Task`.

2. Use the `GetLocationAsync` method on the `Geolocation` class to get the location of the user.

3. Pass the latitude and longitude from the result of the `GetLocationAsync` call and pass it to the `GetForecast` method on the object that implements `IWeatherService` using the following code:

```
public async Task LoadDataAsync()
{
    var location = await Geolocation.GetLocationAsync();
    var forecast = await weatherService.
GetForecastAsync(location.Latitude, location.Longitude);
}
```

Now that we can get data from the service, we need to structure it for our user interface by grouping the individual data items.

Grouping the weather data

When we present the weather data, we will group it by day so that all of the forecasts for one day will be under the same header. To do this, we will create a new model called `ForecastGroup`. To make it possible to use this model with the .NET MAUI `CollectionView`, it has to have an `IEnumerable` type as the base class. Let's set this up:

1. Create a new class called `ForecastGroup` in the `Models` folder.

2. Add `List<ForecastItem>` as the base class for the new model.

3. Add an empty constructor and a constructor that has a list of `ForecastItem` instances as a parameter.

4. Add a `Date` property.

5. Add a property, `DateAsString`, that returns the `Date` property as a short date string.

6. Add a property, `Items`, that returns the list of `ForecastItem` instances, as shown in the following code:

```
using System;

namespace Weather.Models;

public class ForecastGroup : List<ForecastItem>
{
```

```
        public ForecastGroup() { }
        public ForecastGroup(IEnumerable<ForecastItem> items)
        {
            AddRange(items);
        }
        public DateTime Date { get; set; }
        public string DateAsString => Date.ToShortDateString();
        public List<ForecastItem> Items => this;
}
```

When we have done this, we can update `MainViewModel` with two new properties, as follows:

1. Create a private field called `city` with the `ObservableProperty` attribute for the name of the city we are fetching the weather data.

2. Create a private field called `days` with the `ObservableProperty` attribute that will contain the grouped weather data.

3. The `MainViewModel` class should look like the highlighted code shown in the following snippet:

```
    public partial class MainViewModel : ViewModel
    {
        [ObservableProperty]
        private string city;

        [ObservableProperty]
        private ObservableCollection<ForecastGroup> days;
        // Rest of the class is omitted for brevity
    }
```

Now, we are ready to group the data. We will do this in the `LoadDataAsync` method. We will loop through the data from the service and add items to various groups, as follows:

1. Create an `itemGroups` variable of the `List<ForecastGroup>` type.

2. Create a `foreach` loop that loops through all the items in the `forecast` variable.

3. Add an `if` statement that checks whether the `itemGroups` property is empty. If it is empty, add a new `ForecastGroup` to the variable and continue to the next item in the item list.

4. Use the `SingleOrDefault` method (this is an extension method from `System.Linq`) on the `itemGroups` variable to get a group based on the date of the current `ForecastItem`. Add the result to a new variable, `group`.

5. If the `group` property is `null`, then there is no group with the current day in the list of groups. If this is the case, a new `ForecastGroup` should be added to the list in the `itemGroups` variable. The code will continue executing until it gets to the next `forecast` item in the `forecast`. `Items` list. If a group is found, it should be added to the list in the `itemGroups` variable.

6.   After the `foreach` loop, set the `Days` property with a new `ObservableCollection<ForecastGroup>` and use the `itemGroups` variable as an argument in the constructor.

7.   Set the `City` property to the `City` property of the `forecast` variable.

8.   The `LoadDataAsync` method should now look as follows:

```
public async Task LoadDataAsync()
{
    var location = await Geolocation.GetLocationAsync();
    var forecast = await weatherService.
GetForecastAsync(location.Latitude, location.Longitude);

    var itemGroups = new List<ForecastGroup>();
    foreach (var item in forecast.Items)
    {
        if (!itemGroups.Any())
        {
            itemGroups.Add(new ForecastGroup(new
List<ForecastItem>() { item })
            {
                Date = item.DateTime.Date
            });
            continue;
        }
        var group = itemGroups.SingleOrDefault(x => x.Date ==
item.DateTime.Date);

        if (group == null)
        {
            itemGroups.Add(new ForecastGroup(new
List<ForecastItem>() { item })
            {
                Date = item.DateTime.Date
            });
            continue;
        }
        group.Items.Add(item);
    }
    Days = new ObservableCollection<ForecastGroup>(itemGroups);
    City = forecast.City;
}
```

> **Tip**
>
> Don't use the Add method on ObservableCollection when you want to add more than a couple of items. It is better to create a new instance of ObservableCollection and pass a collection to the constructor. The reason for this is that every time you use the Add method, you will have a binding to it from the view, which will cause the view to be rendered. We will get better performance if we avoid using the Add method.

Creating the view for tablets and desktop computers

The next step is to create the view that we will use when the app is running on a tablet or a desktop computer. Let's set this up:

1. Create a new folder in the Weather project called Views.

2. In the Views folder, create a new folder called Desktop.

3. Create a new .NET MAUI ContentPage (XAML) file called MainView in the Views\ Desktop folder:

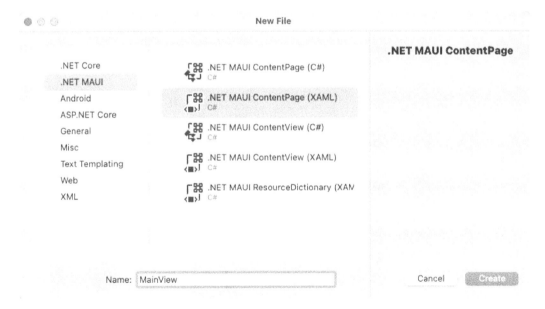

Figure 8.8 – Adding a .NET MAUI XAML ContentPage

4. Pass an instance of MainViewModel in the constructor of the view to set BindingContext, as shown in the following code:

```
public MainView (MainViewModel mainViewModel)
{
```

```
        InitializeComponent ();
        BindingContext = mainViewModel;
}
```

Later, in the *Adding services and ViewModels to dependency injection* section, we will configure dependency injection to provide the instances for us.

To trigger the `LoadDataAsync` method in `MainViewModel`, call the `LoadDataAsync` method by overriding the `OnNavigatedTo` method on the main thread. We need to make sure that the call is executed on the UI thread since it will interact with the user interface.

To do this, perform the following steps:

1. Open the `MainView.xaml.cs` file in the `Views\Desktop` folder.

2. Create an override of the `OnNavigatedTo` method.

3. Add the highlighted code shown in the following snippet to the `OnNavigateTo` method:

```
    protected override void OnNavigatedTo (NavigatedToEventArgs
args)
    {
        base.OnNavigatedTo (args);
        if (BindingContext is MainViewModel viewModel)
        {
            MainThread.BeginInvokeOnMainThread(async () =>
            {
                await viewModel.LoadDataAsync ();
            });
        }
    }
```

In the `MainView` XAML file, add a binding for the `Title` property of `ContentPage` to the `City` property in `ViewModel`, as follows:

1. Open the `MainView.xaml` file in the `Views\Desktop` folder.

2. Add the `Title` binding to the `ContentPage` element, as highlighted in the following code snippet:

```
<ContentPage
    xmlns="http://schemas.microsoft.com/dotnet/2021/maui"
    xmlns:x="http://schemas.microsoft.com/winfx/2009/xaml"
    x:Class="Weather.Views.Desktop.MainView"
    Title="{Binding City}">
```

In this next section, we will use `FlexLayout` to render the data from the ViewModel onto the screen.

Using FlexLayout

In .NET MAUI, we can use `CollectionView` or `ListView` if we want to show a collection of data. Using both `CollectionView` and `ListView` works great in most cases, and we will use `CollectionView` later in this chapter, but `ListView` can only show data vertically. In this app, we want to show data in both directions. In the vertical direction, we will have the days (we group forecasts based on days), while in the horizontal direction, we will have the forecasts within a particular day. We also want the forecasts within a day to wrap if there is not enough space for all of them in one row. `CollectionView` can show data in a horizontal direction, but it will not wrap. With `FlexLayout`, we can add items in both directions and we can use `BindableLayout` to bind items to it. When we use `BindableLayout`, we will use `ItemSource` and `ItemsTemplate` as attached properties.

Perform the following steps to build the view:

1. Add `Grid` as the root view of the page.

2. Add `ScrollView` to `Grid`. We need this to be able to scroll if the content is higher than the height of the page.

3. Add `FlexLayout` to `ScrollView` and set the direction to `Column` so that the content will be in a vertical direction.

4. Add a binding to the `Days` property in `MainViewModel` using `BindableLayout.ItemsSource`.

5. Set `DataTemplate` to the content of `ItemsTemplate`, as shown in the following code:

```
<Grid>
  <ScrollView BackgroundColor="Transparent">
    <FlexLayout BindableLayout.ItemsSource="{Binding Days}"
Direction="Column">
      <BindableLayout.ItemTemplate>
        <DataTemplate>
          <!--Content will be added here -->
        </DataTemplate>
      </BindableLayout.ItemTemplate>
    </FlexLayout>
  </ScrollView>
</Grid>
```

The content for each item will be a header with the date and a horizontal `FlexLayout` with the forecasts for the day. Let's set this up:

1. Open the `MainView.xaml` file.

2. Add `StackLayout` so that the children we add to it will be placed in a vertical direction.

3. Add `ContentView` to `StackLayout` with `Padding` set to `10` and `BackgroundColor` set to `#9F5010`. This will be the header. The reason we need `ContentView` is that we want to have padding around the text.

4. Add `Label` to `ContentView` with `TextColor` set to `White` and `FontAttributes` set to `Bold`.

5. Add a binding to `DateAsString` for the `Text` property of `Label`.

6. The code should be placed at the `<!-- Content will be added here -->` comment and should look as follows:

```
<StackLayout>
  <ContentView Padding="10" BackgroundColor="#9F5010">
    <Label Text="{Binding DateAsString}" TextColor="White"
FontAttributes="Bold" />
  </ContentView>
</StackLayout>
```

Now that we have the date in the user interface, we need to add a `FlexLayout` property, which will repeat through any items in `MainViewModel`. Perform the following steps to do so:

1. Add `FlexLayout` after the `</ContentView>` tag but before the `</StackLayout>` tag.

2. Set `JustifyContent` to `Start` to set the items so that they're added from the left-hand side, without distributing them over the available space.

3. Set `AlignItems` to `Start` to set the content to the left of each item in `FlexLayout`, as shown in the following code:

```
<FlexLayout BindableLayout.ItemsSource="{Binding Items}"
Wrap="Wrap" JustifyContent="Start" AlignItems="Start">
</FlexLayout>
```

After defining `FlexLayout`, we need to provide an `ItemsTemplate` property, which defines how each item in the list should be rendered. Continue adding the XAML directly under the `<FlexLayout>` tag you just added, as follows:

1. Set the `ItemsTemplate` property to `DataTemplate`.

2. `FillDataTemplate` with elements, as shown in the following code:

Tip

If we want to add formatting to a binding, we can use `StringFormat`. In this case, we want to add the degree symbol after the temperature. We can do this by using the `{Binding Temperature, StringFormat='{0}° C'}` phrase. With the `StringFormat` property of the binding, we can format data with the same arguments that we would use if we were to do this in C#. This is the same as `string.Format("{0}° C", Temperature)` in C#. We can also use it to format a date; for example, `{Binding Date, StringFormat='yyyy'}`. In C#, this would look like `Date.ToString("yyyy")`.

```
<BindableLayout.ItemTemplate>
  <DataTemplate>
    <StackLayout Margin="10" Padding="20" WidthRequest="150"
BackgroundColor="#99FFFFFF">
        <Label FontSize="16" FontAttributes="Bold" Text="{Binding
TimeAsString}" HorizontalOptions="Center" />
        <Image WidthRequest="100" HeightRequest="100"
Aspect="AspectFit" HorizontalOptions="Center" Source="{Binding
Icon}" />
        <Label FontSize="14" FontAttributes="Bold" Text="{Binding
Temperature, StringFormat='{0}° C'}" HorizontalOptions="Center"
/>
        <Label FontSize="14" FontAttributes="Bold" Text="{Binding
Description}" HorizontalOptions="Center" />
    </StackLayout>
  </DataTemplate>
</BindableLayout.ItemTemplate>
```

> **Tip**
>
> The AspectFill phrase, as a value of the Aspect property for Image, means that the whole image will always be visible and that the aspects will not be changed. The AspectFit phrase will also keep the aspect of an image, but the image can be zoomed into and out of and cropped so that it fills the whole Image element. The last value that Aspect can be set to, Fill, means that the image can be stretched or compressed to match the Image view to ensure that the aspect ratio is kept.

Adding a toolbar item to refresh the weather data

To be able to refresh the data without restarting the app, we will add a Refresh button to the toolbar. MainViewModel is responsible for handling any logic that we want to perform, and we must expose any action as an ICommand bindable that we can bind to.

Let's start by creating the Refresh command method on MainViewModel:

1. Open the MainViewModel class.

2. Add a using declaration for CommunityToolkit.Mvvm.Input:

   ```
   using System.Collections.ObjectModel;
   using CommunityToolkit.Mvvm.ComponentModel;
   using CommunityToolkit.Mvvm.Input;
   using Weather.Models;
   using Weather.Services;
   ```

3. Add a method called `RefreshAsync` that calls the `LoadDataAsync` method, as shown in the following code:

```
public async Task RefreshAsync()
{
    await LoadDataAsync();
}
```

4. Since these methods are asynchronous, `Refresh` will return `Task`, and we can use `async` and `await` to call `LoadDataAsync` without blocking the UI thread.

5. Add a `RelayCommand` attribute to the `RefreshAsync` method to auto-generate the `ICommand` bindable property to the method:

```
[RelayCommand]
public async Task RefreshAsync()
{
    await LoadDataAsync();
}
```

Now that we have defined the `Refresh` command, we need to bind it to the user interface so that when the user clicks the toolbar button, the action will be executed.

To do this, perform the following steps:

1. Open the `MainView.xaml` file.

2. Download the `refresh.png` file from `https://raw.githubusercontent.com/PacktPublishing/MAUI-Projects-3rd-Edition/main/Chapter08/Weather/Resources/Images/refresh.png` and save it to the `Resources/Images` folder of the project.

3. Add a new `ToolbarItem` with the `Text` property set to `Refresh` to the `ToolbarItems` property of `ContentPage` and set the `IconImageSource` property to `refresh.png` (alternatively, you can set the `IconImageSource` property to the URL of the image and .NET MAUI will download the image).

4. Bind the `Command` property to the `Refresh` property in `MainViewModel`, as shown in the following code:

```
<ContentPage.ToolbarItems>
    <ToolbarItem IconImageSource="refresh.png" Text="Refresh"
Command="{Binding RefreshCommand}" />
</ContentPage.ToolbarItems>
```

That's all for refreshing the data. Now, we need some kind of indicator that the data is loading.

Adding a loading indicator

When we refresh the data, we want to show a loading indicator so that the user knows that something is happening. To do this, we will add `ActivityIndicator`, which is what this control is called in .NET MAUI. Let's set this up:

1. Open the `MainViewModel` class.

2. Add a `Boolean` field called `isRefreshing` to `MainViewModel`.

3. Add the `ObservableProperty` attribute to `isRefreshingField` to generate the `IPropertyChanged` implementation.

4. Set the `IsRefreshing` property to `true` at the beginning of the `LoadDataAsync` method.

5. At the end of the `LoadDataAsync` method, set the `IsRefreshing` property to `false`, as shown in the following code:

```
[ObservableProperty]
private bool isRefreshing;

....// The rest of the code is omitted for brevity

public async Task LoadData()
{
    IsRefreshing = true;
....// The rest of the code is omitted for brevity
    IsRefreshing = false;
}
```

Now that we have added some code to `MainViewModel`, we need to bind the `IsRefreshing` property to a user interface element that will be displayed when the `IsRefreshing` property is `true`, as shown in the following steps:

1. In `MainView.xaml`, add `Frame` after `ScrollView` as the last element in `Grid`.

2. Bind the `IsVisible` property to the `IsRefreshing` method that we created in `MainViewModel`.

3. Set `HeightRequest` and `WidthRequest` to `100`.

4. Set `VerticalOptions` and `HorizontalOptions` to `Center` so that `Frame` will be in the middle of the view.

5. Set `BackgroundColor` to `#99000000` to set the background to white with a little bit of transparency.

6. Add `ActivityIndicator` to `Frame` with `Color` set to `Black` and `IsRunning` set to True, as shown in the following code:

```
<Frame IsVisible="{Binding IsRefreshing}"
BackgroundColor="#99FFFFFF" WidthRequest="100"
HeightRequest="100" VerticalOptions="Center"
HorizontalOptions="Center">
  <ActivityIndicator Color="Black" IsRunning="True" />
</Frame>
```

This will create a spinner that will be visible while data is loading, which is a really good practice when creating any user interface. Now, we'll add a background image to make the app look a bit nicer.

Setting a background image

The last thing we will do to this view, for the moment, is add a background image. The image we will be using in this example is a result of a Google search for images that are free to use. Let's set this up:

1. Open the `MainView.xaml` file.

2. Set the `Background` property of `ScrollView` to `Transparent`.

3. Add an `Image` element in `Grid` with `UriImageSource` as the value of the `Source` property.

4. Set the `CachingEnabled` property to true and `CacheValidity` to 5. This means that the image will be cached for 5 days.

> **Note**
>
> You could also set these properties if you used a URL for the `Refresh IconImageSource` property to avoid downloading the image on every run of the app.

5. The XAML should now look as follows:

```
<ContentPage xmlns="http://schemas.microsoft.com/dotnet/2021/
maui"
              xmlns:x="http://schemas.microsoft.com/winfx/2009/
xaml"
   x:Class="Weather.Views.Desktop.MainView"
   Title="{Binding City}">
  <ContentPage.ToolbarItems>
    <ToolbarItem Icon="refresh.png" Text="Refresh"
Command="{Binding RefreshCommand}" />
  </ContentPage.ToolbarItems>

<Grid>
  <Image Aspect="AspectFill">
    <Image.Source>
```

```
        <UriImageSource Uri="https://upload.wikimedia.org/
    wikipedia/commons/7/79/Solnedg%C3%A5ng_%C3%B6ver_Laholmsbukten_
    augusti_2011.jpg" CachingEnabled="true" CacheValidity="5" />
        </Image.Source>
    </Image>
    <ScrollView BackgroundColor="Transparent">

    <!-- The rest of the code is omitted for brevity -->
```

We can also set the URL directly in the `Source` property by using `<Image Source="https://ourgreatimage.url" />`. However, if we do this, we can't specify caching for the image.

With the desktop view complete, we need to consider how this page will look when we are running the app on a phone or tablet.

Creating the view for phones

Structuring content on a tablet and a desktop computer is very similar in many ways. On phones, however, we are much more limited in what we can do. Therefore, in this section, we will create a specific view for this app when it's used on phones. To do so, perform the following steps:

1. Create a new XAML-based **ContentPage** in the `Views` folder.

2. In the `Views` folder, create a new folder called `Mobile`.

3. Create a new `.NET MAUI ContentPage (XAML)` file called `MainView` in the `Views\Mobile` folder:

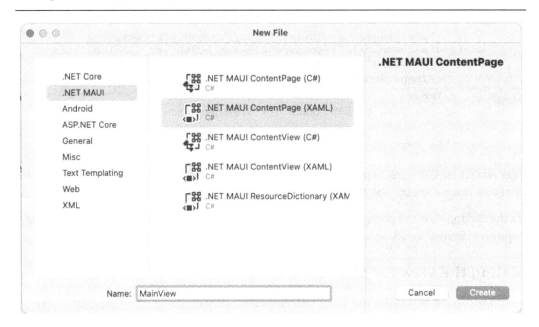

Figure 8.9 – Adding a .NET MAUI XAML ContentPage

4.   Pass an instance of `MainViewModel` in the constructor of the view to set `BindingContext`, as shown in the following code:

```
public MainView (MainViewModel mainViewModel)
{
    InitializeComponent();
    BindingContext = mainViewModel;
}
```

Later, in the *Adding services and ViewModels to dependency injection* section, we will configure dependency injection to provide the instances for us.

To trigger the `LoadDataAsync` method in `MainViewModel`, call the `LoadDataAsync` method by overriding the `OnNavigatedTo` method on the main thread. We need to make sure that the call is executed on the UI thread since it will interact with the user interface.

To do this, perform the following steps:

1.   Open the `MainView.xaml.cs` file in the `Views\Mobile` folder.

2.   Create an override of the `OnNavigatedTo` method.

3.   Add the highlighted code in the following snippet to the `OnNavigateTo` method:

```
protected override void OnNavigatedTo(NavigatedToEventArgs args)
{
```

```
        base.OnNavigatedTo(args);
        if (BindingContext is MainViewModel viewModel)
        {
            MainThread.BeginInvokeOnMainThread(async () =>
            {
                await viewModel.LoadDataAsync();
            });
        }
    }
```

In the MainView XAML file, add a binding for the Title property of ContentPage to the City property in ViewModel, as follows:

1. Open the MainView.xaml file in the Views\Mobile folder.

2. Add the Title binding to the ContentPage element, as highlighted in the following code snippet:

    ```
    <ContentPage xmlns="http://schemas.microsoft.com/dotnet/2021/
    maui"
                     xmlns:x="http://schemas.microsoft.com/winfx/2009/
    xaml"
        x:Class="Weather.Views.Desktop.MainView"
        Title="{Binding City}">
    ```

In the next section, we will use CollectionView to display the weather data instead of using FlexView, as we did for the desktop view.

Using a grouped CollectionView

We could use FlexLayout for the phone's view, but because we want our user experience to be as good as possible, we will use CollectionView instead. To get the headers for each day, we will use grouping for CollectionView. For FlexLayout, we had ScrollView, but for CollectionView, we don't need this because CollectionView can handle scrolling by default.

Let's continue creating the user interface for the phone's view:

1. Open the MainView.xaml file in the Views\Mobile folder.

2. Add CollectionView to the root of the page.

3. Set a binding to the Days property in MainViewModel for the ItemSource property.

4. Set IsGrouped to True to enable grouping in CollectionView.

5. Set BackgroundColor to Transparent, as shown in the following code:

    ```
    <CollectionView ItemsSource="{Binding Days}" IsGrouped="True"
    BackgroundColor="Transparent">
    </CollectionView>
    ```

To format how each header will look, we will create a `DataTemplate` property, as follows:

1. Add a `DataTemplate` property to the `GroupHeaderTemplate` property of `CollectionView`.

2. Add the content for the row to `DataTemplate`, as shown in the following code:

```
<CollectionView ItemsSource="{Binding Days}" IsGrouped="True"
BackgroundColor="Transparent">
  <CollectionView.GroupHeaderTemplate>
    <DataTemplate>
      <ContentView Padding="15,5" BackgroundColor="#9F5010">
        <Label FontAttributes="Bold" TextColor="White"
Text="{Binding DateAsString}" VerticalOptions="Center"/>
      </ContentView>
    </DataTemplate>
  </CollectionView.GroupHeaderTemplate>
</CollectionView>
```

To format how each forecast will look, we will create a `DataTemplate` property, as we did with the group header. Let's set this up:

1. Add a `DataTemplate` property to the `ItemTemplate` property of `CollectionView`.

2. In `DataTemplate`, add a `Grid` property that contains four columns. Use the `ColumnDefinition` property to specify the width of the columns. The second column should be 50; the other three will share the rest of the space. We will do this by setting `Width` to `*`.

3. Add the following content to `Grid`:

```
<CollectionView.ItemTemplate>
  <DataTemplate>
    <Grid Padding="15,10" ColumnSpacing="10"
BackgroundColor="#99FFFFFF">
      <Grid.ColumnDefinitions>
        <ColumnDefinition Width="*" />
        <ColumnDefinition Width="50" />
        <ColumnDefinition Width="*" />
        <ColumnDefinition Width="*" />
      </Grid.ColumnDefinitions>
      <Label FontAttributes="Bold" Text="{Binding TimeAsString}"
VerticalOptions="Center" />
      <Image Grid.Column="1" HeightRequest="50"
WidthRequest="50" Source="{Binding Icon}" Aspect="AspectFit"
VerticalOptions="Center" />
      <Label Grid.Column="2" Text="{Binding Temperature,
StringFormat='{0}° C'}"
VerticalOptions="Center" />
```

```
          <Label Grid.Column="3" Text="{Binding Description}"
VerticalOptions="Center" />

      </Grid>
    </DataTemplate>
  </CollectionView.ItemTemplate>
```

Adding pull-to-refresh functionality

For the tablet and desktop versions of the view, we added a button to the toolbar to refresh the weather forecast. In the phone version of the view, however, we will add pull-to-refresh functionality, which is a common way to refresh content in a list of data. CollectionView in .NET MAUI has no built-in support for pull-to-refresh as ListView has.

Instead, we can use RefreshView. RefreshView can be used to add pull-to-refresh behavior to any control. Let's set this up:

1. Go to Views\Mobile\MainView.xaml.

2. Wrap CollectionView inside RefreshView.

3. Bind the RefreshCommand property in MainViewModel to the Command property of RefreshView to trigger a refresh when the user performs a pull-to-refresh gesture.

4. To show a loading icon when the refresh is in progress, bind the IsRefreshing property in MainViewModel to the IsRefreshing property of RefreshView. When we are setting this up, we will also get a loading indicator when the initial load is running, as shown in the following code:

```
<RefreshView Command="{Binding Refresh}" IsRefreshing="{Binding
IsRefreshing}">
  <CollectionView ItemsSource="{Binding Days}" IsGrouped="True"
BackgroundColor="Transparent">
. . . .
  </CollectionView>
</RefreshView>
```

That concludes the views for the moment. Now, let's wire them up into dependency injection so that we can see our work.

Adding services and ViewModels to dependency injection

For our views to get an instance of `MainViewModel` and `MainViewModel` to get an instance of `OpenWeatherMapWeatherService`, we need to add them to dependency injection. Let's set this up:

1. Open `MauiProgram.cs`.

2. Add the following highlighted code:

```
#if DEBUG
    builder.Logging.AddDebug();
#endif
    builder.Services.AddSingleton<IWeatherService,
OpenWeatherMapWeatherService>();

    builder.Services.AddTransient<MainViewModel,
MainViewModel>();

    return builder.Build();
```

In the next section, we will add the navigation to the views based on the device's form factor.

Navigating to different views based on the form factor

We now have two different views that should be loaded in the same place in the app. `Weather.Views.Desktop.MainView` should be loaded if the app is running on a tablet or a desktop, while `Weather.Views.Mobile.MainView` should be loaded if the app is running on a phone.

The `Device` class in .NET MAUI has a static `Idiom` property that we can use to check which form factor the app is running on. The value of `Idiom` can be `Phone`, `Tablet`, `Desktop`, `Watch`, or `TV`. Because we only have one view in this app, we could have used an `if` statement when we were setting `MainPage` in `App.xaml.cs` and checked what the `Idiom` value was.

Since we are only ever going to need one view, we can register just the view that we need in dependency injection – all we need is a common type to register the views with. Let's create a new interface that our views will implement:

1. Create a new interface in the `Views` folder named `IMainView`.

    We won't add any additional properties or methods to the interface – we'll just use it as a marker.

2. Open `Views\Desktop\MainView.xaml.cs` and add the `IMainView` interface to the class:

```
public partial class MainView : ContentPage, IMainView
```

3. Open `Views\Mobile\MainView.xaml.cs` and add the `IMainView` interface to the class:

```
public partial class MainView : ContentPage, IMainView
```

Now that we have a common interface, we can register the views with dependency injection:

1. Open the `MauiProgram.cs` file.

2. Add the following code:

```
#if DEBUG
    builder.Logging.AddDebug();
#endif
    builder.Services.AddSingleton<IWeatherService,
OpenWeatherMapWeatherService>();

    builder.Services.AddTransient<MainViewModel,
MainViewModel>();
    if (DeviceInfo.Idiom == DeviceIdiom.Phone)
    {
        builder.Services.AddTransient<IMainView, Views.Mobile.
MainView>();
    }
    else
    {
        builder.Services.AddTransient<IMainView, Views.Desktop.
MainView>();
    }

    return builder.Build();
```

With these changes, we can now test our application. If you run your app, you should see something like the following:

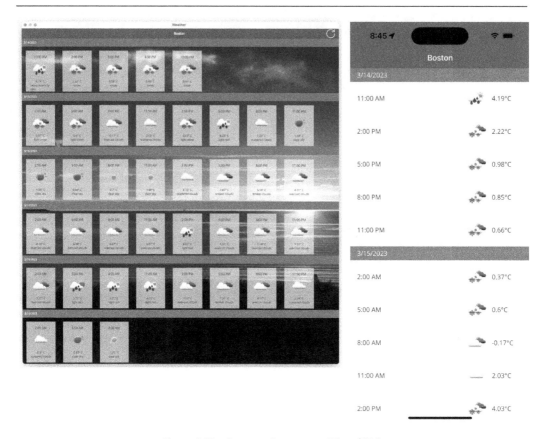

Figure 8.10 – App running on macOS and iOS

Next, let's update the desktop views to handle resizing properly by using `VisualStateManager`.

Handling states with VisualStateManager

`VisualStateManager` is a way to make changes in the UI from the code. We can define states and set values for selected properties to apply to a specific state. `VisualStateManager` can be useful in cases where we want to use the same view for devices with different screen resolutions. It was first introduced in **UWP** to make it easier to create Windows 10 applications for multiple platforms. This was because Windows 10 could run on Windows Phone, as well as on desktops and tablets (the operating system was called Windows 10 Mobile). However, Windows Phone has now been deprecated. `VisualStateManager` is interesting for us as .NET MAUI developers, especially when both iOS and Android can run on both phones and tablets.

In this project, we will use it to make a `forecast` item bigger when the app is running in landscape mode on a tablet or a desktop. We will also make the weather icon bigger. Let's set this up:

1. Open the `Views\Desktop\MainView.xaml` file.

2. In the first `FlexLayout` and `DataTemplate`, insert a `VisualStateManager`. `VisualStateGroups` element into the first `StackLayout`:

    ```
    <StackLayout Margin="10" Padding="20" WidthRequest="150"
    BackgroundColor="#99FFFFFF">
      <VisualStateManager.VisualStateGroups>
        <VisualStateGroup>
        </VisualStateGroup>
      </VisualStateManager.VisualStateGroups>
    ......
    </StackLayout>
    ```

Regarding `VisualStateGroup`, we should add two states, as follows:

1. Add a new `VisualState` called `Portrait` to `VisualStateGroup`.

2. Create a setter in `VisualState` and set `WidthRequest` to `150`.

3. Add another `VisualState` called `Landscape` to `VisualStateGroup`.

4. Create a setter in `VisualState` and set `WidthRequest` to `200`, as shown in the following code:

    ```
    <VisualStateGroup>
      <VisualState Name="Portrait">
        <VisualState.Setters>
          <Setter Property="WidthRequest" Value="150" />
        </VisualState.Setters>
      </VisualState>
      <VisualState Name="Landscape">
        <VisualState.Setters>
          <Setter Property="WidthRequest" Value="200" />
        </VisualState.Setters>
      </VisualState>
    </VisualStateGroup>
    ```

We also want the icons in a forecast item to be bigger when the item itself is bigger. To do this, we will use `VisualStateManager` again. Let's set this up:

1. Insert a `VisualStateManager.VisualStateGroups` element into the second `FlexLayout` and in the `Image` element in `DataTemplate`.

2. Add `VisualState` for both `Portrait` and `Landscape`.

3. Add setters to the states to set `WidthRequest` and `HeightRequest`. The value should be 100 in the `Portrait` state and 150 in the `Landscape` state, as shown in the following code:

```
<Image WidthRequest="100" HeightRequest="100" Aspect="AspectFit"
HorizontalOptions="Center" Source="{Binding Icon}">
    <VisualStateManager.VisualStateGroups>
      <VisualStateGroup>
        <VisualState Name="Portrait">
          <VisualState.Setters>
            <Setter Property="WidthRequest" Value="100" />
            <Setter Property="HeightRequest" Value="100" />
          </VisualState.Setters>
        </VisualState>
        <VisualState Name="Landscape">
          <VisualState.Setters>
            <Setter Property="WidthRequest" Value="150" />
            <Setter Property="HeightRequest" Value="150" />
          </VisualState.Setters>
        </VisualState>
      </VisualStateGroup>
    </VisualStateManager.VisualStateGroups>
</Image>
```

Creating a behavior to set state changes

With `Behavior`, we can add functionality to controls without having to subclass them. With behaviors, we can also create more reusable code than we could if we subclassed a control. The more specific `Behavior` we create, the more reusable it will be. For example, `Behavior` that inherits from `Behavior<View>` could be used on all controls, but `Behavior` that inherits from `Button` can only be used for buttons. Because of this, we always want to create behaviors with a less specific base class.

When we create `Behavior`, we need to override two methods: `OnAttached` and `OnDetachingFrom`. It is really important to remove event listeners in the `OnDeattached` method if we have added them to the `OnAttached` method. This will make the app use less memory. It is also important to set values back to the values that they had before the `OnAppearing` method ran; otherwise, we might see some strange behavior, especially if the behavior is in a `CollectionView` or `ListView` view that is reusing cells.

In this app, we will create a behavior for `FlexLayout`. This is because we can't set the state of an item in `FlexLayout` from the code-behind. We could have added some code to check whether the app runs in portrait or landscape in `FlexLayout`, but if we use `Behavior` instead, we can separate that code from `FlexLayout` so that it will be more reusable. Perform the following steps to do so:

1. Create a new folder called `Behaviors`.

2. Create a new class called `FlexLayoutBehavior`.

3. Add `Behavior<FlexLayoutView>` as a base class.

4. Create a `private` field of the `FlexLayout` type called `view`.

5. The code should look as follows:

```
using System;
namespace Weather.Behaviors;

public class FlexLayoutBehavior : Behavior<FlexLayout>
{
    private FlexLayout view;
}
```

`FlexLayout` is a class that inherits from the `Behavior<FlexLayout>` base class. This will give us the ability to override some virtual methods that will be called when we attach and detach the behavior from a `FlexLayout` class.

But first, we need to create a method that will handle the change in state. Perform the following steps to do so:

1. Open the `FlexlayoutBehavior.cs` file.

2. Create a `private` method called `SetState`. This method will have a `VisualElement` value and a `string` argument.

3. Call `VisualStateManager.GoToState` and pass the parameters to it.

4. If the view is of the `Layout` type, there might be child elements that also need to get the new state. To do that, we will loop through all the children of the layout. Instead of just setting the state directly to the children, we will call the `SetState` method, which is the method that we are already inside. The reason for this is that some of the children may have their own children:

```
private void SetState(VisualElement view, string state)
{
    VisualStateManager.GoToState(view, state);
    if (view is Layout layout)
    {
        foreach (VisualElement child in layout.Children)
        {
```

```
                    SetState(child, state);
            }
        }
    }
```

Now that we have created the SetState method, we need to write a method that uses it and determines what state to set. Perform the following steps to do so:

1. Create a private method called UpdateState.

2. Run the code on MainThread to check whether the app is running in portrait or landscape mode.

3. Create a variable called page and set its value to Application.Current.MainPage.

4. Check whether Width is larger than Height. If this is true, set the VisualState property on the view variable to Landscape. If this is false, set the VisualState property on the view variable to Portrait, as shown in the following code:

```
private void UpdateState()
{
    MainThread.BeginInvokeOnMainThread(() =>
    {
        var page = Application.Current.MainPage;
        if (page.Width > page.Height)
        {
            SetState(view, "Landscape");
            return;
        }
        SetState(view, "Portrait");
    });
}
```

With that, the UpdateState method has been added. Now, we need to override the OnAttachedTo method, which will be called when the behavior is added to FlexLayout. When it is, we want to update the state by calling this method and hook it up to the SizeChanged event of MainPage so that when the size changes, we will update the state again.

Let's set this up:

1. Open the FlexLayoutBehavior file.

2. Override the OnAttachedTo method from the base class.

3. Set the view property to the parameter from the OnAttachedTo method.

4. Add an event listener to Application.Current.MainPage.SizeChanged. In the event listener, add a call to the UpdateState method, as shown in the following code:

```
protected override void OnAttachedTo(FlexLayout view)
{
```

```
    this.view = view;
    base.OnAttachedTo(view);
    UpdateState();
    Application.Current.MainPage.SizeChanged += MainPage_
SizeChanged;
}

void MainPage_SizeChanged(object sender, EventArgs e)
{
    UpdateState();
}
```

When we remove behaviors from a control, it's very important to also remove any event handlers from it to avoid memory leaks, and in the worst case, the app crashing. Let's do this:

1. Open the `FlexLayoutBehavior.cs` file.

2. Override `OnDetachingFrom` from the base class.

3. Remove the event listener from `Application.Current.MainPage.SizeChanged`.

4. Set the `view` field to `null`, as shown in the following code:

```
protected override void OnDetachingFrom(FlexLayout view)
{
    base.OnDetachingFrom(view);
    Application.Current.MainPage.SizeChanged -= MainPage_
SizeChanged;
    this.view = null;
}
```

Perform the following steps to add `behavior` to the view:

1. Open the `MainView.xaml` file inside the `Views/Desktop` folder.

2. Import the `Weather.Behaviors` namespace, as shown in the following code:

```
<ContentPage xmlns="http://schemas.microsoft.com/dotnet/2021/
maui"
             xmlns:x="http://schemas.microsoft.com/winfx/2009/
xaml"
    xmlns:behaviors="clr-namespace:Weather.Behaviors"
    x:Class="Weather.Views.Desktop.MainView"
    Title="{Binding City}">
```

The last thing we will do is add `FlexLayoutBehavior` to the second `FlexLayout`, as shown in the following code:

```
<FlexLayout ItemsSource="{Binding Items}" Wrap="Wrap"
JustifyContent="Start" AlignItems="Start">
    <FlexLayout.Behaviors>
```

```
    <behaviors:FlexLayoutBehavior />
  </FlexLayout.Behaviors>
<FlexLayout.ItemsTemplate>
```

Congratulations – that is a wrap on the weather app!

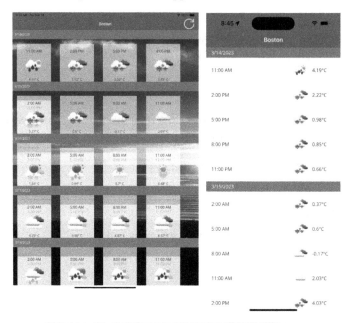

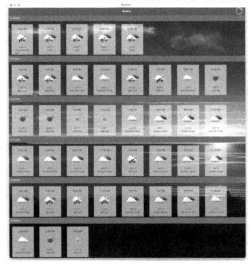

Figure 8.11 – Completed app on tablet, phone, and desktop

Summary

In this chapter, we successfully created an app for four different operating systems – iOS, macOS, Android, and Windows – and three different form factors – phones, tablets, and desktop computers. To create a good user experience on all platforms and form factors, we used `FlexLayout` and `VisualStateManager`. We also learned how to handle different views for different form factors, as well as how to use `Behaviors`.

The next app we will build will be a game with real-time communication. In the next chapter, we will take a look at how we can use the **SignalR** service in **Azure** as the backend game service.

Part 3:
Advanced Projects

In this part, you will work with more advanced topics and complicated projects. You will learn how to create and deploy a service in Azure. Additionally, you will work with Azure Storage and SignalR services. You will learn how to call your service from a .NET MAUI application, properly handle error conditions, and integrate the camera into your app. You will explore a project using Blazor embedded into a .NET MAUI app, and learn how to integrate artificial intelligence services into a .NET MAUI app.

This part has the following chapters:

- *Chapter 9, Setting Up a Backend for a Game Using Azure Services*
- *Chapter 10, Building a Real-Time Game*
- *Chapter 11, Building a Calculator Using .NET MAUI Blazor*
- *Chapter 12, Hot Dog or Not Hot Dog Using Machine Learning*

Setting Up a Backend for a Game Using Azure Services

In this chapter, we will set up a backend for a game app with real-time communication. We will not only create a backend that can scale up to handle a large number of users but also scale down when the number of users is reduced. To build that backend, we will use a serverless architecture based on services in **Microsoft Azure**.

The following topics will be covered in this chapter:

- Understanding the different Azure serverless services
- Creating a **SignalR** service in Microsoft Azure
- Using Azure Functions as an **application programming interface** (**API**)

Technical requirements

To be able to complete this project, you need to have Visual Studio for Mac or PC installed, as well as the necessary .NET MAUI components. See *Chapter 1*, *Introduction to .NET MAUI*, for more details on how to set up your environment.

You also need an Azure account. If you have a Visual Studio subscription, there are a specific amount of Azure credits included each month. To activate your Azure benefits, go to `https://my.visualstudio.com`.

You can also create a free account, where you can use selected services for free over 12 months. You will get $200 worth of credit to explore any Azure service for 30 days, and you can also use the free services at any time. Read more at `https://azure.microsoft.com/en-us/free/`.

If you do not have and do not want to sign up for a free Azure account, you can use local development tools to run the services without Azure.

You can find the full source for the code in this chapter at `https://github.com/PacktPublishing/MAUI-Projects-3rd-Edition`.

Project overview

The main aim of this project will be to set up the backend for a game. A large part of the project will be the configuration that we will carry out in the Azure portal. We will also write some code for the Azure functions that will handle the SignalR connections and a bit of the game logic and state. SignalR is a library that makes real-time communication in applications easier. Azure SignalR is a service that makes it easier to connect multiple clients to send messages via the SignalR library. SignalR is described in more detail later. There will be functions to return information about the SignalR connection, manage matching players to play against each other, and post the result of each player's turn to the SignalR service.

The following diagram shows an overview of the architecture of this application:

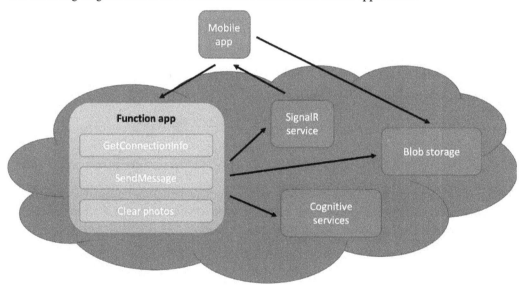

Figure 9.1 – Application architecture

The estimated time to complete this part of the project is about 2 hours.

An overview of the game

Sticks & Stones is a turn-based social game based on the concepts of two childhood games mashed into one, Dots and Boxes and Tic-Tac-Toe. The game board is laid out in a 9x9 grid. Each player will take a turn placing a stick along the side of a box, earning one point. If a stick completes a box, then the player takes ownership of the box, earning five points. The game is won when a player owns three boxes in a row, horizontally, vertically, or diagonally. If no player can own three boxes in a row, the winner of the game is determined by the player with the highest score.

To keep the app and the service side relatively simple, we will eliminate a lot of state management. When the player opens the app, they will have to connect to the game service. They will have to provide a gamer tag or username and an email address. Optionally, they can upload a picture of themselves to use as a profile picture.

Once connected, the player will see a list of all the other players connected to the same game service; this is called the lobby. The player's status, either "Ready to play" or "In a match," will be displayed along with the player's gamer tag and profile picture. If the player is not in a match, then there will also be a button to challenge the player to a match.

Challenging a player to a match will cause the app to prompt the opponent to respond to the challenge, either accept or decline. If the opponent accepts the challenge, then both players are navigated to a new game board where the player who received the challenge will have the first turn. Both players' statuses will update to "In a match" in all the other players' lobbies. Play will alternate between players as they choose a location to place a single stick. Each time a stick is placed by a player, the game board and score will update on both players' devices. When a stick is placed that completes one or more squares, the player then "owns" that square, and a pile of stones is placed in the center of the square. When all sticks have been placed, or a player owns three stones in a row, the game is over, the players navigate back to the lobby, and their status is updated to "Ready to play."

If a player leaves the app during a game, then they will have forfeited the game and the remaining opponent will be credited with the win and navigated back to the lobby.

The following screenshot should give you an idea of what the app will look like when it's completed in *Chapter 10*:

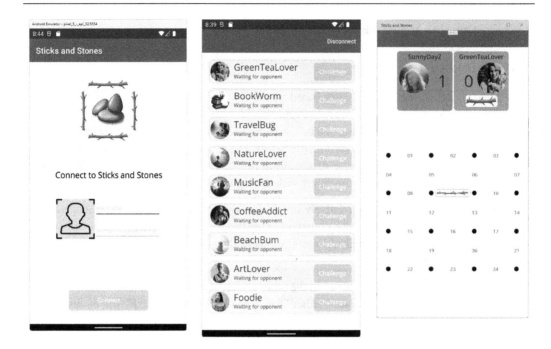

Figure 9.2 – The main game screens

Understanding the different Azure serverless services

Before we start to build a backend with a serverless architecture, we need to define what **serverless** means. In a serverless architecture, the code will run on a server, but we don't need to worry about that; the only thing we need to focus on is building our software. We let someone else handle everything to do with servers. We don't need to think about how much memory or **central processing units** (**CPUs**) the server needs, or even how many servers we need. When we use services in Azure, Microsoft takes care of this for us.

Azure SignalR Service

Azure SignalR Service is a service in **Microsoft Azure** for real-time communication between a server and clients. The service will push content to the clients without them having to poll the server to get content updates. SignalR can be used for multiple types of applications, including mobile applications, web applications, and desktop applications.

SignalR will use **WebSockets** if that option is available. If it is not, SignalR will use other techniques for communication, such as **Server-Sent Events** (**SSEs**) or **long polling**. SignalR will detect which transport technology is available and use it without the developer having to think about it at all.

SignalR can be used in the following examples:

- **Chat applications**: Where the application requires updates from the server as soon as new messages are available
- **Collaborative applications**: For example, meeting applications or when users on multiple devices are working with the same document
- **Multiplayer games**: Where all users need live updates about other users
- **Dashboard applications**: Where users need live updates

Azure Functions

Azure Functions is a Microsoft Azure service that allows us to run code in a serverless way. We will deploy small pieces of code called **functions**. Functions are deployed in groups, called **function apps**. When we are creating a function app, we need to select whether we want it to run on a Consumption plan or an App Service plan. We select a Consumption plan if we want the application to be completely serverless, while with an App Service plan, we have to specify the requirements of the server. With a Consumption plan, we pay for the execution time and for how much memory the function uses. One benefit of an App Service plan is that you can configure it to be **Always-On**, and you won't have any cold starts, so long as you don't have to scale up to more instances. The big benefit of a Consumption plan is that it will always scale according to which resources are needed at that time.

There are several ways in which a function can be triggered to run. Two examples are `HttpTrigger` and `TimeTrigger`. `HttpTrigger` will trigger the function to run when an HTTP request is calling the function. With `TimeTrigger`, functions will run at an interval that we specify. There are also triggers for other Azure services. For example, we can configure a function to run when a file is uploaded to Azure Blob storage, when a new message is posted to an event hub or service bus, or when data is changed in an Azure Cosmos DB service.

Now that we understand what features Azure SignalR Service and Functions offer, let's use them to build our game backend.

Building the serverless backend

In this section, we will set up the backend based on the services described in the preceding section.

Creating a SignalR service

The first service that we will set up is the one for SignalR. To create such a service, proceed as follows:

1. Go to the Azure portal at `https://portal.azure.com`.
2. Create a new resource. The **SignalR Service** resource is in the **Web & Mobile** category.

3. Provide a name for the resource in the form.

4. Select the subscription you want to use for this project.

    We recommend that you create a new **Resource group** and use it for all the resources that we will create for this project. The reason that we want one resource group is that it is easier to track which resources are related to this project, and it is also easier to delete all the resources together.

5. Select a location that is close to your users.

6. Select a pricing tier. For this project, we will use the **Free** tier. We can always use the **Free** tier for development and later scale up to a tier that can handle more connections.

7. Set **Service mode** to **Serverless**.

8. Click **Review + create** to review the settings before creating the SignalR service.

9. Click **Create** to create the storage account.

Refer to the following screenshot to view the preceding information:

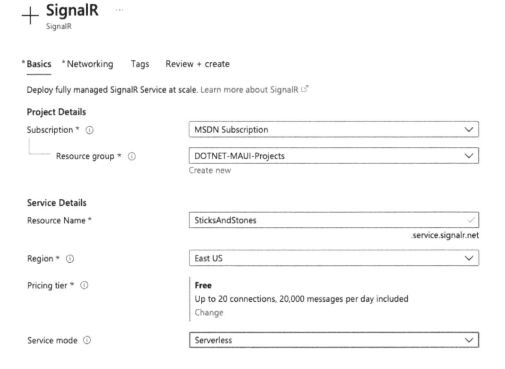

Figure 9.3 – Creating the SignalR service

This is all we need to do to set up a SignalR service. We will return to it in the Azure portal later to grab a connection string to it.

The next step is to set up a storage account in which we can store the images that are uploaded by the users.

With the Computer Vision service created, we can now create the Azure Functions service, which will run our game logic and use SignalR, Blob storage, and Cognitive Services, which we just created.

Using Azure Functions as an API

All the code we will write for the backend will be in Azure Functions. We will use a Visual Studio project to write, debug, and deploy our functions. Before we create the project, we will have to set up and configure the Azure Functions service. Then, we will implement the function to connect the player to the game and provide the client with a list of current players. Next, we will write the functions that allow one player to challenge another to a game. Finally, we will wrap up by writing the function that allows players to take turns placing sticks on the board.

Let's begin by creating the Azure Functions service.

Creating the Azure Functions service

Before we write any code, we will create the function app. This will contain the functions in the Azure portal. Proceed as follows:

1. Create a new **Function App** resource. **Function App** can be found under the **Compute** category.
2. Select a subscription for the function app.
3. Select a resource group for the function app. This should be the same as the other resources we have created in this chapter.
4. Give the function app a name. The name will also be the start of the URL of the function.
5. Select **Code** as the deployment mechanism.
6. Select **.NET** as the runtime stack for the functions.
7. Select **.NET 6.0 (Long Term Support)** for the version.
8. Select a location that is closest to your users.
9. Select **Windows** for **Operating System**.
10. We will use the **Consumption** plan as our **Hosting** plan, so we'll only pay for what we use. **Function app** will scale both up and down according to our requirements – without us having to think about it at all – if we select a **Consumption** plan.

Refer to the following screenshot to view the preceding information:

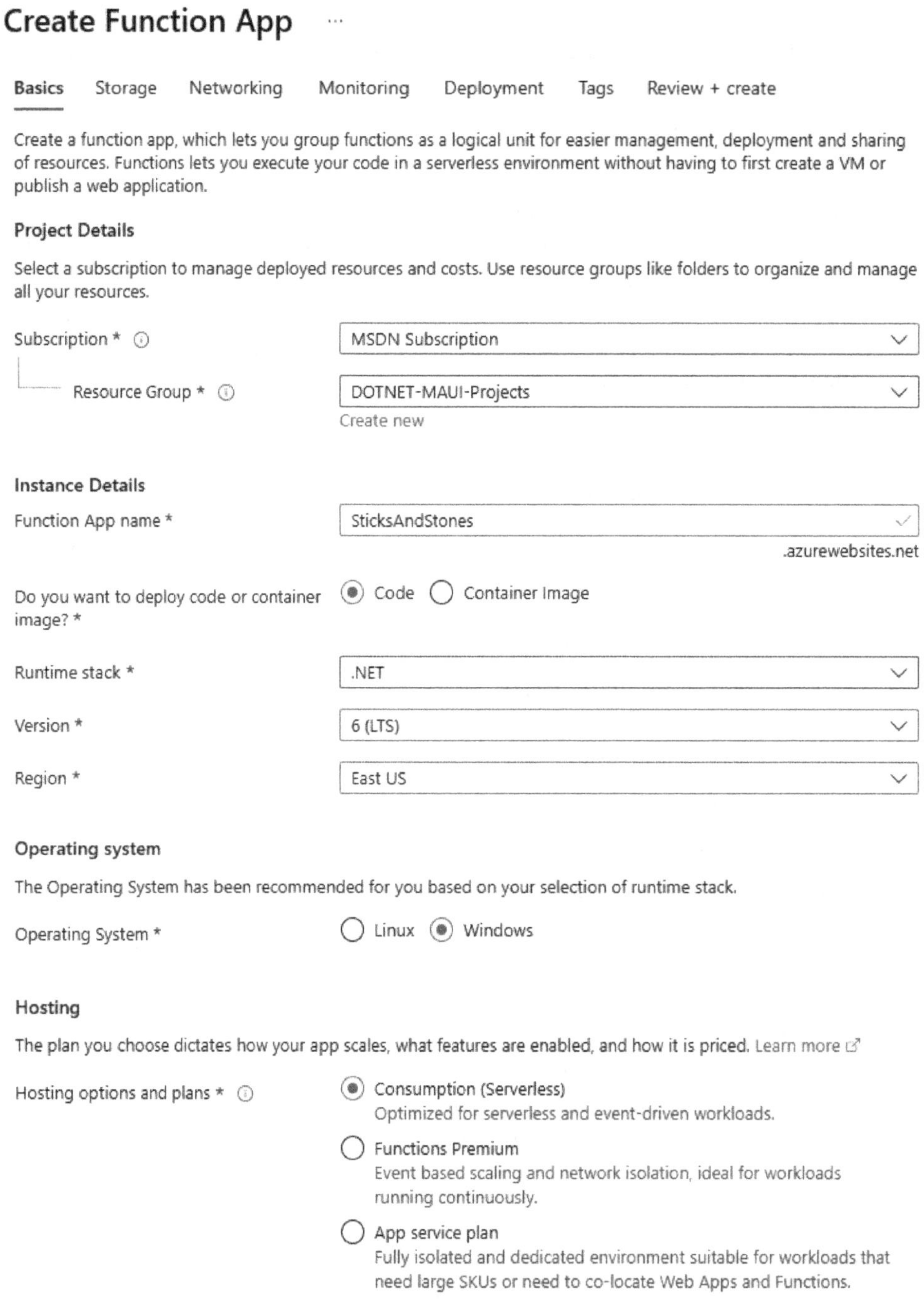

Create Function App   ···

Basics   Storage   Networking   Monitoring   Deployment   Tags   Review + create

Create a function app, which lets you group functions as a logical unit for easier management, deployment and sharing of resources. Functions lets you execute your code in a serverless environment without having to first create a VM or publish a web application.

Project Details

Select a subscription to manage deployed resources and costs. Use resource groups like folders to organize and manage all your resources.

Subscription * ⓘ	MSDN Subscription ∨
Resource Group * ⓘ	DOTNET-MAUI-Projects ∨
	Create new

Instance Details

Function App name *	SticksAndStones ✓
	.azurewebsites.net

Do you want to deploy code or container image? *	◉ Code ◯ Container Image

Runtime stack *	.NET ∨
Version *	6 (LTS) ∨
Region *	East US ∨

Operating system

The Operating System has been recommended for you based on your selection of runtime stack.

Operating System *	◯ Linux ◉ Windows

Hosting

The plan you choose dictates how your app scales, what features are enabled, and how it is priced. Learn more ↗

Hosting options and plans * ⓘ	◉ Consumption (Serverless) Optimized for serverless and event-driven workloads.
	◯ Functions Premium Event based scaling and network isolation, ideal for workloads running continuously.
	◯ App service plan Fully isolated and dedicated environment suitable for workloads that need large SKUs or need to co-locate Web Apps and Functions.

Figure 9.4 – Create Function App – Basics

11. Click **Review + create** to review the settings before creating the function app.

12. Click **Create** to create the function app.

Creating the projects

If you want, you can create functions in the Azure portal. I prefer to use Visual Studio, however, because the code editing experience is better, and you can use source control. For this project, we will need to separate projects as part of our solution – an Azure Functions project and a class library for shared code between the functions and the .NET MAUI app that will be built in *Chapter 10*. To create and configure the projects, proceed as follows:

1. Create a new project in Visual Studio.

2. Enter `function` in the search field to find the template for Azure Functions.

3. Click the **Azure Functions** template to continue, as illustrated in the following screenshot:

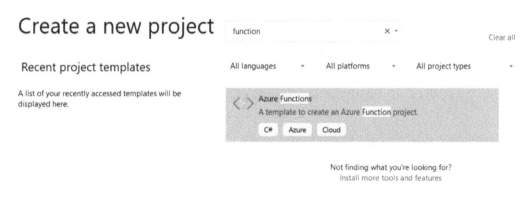

Figure 9.5 – Create a new project

4. Name the project `SticksAndStones.Functions`.

5. Name the solution `SticksAndStones.Functions`, as illustrated in the following screenshot, and click **Next**:

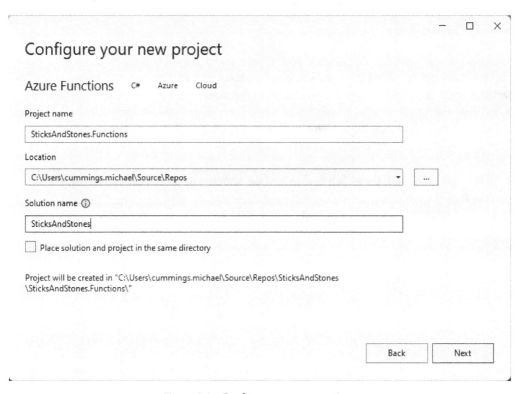

Figure 9.6 – Configure your new project

The next step is to create our first function, as follows:

1. Select **.Net 6.0 (Long Term Support)** for **Functions worker** at the top of the dialog box.

2. Select **Http trigger** as the trigger for our first function.

3. Click **Create** to continue; our functions project will be created.

Refer to the following screenshot to view the preceding information:

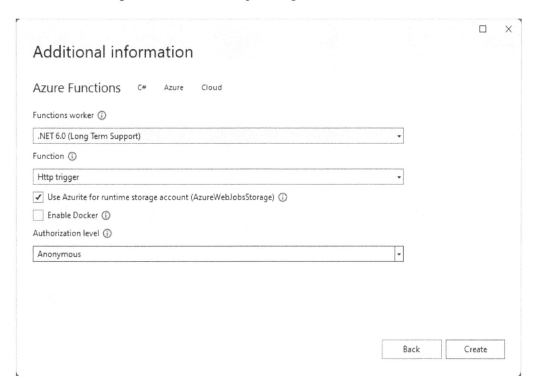

Figure 9.7 – Creating a new Azure Functions application – Additional information

Our first function will return the connection information for the SignalR service. To do that, we need to connect the function by adding a connection string to the SignalR service, as follows:

1. Go to the **SignalR Service** resource in the Azure portal.

2. Go to the **Keys** tab on the left and copy the connection string.

3. Go to the **Function App** resource and add the connection string under **Application Settings**. Use `AzureSignalRConnectionString` as the name for the setting.

4. Add the connection string to the **Values** array in the `local.settings.json` file in the Visual Studio project to be able to run the function locally on the development machine, as illustrated in the following code block:

```
{
  "IsEncrypted": false,
  "Values": {
    "AzureWebJobsStorage": "UseDevelopmentStorage=true",
    "FUNCTIONS_WORKER_RUNTIME": "dotnet",
    "AzureSignalRConnectionString":
"{EnterTheConnectingStringHere}"
  }
}
```

If the file doesn't exist in your project, create a new file in the root of the `SticksAndStones.Functions` project, and add the code listed previously with your connection string.

Next, in the `SticksAndStones.Functions` project, we need to reference the `Microsoft.Azure.WebJobs.Extensions.SignalRService` NuGet package. This package contains the classes we need to communicate with the SignalR service. If an error occurs during this and you are not able to install the package, make sure that you have the latest version of all other packages in the project and try again.

The last change we need to make is to adjust the automatic namespace generation. By default, the default namespace is the name of the project, which means all types in this project would have a root namespace of `SticksAndStones.Functions`. We don't need the `Functions` part of that, so let's remove it:

1. Right-click the `SticksAndStones.Functions` project in **Solution Explorer** and select **Properties**.

2. In the **Properties** window, use the search box at the top to search for `Default namespace`.

3. Change the **Default namespace** property value to `$(MSBuildProjectName.Split(".")[0].Replace(" ", "_"))`.

    This will split the project name on `.`, using only the first part and replacing any spaces with underscores.

Now, when we create a new class, the namespace will start with just `SticksAndStones`. It's time to create a shared project so that we can reuse code in both the .NET MAUI client and the Azure Functions service.

The shared code will go into a class library project. To create the project and reference it from the `SticksAndStones.Functions` project, follow these steps:

1. Right-click on the `SticksAndStones` solution node in **Solution Explorer** and select **Add**, then **New Project**.

2. Search for `Class Library` in the **Add a new project** dialog box, as shown here:

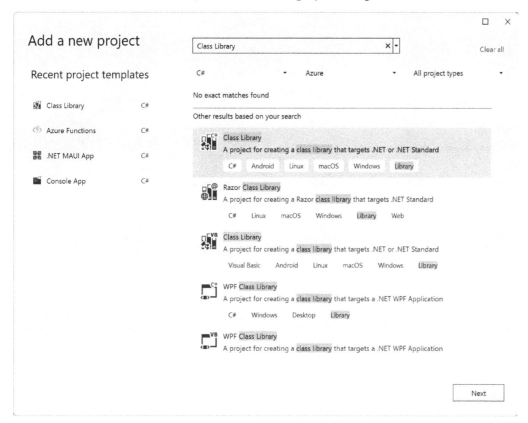

Figure 9.8 – Add a new project

3. Select the **Class Library** template, then click **Next**.

4. In the **Configure your new project** dialog box, enter `StickAndStones.Shared` for the name, as shown in the following screenshot, and click **Next**:

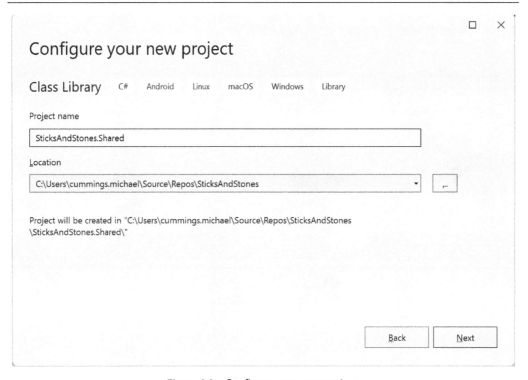

Figure 9.9 – Configure your new project

5. Select the **.NET 6.0 (Long Term Support)** framework project in the **Additional information** dialog, then click **Create**.

6. Delete the `Class1.cs` file that is created as part of the project template.

7. Add a reference to `SticksAndStones.Shared` in the `SticksAndStones.Functions` project.

As we did for the `SticksAndStones.Functions` project, we will change the default namespace by following these steps:

1. Right-click the `SticksAndStones.Functions` project in **Solution Explorer** and select **Properties**.

2. In the **Properties** window, use the search box at the top to search for `Default namespace`.

3. Change the **Default namespace** property value to `$(MSBuildProjectName.Split("."))[0].Replace(" ", "_"))`.

    This will split the project name on `.`, using only the first part and replacing any spaces with underscores.

Now, we can write the code for the function that will return the connection information.

Connecting a player to the game

The first step in the game is to get connected. Being connected adds you to the list of available players so that you or other players can then join a game. As we have done in other projects in this book, first, we will create the models that we need to store or transfer data between the service and the clients. Then, we will implement the `Connect` function itself.

Creating the models

We will need a few functions for the app in *Chapter 10* to call during the life cycle of the app. The first is to establish a connection with the game services, called `Connect`. Essentially, this tells the service that a new or existing player is active and ready for play. The `Connect` function will register the player details and return the connection string to the SignalR hub so that the app can receive messages. We will need a few models before we can complete the function. There needs to be a `Player` model, a `Game` model, and models to assist with passing data between Azure Functions and the SignalR service to the app.

Before we dive into creating the library, we should discuss the naming convention used in this chapter. Having a convention for how you name things will make it easier to determine how the class is used. When the app calls any Azure function, if it needs to send any data, it will do so using a class that has a suffix of `-Request`, and any Azure function that returns data will do so using a class that ends with `-Response`. For any data that is sent via the SignalR hub, we will use a class that has an `EventArgs` suffix. These classes will contain references to our actual models, and just act as a container for the data. Having these classes in place means that you can modify the data being sent or received without affecting the models themselves.

Since this is a two-player game, we need to track a little bit of the state so that we know who is online and what matches are in play. For this project, we will keep the state simple and not involve an actual database, but we will still use Entity Framework to do most of the work for us.

Now that we have created and referenced the new project, and we have a naming convention in place, we can start creating the classes we will need. We will start with the two models, `Player` and `Match`. `Player` represents each person, while `Game` is a match between two `Player instances` and the state of play. To create the two models, follow these steps:

1. Create a new folder in the `SticksAndStones.Shared` project called `Models`.

2. Create a new class in the `Models` folder called `Player`.

3. Create a `public` property in the `Player` class called `Id` as `Guid` and initialize it to `Guid.Empty`.

4. Create another `public` property called `GamerTag` as `string` and initialize it to `string.Empty`.

5. Create a `public` property named `GameId` as `Guid` and initialize it to `Guid.Empty`.

6. Your `Player` class should now resemble the following code block:

```
namespace SticksAndStones.Models;

public class Player
{
    public Guid Id { get; set; } = Guid.Empty;

    public string GamerTag { get; set; } = string.Empty;
    public string EmailAddress { get; set; } = string.Empty;
    public Guid MatchId { get; set; } = Guid.Empty;
}
```

Our model classes will use the `Id` field as a unique identifier so that we can locate each one individually. It will be used to locate specific players for messaging and relating `Match` instances to `Player` instances. `GamerTag` will be the display name for the player, and `EmailAddress` is how we can correlate players if they leave the app and then log back in again. Finally, the `MatchId` property will track whether the player is actively in a game.

Now that we have defined the `Player` class, it's time to define the `Match` class:

1. Create a new class in the `Models` folder called `Match`.

2. Create a `public` property in the Game class called `Id` as `Guid` and initialize it to `Guid.Empty`.

3. Create a `public` property called `PlayerOneId` as `Guid`.

4. Add a `public` property named `PlayerOneScore` as `int`.

5. Create another `public` property called `PlayerTwoId` as `Guid`.

6. Add a `public` property named `PlayerTwoScore` as `int`.

7. Create a `public` property called `NextPlayerId` as `Guid`.

8. Create a `public` property called `Sticks` as `List<int>` and initialize it to new `List<int>(24)`.

9. Create a `public` property called `Stones` as `List<int>` and initialize it to new `List<int>(9)`.

10. Create a `public` property called `Scores` as `List<int>` and initialize it to new `List<int>(2)`.

11. Create a `public` property called `Completed` as `bool` and initialize it to `false`.

12. Create a `public` property called `WinnerId` as `Guid` and initialize it to `Guid.Empty`.

13. Add a `public static` method named `New` that takes two parameters, both of the `Guid` type, called `challengerId` and `opponentId`. The method returns an object of the Game type. The method should return a new instance of Game and assign the `Id` property to `Guid.NewGuid()`, `PlayerOneId` and `NextPlayerId` to `opponentId`, and `PlayerTwoId` to `challengerId`.

14. The `Player` class should now resemble the following code block:

```
using System;
using System.Collections.Generic;

namespace SticksAndStones.Models;

public class Match
{
    public Guid Id { get; set; } = Guid.Empty;

    public Guid PlayerOneId { get; set; }
    public int PlayerOneScore { get; set; }

    public Guid PlayerTwoId { get; set; }
    public int PlayerTwoScore { get; set; }

    public Guid NextPlayerId { get; set; }

    public List<int> Sticks {get; set; } = new(new int[24]);
    public List<int> Stones {get; set;} = new(new int[9]);
    public List<int> Score = new(new int[2]);

    public bool Completed { get; set; } = false;
    public Guid WinnerId { get; set; } = Guid.Empty;

    public static Game New(Guid challengerId, Guid opponentId)
    {
        return new()
        {
            Id = Guid.NewGuid(),
            PlayerOne = opponent,
            PlayerTwo = challenger,
            NextPlayer = opponent
        };
    }
}
```

The `Player` and `Match` classes will be used for data storage and data transfer between the client and server. Before we go any further with creating our models, let's add the database using Entity Framework. Perform the following steps to add a reference to Entity Framework and create the database context so that the `Player` and `Match` classes can be stored in an `InMemory` database:

1. Add a package reference to `Microsoft.EntityFrameworkCore.InMemory` to the `SticksAndStones.Functions` project.

2. Create a new folder called `Repository` in the `SticksAndStones.Functions` project.

3. Create a class named `GameDbContext` in the `Repository` folder.

4. Modify the constructor for the class to set database options:

```
public GameDbContext(DbContextOptions<GameDbContext> options) :
base(options) { }
```

5. Add a public `Players` property to store the `Player` objects:

    ```
    public DbSet<Player> Players { get; set; }
    ```

6. Add a public `Matches` property to store the `Match` objects:

    ```
    public DbSet<Match Matches { get; set; }
    ```

7. Add an override to the `OnModelCreating` method:

    ```
    protected override void OnModelCreating(ModelBuilder
    modelBuilder)
    {
    }
    ```

    This method is where we specify to Entity Framework how to relate our classes together in a relational database.

8. Start by declaring the identifiers for each class in the `OnModelCreating` method, as shown here:

    ```
    modelBuilder.Entity<Player>()
        .HasKey<Player>(p => p.Id);

    modelBuilder.Entity<Match>()
        .HasKey<Match>(g => g.Id);

    base.OnModelCreating(modelBuilder);
    ```

9. Entity Framework does not handle our `List<int>` properties properly. It assumes that since it is a list, they are related instances. To change the default behavior in Entity Framework, we can use the following highlighted code:

    ```
    modelBuilder.Entity<Game>()
        .HasKey(g => g.Id);

    modelBuilder.Entity<Match>()
        .Property(p => p.Sticks)
        .HasConversion(
            toDb => string.Join(",", toDb),
            fromDb => fromDb.Split(',', StringSplitOptions.None).
    Select(int.Parse).ToList() ?? new(new int[24]));

    modelBuilder.Entity<Match>()
        .Property(p => p.Stones)
        .HasConversion(
            toDb => string.Join(",", toDb),
            fromDb => fromDb.Split(',', StringSplitOptions.None).
    Select(int.Parse).ToList() ?? new(new int[9]));

    base.OnModelCreating(modelBuilder);
    ```

What each block does is define a conversion for the property. A conversion has two Lambda expressions – one from the C# object to the database and the other from the database to the C# object. For our List<int> properties, we want to convert a C# List<int> into a comma-separated string of integers, and then a comma-separated string of integers to List<int>.

toDB is an instance of List<int>, so to convert that into a comma-separated list of numbers, we can use the String.Join function to join each element of the list with , between them.

fromDb is a string value containing numbers separated by commas. To convert that into List<int>, we can use the String.Split method to isolate each number, then pass each number into the Int.Parse method to convert the number into an int value. Select will produce IEnumberable<int>; we can use the **LINQ** extension, ToList, to convert that into List<int>. If it doesn't create a list, we can supply a default list of values, just like we did in the Match class itself.

To initialize Entity Framework to use an in-memory database, we need to create a Startup method. To create the method and initialize the database, follow these steps:

1. Create a new class named Startup in the root of the SticksAndStones.Functions project.

2. Modify the class file with the following highlighted code:

```
using Microsoft.Azure.Functions.Extensions.DependencyInjection;
using Microsoft.EntityFrameworkCore;
using Microsoft.Extensions.DependencyInjection;
using SticksAndStones.Repository;

[assembly: FunctionsStartup(typeof(SticksAndStones.Startup))]

namespace SticksAndStones;

public class Startup : FunctionsStartup
{
    public override void Configure(IFunctionsHostBuilder
builder)
    {
        string SqlConnection = Environment.
GetEnvironmentVariable("SqlConnectionString");

        builder.Services.AddDbContextFactory<GameDbContext>(
            options =>
            {
                options.UseInMemoryDatabase("SticksAndStones");
            });
    }
}
```

The Startup method will now be called when the SticksAndStones.Functions project is loaded at runtime. It will then create a factory for creating instances of the GameDbContext class we created previously and initialize it with an in-memory database.

That concludes our setup for Entity Framework and our basic models, `Player` and `Game`. There is one final model that we need to send the SignalR connection information to the client. To create this model, follow these steps:

1. Create a new class in the `Models` folder named `ConnectionInfo`.

2. Add a public property `Url` that is a `string` value.

3. Add another public property named `AccessToken` that is also a `string` value.

4. The `ConnectionInfo` class should look like this:

    ```
    namespace SticksAndStones.Models;

    public class ConnectionInfo
    {
        public string Url { get; set; }
        public string AccessToken { get; set; }
    }
    ```

With the models now created, we can start creating the `Connect` function.

Creating the Connect function

We will start with a function to connect our player to the game, aptly named `Connect`. This function will expect a partially filled `Player` object to be sent in the request body. The function will return a fully populated `Player` object, a list of the currently connected players, and the connection information needed by the client to connect to the SignalR hub. To make the inputs and outputs cleaner, we will wrap them.

To create the input and output classes, proceed as follows:

1. Create a new class in the `Messages` folder called `ConnectMessages`.

2. Modify `ConnectMessages.cs` so that it looks like this:

    ```
    using SticksAndStones.Models;

    namespace SticksAndStones.Messages;

    public record struct ConnectRequest(Player Player);
    public record struct ConnectResponse(Player Player, List<Player>
    Players, ConnectionInfo ConnectionInfo);
    ```

For all the classes that will be used to transfer data between the client and Azure Functions or the SignalR service, we will use the `record` syntax. Since these classes will not have any real functionality, their sole purpose is to contain our models. By using a `record` struct, we also improve the memory usage of our functions since a new instance would be created in local memory and not global memory, which requires additional handling. The `record` syntax combines the constructor and property declarations into a single line of code, eliminating a lot of boilerplate code that adds no real benefit.

You will notice that we are using the conventions we discussed in the *Creating the models* section. Classes that have a suffix of `Request` or `Response` are used as input and output for any Azure function. For any data that is sent via the SignalR service, the class will use a suffix of `EventArgs`.

When a new client is connecting, a message will be sent to other users via the SignalR service to indicate that they have connected. This message will also be used to notify when players start or end a game. To create such a message, proceed as follows:

1. Create a new class called `PlayerUpdatedEventArgs` in the `Messages` folder of the `SticksAndStones.Shared` project.

2. Modify the class so that it's a `record` struct with a single `Player` parameter, as shown in the following code snippet:

    ```
    using SticksAndStones.Models;

    namespace SticksAndStones.Messages;

    public record struct PlayerUpdatedEventArgs(Player Player);
    ```

Now that we have created the structures needed for the `Connect` function, we can start writing the function itself:

1. Create a new folder called `Hubs`. We will put our service class into this folder.

2. Move the `Function1.cs` file into the `Hubs` folder.

3. Respond **Yes** to the next two prompts for moving the file and adjusting the namespace.

4. Rename the `Function1.cs` file `GameHub.cs` and click **Yes** in the rename prompt.

5. Open the `GameHub.cs` file and rename the class `GameHub`, replace the `internal static` access modifiers and replace them with `public`, and derive it from the `ServerlessHub` base class, as highlighted here:

    ```
    public class GameHub : ServerlessHub
    {
        [FunctionName("Function1")]
        public static async Task<IActionResult> Run(
        [HttpTrigger(AuthorizationLevel.Function, "get", "post",
    Route = null)] HttpRequest req,
        ILogger log)
        {
            log.LogInformation("C# HTTP trigger function processed a
    request.");

            string name = req.Query["name"];

            string requestBody = await new StreamReader(req.Body).
    ReadToEndAsync();
            dynamic data = JsonConvert.
    DeserializeObject(requestBody);
            name = name ?? data?.name;
    ```

```
            string responseMessage = string.IsNullOrEmpty(name)
            ? "This HTTP triggered function executed successfully.
Pass a name in the query string or in the request body for a
personalized response."
                    : $"Hello, {name}. This HTTP triggered function
executed successfully.";

            return new OkObjectResult(responseMessage);
        }
    }
```

6. Rename the default `Function1` function `Connect`, removing the `static` modifier as well. The method signature should look like the following highlighted code snippet:

```
[FunctionName("Connect")]
public async Task<IActionResult> Connect(
  [HttpTrigger(AuthorizationLevel.Function, "post", Route =
null)] HttpRequest req,
    ILogger log)
```

The `HttpTrigger` attribute indicates that this function is called by using the HTTP protocol and not by some other means, such as a SignalR message or a timer. The function is only called using the HTTP `POST` method, not `GET`.

7. To send a message to all clients connected to the SignalR hub, we will need another SignalR binding. This time, it is an **output binding**. The output binding is decorated with a SignalR attribute with `HubName` and is of the `IAsyncCollector<SignalRMessage>` type, as follows:

```
[HttpTrigger(AuthorizationLevel.Anonymous, "post", Route =
null)] HttpRequest req,
[SignalRConnectionInfo(HubName = "GameHub")]
SignalRConnectionInfo connectionInfo,
[SignalR(HubName = "GameHub")] IAsyncCollector<SignalRMessage>
signalRMessages,
ILogger log)
```

Remove the contents of the `Connect` method and proceed as follows to implement the function:

1. Logging is important in Azure Functions as it helps when debugging in production environments. So, let's add a `log` message:

```
log.LogInformation("A new client is requesting connection");
```

2. The client, a .NET MAUI app, will send `ConnectRequest` in the body of the HTTP request as JSON. To get an instance of `ConnectRequest` from the request body, use the following lines of code:

```
var result = await JsonSerializer.
DeserializeAsync<ConnectRequest>(req.Body, jsonOptions);
var newPlayer = result.Player;
```

You will have to add using System.Text.Json to the namespace declarations as well. This uses the System.Text.Json.JsonSerializer class to read the contents of the request body and create a ConnectRequest object from it. It uses jsonOptions to properly deserialize the object.

3. Now, we need to define the jsonOptions field. Add the following line of code above the Connect method:

```
internal class GameHub : ServerlessHub
{
    private readonly JsonSerializerOptions jsonOptions =
new(JsonSerializerDefaults.Web);

    [FunctionName("Connect")]
    public async Task<IActionResult> Connect(
```

JsonSerializerDefaults.Web ensures that the JSON is formatted properly so that Azure Functions and the SignalR service will properly serialize and deserialize the objects. Mainly, it will enforce the following:

*   Property names are case-insensitive

*   All property and object names will be formatted as camelCase

*   Numbers can be quoted

4. If we receive bad player data, return ArgumentException to the client, as follows:

```
if (newPlayer is null)
{
    var error = new ArgumentException("No player data.",
"Player");
    log.LogError(error, "Failure to deserialize arguments");
    return new BadRequestObjectResult(error);
}

if (string.IsNullOrEmpty(newPlayer.GamerTag))
{
    var error = new ArgumentException("A GamerTag is required
for all players.", "GamerTag");
    log.LogError(error, "Invalid value for GamerTag");
    return new BadRequestObjectResult(error);
}

if (string.IsNullOrEmpty(newPlayer.EmailAddress))
{
    var error = new ArgumentException("An Email Address is
required for all players.", "EmailAddress");
    log.LogError(error, "Invalid value for EmailAddress");
    return new BadRequestObjectResult(error);
}
```

Since the return type of the function is `IActionResult`, we can't simply return our custom objects. Instead, we need to create an object that derives or implements `IActionResult` and pass in our result. In the case of errors, we will use `BadRequestObjectResult`, which will accept `Exception` as a parameter in the constructor. `BadRequestObjectResult` will set the HTTP status code to `400`, indicating an error. This status code can then be checked by the client to know whether the request succeeded or not before parsing the body of the response.

5. The next couple of steps will require us to query the database, so we need to add the database context factory to the class. Add the `Microsoft.EntityFrameworkCore` namespace declaration:

```
using Microsoft.Azure.WebJobs.Extensions.SignalRService;
using Microsoft.EntityFrameworkCore;
using Microsoft.Extensions.Logging;
```

6. Add a `private` field to store the context factory and a constructor with an argument that will be fulfilled by dependency injection, as follows:

```
private readonly JsonSerializerOptions jsonOptions = new
JsonSerializerOptions(JsonSerializerDefaults.Web);

private readonly IDbContextFactory <GameDbContext>
dbContextFactory;

public GameHub(IDbContextFactory<GameDbContext> dbcontext)
{
    contextFactory = dbContextFactory;
}

[FunctionName("Connect")]
public async Task<IActionResult> Connect(
```

7. Add a namespace declaration for `System.Linq` to allow the use of **Linq** queries:

```
using System;
using System.Linq;
using System.Text.Json;
```

8. Now that we can access the database context and have determined that the input is valid, we need to make sure that the data is unique, so we will look for `GamerTag` in the database to ensure it isn't in use already by another player, as follows:

```
using var context = contextFactory.CreateDbContext();

log.LogInformation("Checking for GamerTag usage");
var gamerTagInUse = (from p in context.Players
                where string.Equals(p.GamerTag, newPlayer.
GamerTag, StringComparison.InvariantCultureIgnoreCase)
                    && !string.Equals(p.EmailAddress,
newPlayer.EmailAddress, StringComparison.OrdinalIgnoreCase)
                    select p).Any();
if (gamerTagInUse)
{
```

```
        var error = new ArgumentException($"The GamerTag {newPlayer.
    GamerTag} is in use, please choose another.", "GamerTag");
        log.LogError(error, "GamerTag in use.");
        return new BadRequestObjectResult(error);
    }
```

The first step is to get a new database context from the factory. The **Linq** query is simple in this case – it uses the `Players` list from the database context to compare `GamerTag` value to other GamerTagsvalues. However, we want to exclude a result if it matches `EmailAddress` since that would indicate the records are identical, and this user is just signing back in again.

9. Now, query the `Players` dataset for a player with a matching email:

```
        log.LogInformation("Locating Player record.");
        var thisPlayer = (from p in context.Players where string.
    Equals(p.EmailAddress, newPlayer.EmailAddress, StringComparison.
    OrdinalIgnoreCase) select p).FirstOrDefault();
```

10. If there is no `Player` in `Players` that matches, then add `Player` to the dataset:

```
    if (thisPlayer is null)
    {
        log.LogInformation("Player not found, creating.");
        thisPlayer = newPlayer;
        thisPlayer.Id = Guid.NewGuid();
        context.Add(thisPlayer);
        await context.SaveChangesAsync();
    }
```

We assign the `Player` object a new `Guid` so that each `Player` has a unique identifier. This could also be done by Entity Framework; however, we will take care of it here. The context is then used to add the `Player` instance so that it is tracked for any changes. After this, `SaveChangesAsync` will commit all changes to the database.

11. The next step in the `Connect` function is to send a message to all the connected players that a new player has joined. We can do that using the `SendAsync` method. The `SendAsync` method takes two parameters – the method name as a `string` value that the message is intended for, and the message as an `object` value. To ensure we are sending and receiving the right method, we will create a constant value. Create a new class named `Constants` in the root of the `SticksAndStones.Shared` project, then update it so that it looks like this:

```
    namespace SticksAndStones;

    public static class Constants
    {
        public static class Events
        {
            public static readonly string PlayerUpdated =
    nameof(PlayerUpdated);
        }
    }
```

12. Now, we can notify other connected players that a new player has connected. Open the `GameHub` class and, at the end of the `Connect` method, add the following code:

```
        log.LogInformation("Notifying connected players of new
player.");
        await Clients.All.SendAsync(Constants.Events.PlayerUpdated,
new PlayerUpdatedEventArgs(thisPlayer));
```

This code uses the `SendAsync` method from the `Clients.All` collection in the `ServerlessHub` base class to send a message to all connected clients. We pass `Constants.Events.PlayerUpdated`, which is the `"PlayerUpdated"` string, as the method name. As arguments, we are sending the `Player` instance wrapped in `PlayerUpdatedEventArgs`. We will handle this message in *Chapter 10*.

13. Now, get the set of available players from the database to send back to the client:

```
// Get the set of available players
log.LogInformation("Getting the set of available players.");
        var players = (from player in context.Players
            where player.Id != thisPlayer.Id
            select player).ToList();
```

Using Linq, we can easily query the `Players` collection and exclude the current player.

14. At this point, we need to get the SignalR connection information from the SignalR service. This can be accomplished by calling the `NegotiateAsync` method of the `ServerlessHub` base class. Additionally, so that we can send directed messages to individual users, we will set the `UserId` value for the connection to the player ID value. Add the following line of code to configure and retrieve the SignalR connection information:

```
var connectionInfo = await NegotiateAsync(new
NegotiationOptions() { UserId = thisPlayer.Id.ToString() });
```

15. Now that we have all the information that we need to return to the client, we can construct the `ConnectResponse` object. We will use the `ConnectionInfo` class and map the `SignalRConnection` properties to it so that we avoid having to reference the SignalR service in the shared library:

```
log.LogInformation("Creating response.");
var connectResponse = new ConnectResponse()
{
    Player = thisPlayer,
    Players = players,
    ConnectionInfo = new Models.ConnectionInfo { Url =
connectionInfo.Url, AccessToken = connectionInfo.AccessToken }
};
```

16. Once `ConnectResponse` has been initialized, we can return it by using `OkObjectResult`, which will use an HTTP response code of **200 OK**:

```
log.LogInformation("Sending response.");
return new OkObjectResult(connectResponse);
```

To test the function we just wrote, you can use a PowerShell command prompt and the following command, after pressing *F5* in Visual Studio:

```
Invoke-WebRequest -Headers @{ ContentType = "application/json" } -Uri
http://localhost:7024/api/Connect -Method Post -Body ''
```

The port number used in the `Uri` parameter may be different for your project. You can get the correct port number by opening the `launchSettings.json` file in the `Properties` folder of the `SticksAndStones.Functions` project. The port number is set in the `commandLineArgs` property, as highlighted here:

```
{
  "profiles": {
    "SticksAndStones.Functions": {
      "commandName": "Project",
      "commandLineArgs": "--port 7024",
      "launchBrowser": false
    }
  }
}
```

Replace your port number in the `Uri` parameter after `localhost:`.

In the `Body` parameter, you can add the JSON that the command is expecting. For the `Connect` function, this would be `ConnectRequest` and would look like this:

```
'{
    "player": {
        "gamerTag": "NewPlayer2",
        "emailAddress": "newplayer2@gmail.com",
    }
}'
```

The full command will look like this:

```
Invoke-WebRequest -Headers @{ ContentType = "application/json" } -Uri
http://localhost:7024/api/Connect -Method Post -Body '{
    "player": {
        "gamerTag": "NewPlayer2",
        "emailAddress": "newplayer2@gmail.com",
    }
}'
```

Go ahead and try out various versions of the command to see how the function reacts.

Now that we can connect players to the game server, let's look at what is needed for the lobby.

Refreshing the lobby

In the Sticks and Stones App, which you will create in *Chapter 10*, once a player has connected, they will move to the lobby page. Initially, the lobby will be populated from the list of players sent in the response from the Connect function. Additionally, as each player connects, the lobby will be updated through a SignalR event.

But we all get impatient and want a way to refresh the list immediately. So, the lobby page has a way to refresh the list; to do so, it will call the GetAllPlayers function.

Let's start by creating the messages needed for GetAllPlayers.

Creating the messages

GetAllPlayers takes no parameters, so we only need to create the GetAllPlayersResponse type. Follow these steps to add GetAllPlayersResponse:

1. In the SticksAndStones.Shared project, create a new file named GetAllPlayers Messages.cs in the Messages folder.

2. Modify the contents of the file so that it looks as follows:

    ```
    using SticksAndStones.Models;

    namespace SticksAndStones.Messages;

    public record struct GetAllPlayersResponse(List<Player>
    Players);
    ```

With the messages created, we can move on to the GetAllPlayers function.

Getting all the players

GetAllPlayers is called using the Http GET method and has an optional parameter that is passed through QueryString using the id key. The optional parameter is used to exclude a specific id from the returned list. This makes it so that the app can send the current player's id and not have it returned in the list. To create the GetAllPlayers function, follow these steps:

1. In the GameHub class, after the Connect method, add the following method declaration:

    ```
    [FunctionName("GetAllPlayers")]
    public IActionResult GetAllPlayers(
    [HttpTrigger(AuthorizationLevel.Function, "get", Route =
    "Players/GetAll")] HttpRequest req,
    ILogger log)
    {
    }
    ```

Not much is new here other than using "get" instead of "post" for the Http method, and Route is set to "Players/GetAll", which would make the URL for the function http://localhost:7024/api/Players/GetAll.

2. In the method, we will process the id option parameter. To do so, add the following code:

```
// Exclude the playerId if provided
Guid playerId = Guid.Empty;
if (req.Query.ContainsKey("id"))
{
    string id = req.Query["id"];
    if (!string.IsNullOrEmpty(id))
    {
        playerId = new Guid(id);
    }
}
```

In this code, we check for the existence of a key named id. If it exists, then its value is retrieved and converted into a Guid value and assigned to the playerId variable.

3. Next, we can query the database for all players, and exclude player.Id using the following code:

```
using var context = contextFactory.CreateDbContext();

// Get the set of available players
log.LogInformation("Getting the set of available players.");
var players = (from player in context.Players
               where player.Id != playerId
               select player).ToList();
```

4. Finally, return OkObjectResult with a new GetAllPlayersResponse object initialized with the list of players:

```
return new OkObjectResult(new GetAllPlayersResponse(players));
```

Now that we can refresh the lobby with a list of all the players, it's time to match them up for a game.

Challenging another player to a game

To test your skills at this game, you'll need an opponent – someone who would also like to test their skills against yours. This section will build the functionality needed in the SticksAndStones.Function project to have one player – the challenger – challenge another player – the opponent – to a game. The opponent has the option to accept the challenge or deny it. We will also handle the case where the opponent does not respond since they might have put their phone down; this is an edge case.

The interactions in this use case can get tricky, so let's review the following diagram to get a better understanding of what we are building:

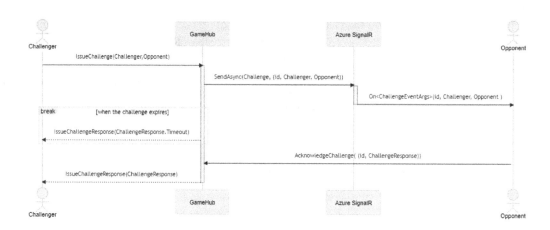

Figure 9.10 – Challenge diagram

The process starts with a user interaction that results in the client making an HTTP request to the GameHub instance via the IssueChallenge function. The client will pass the challenger and opponent details when making the HTTP call. IssueChallenge will create an Id value to track this process. IssueChallenge will then send a direct message to the opponent via the SignalR hub using the SendAsync method. The message will include the Id value that was created earlier, the challenger, and the opponent details as an instance of ChallengeEventArgs. The opponent's client will receive the message via an On<ChallengeEventArgs> event handler. The opponent will then have the choice of accepting or declining the challenge. The response is sent back to the GameHub instance using the AcknowledgeChallenge function. The Id value of the challenge is sent along with ChallengeResponse, either Accept or Decline. A third possibility is Timeout. If the opponent never responds, then after a certain amount of time has passed, Challenge will Timeout. In either event, the result is then returned to the challenger using the response of the IssueChallenge call.

Let's get started by defining the messages.

Creating the messages and models

We will start with IssueChallengeRequest since that is the first message that is being sent. Follow these steps to create the class:

1. In the SticksAndStones.Shared project, create a new file named ChallengeMessages.cs under the Message folder.

2. Modify the file so that it looks as follows:

    ```
    using SticksAndStones.Models;
    using System;

    namespace SticksAndStones.Messages;
    ```

```
public record struct IssueChallengeRequest(Player Challenger,
Player Opponent);
```

As we have in the `Connect` function, we use a `record` struct to eliminate a lot of the boilerplate code needed to define a `struct` value. Our message only needs two `Player` objects – `Challenger` and `Opponent`. The client will have both available when it makes the call to this function. `Challenger` will be the client making the call to `IssueChallenge` and `Opponent` will be the opposing side.

The `IssueChallenge` function will return `IssueChallengeResponse`. This `Issue ChallengeResponse` will have just one field, `Response`, which will be an enum value called `ChallengeResponse`. Follow these steps to create `ChallengeResponse`:

1. In the `SticksAndStones.Shared` project, in the `Models` folder, create a new enum named `ChallengeResponse`.

2. Add the following values to the enum value:

    *   `None`

    *   `Accepted`

    *   `Declined`

    *   `Timeout`

    Your code should look like this:

    ```
    namespace SticksAndStones.Models;

    public enum ChallengeResponse
    {
        None,
        Accepted,
        Declined,
        TimeOut
    }
    ```

To create the remaining messages for the `IssueChallenge` and `AcknowledgeChallenge` functions, follow these steps:

1. Open the `ChallengeMessages.cs` file and add the following declaration at the end of the file:

    ```
    public record struct IssueChallengeResponse(ChallengeResponse
    Response);
    ```

2. When the opponent responds to a challenge, they will call the `AcknowledgeChallenge` function and pass an `AcknowledgeChallengeRequest` object as an argument. In the `ChallengeMessages.cs` file, add the following declaration to create `AcknowledgeChallengeRequest`:

    ```
    public record struct AcknowledgeChallengeRequest(Guid Id,
    ChallengeResponse Response);
    ```

That completes the messages that are sent or received from the GameHub functions for a challenge. That just leaves `ChallengeEventArgs`, which is sent from GameHub to the opponent. To create the `ChallengeEventArgs` class, follow these steps:

1. In the `Messages` folder of the `SticksAndStones.Shared` project, create a new file named `ChallengeEventArgs.cs`.

2. Replace the contents of the `ChallengeEventArgs.cs` file with the following:

    ```
    using SticksAndStones.Models;
    using System;

    namespace SticksAndStones.Messages;

    public record struct ChallengeEventArgs(Guid Id, Player
    Challenger, Player Opponent);
    ```

3. To add the method name constant for the `SendAsync` method, open the `Constants.cs` file in the `SticksAndStones.Shared` project and add the following highlighted field to the `Events` class:

    ```
    public static class Events
    {
        public static readonly string PlayerUpdated =
    nameof(PlayerUpdated);
        public static readonly string Challenge = nameof(Challenge);
    }
    ```

As with the previous message definitions, `ChallengeEventArgs` is also defined as `public record struct`. The parameters are an `Id` value of the `Guid` type and two `Player` objects – one for `Challenger` or the initiator, and one for `Opponent` or the receiver. `Id` will be created in the `IssueChallenge` function and is used to correlate the challenge with the response. This is needed because we are tracking the challenge and if a certain amount of time has passed, we expire the challenge. `Id` is used to track that state and check whether the challenge is still valid if the client responds.

What was not included in the diagram in *Figure 9.10* is a structure that is used to track the challenge in the GameHub class. It is only needed while a challenge is active and hasn't timed out. To create the Challenge class, follow these steps:

1. Create a file named Challenge.cs in the Models folder of the SticksAndStones. Shared project.

2. Replace the contents of the Challenge.cs file with the following code:

```
using System;

namespace SticksAndStones.Models;

public record struct Challenge(Guid Id, Player Challenger,
Player Opponent, ChallengeResponse Response);
```

As with previous models, we use a record struct. The Challenge class has various properties – Id, Challenger as a Player type, Opponent as a Player type, and ChallengeResponse, which is called Response.

If a player accepts the challenge, then the two players will start a match with each other. Each player will be notified that the match has begun by receiving a MatchStarted SignalR event. To create the event and its arguments, follow these steps:

1. Open the Constants.cs file in the SticksAndStones.Shared project and add the following highlighted field to the Events class:

```
public static class Events
{
    public static readonly string PlayerUpdated =
nameof(PlayerUpdated);
    public static readonly string Challenge = nameof(Challenge);
    public static readonly string GameStarted =
nameof(MatchStarted);
}
```

2. In the Messages folder of the SticksAndStones.Shared project, create a new file named MatchStartedEventArgs.cs.

3. Replace the contents of the MatchStartedEventArgs.cs file with the following:

```
using SticksAndStones.Models;

namespace SticksAndStones.Messages;

public record struct MatchStartedEventArgs(Match Match);
```

That concludes the new messages and models that are needed to allow one player to challenge another to a game of SticksAndStones. Next, we will create the first of two functions that will handle the process in the GameHub class.

Creating the IssueChallenge function

We'll start with the `IssueChallenge` function. This function is called from the `Challenger` client to start the challenge process. The client will pass their `Player` object, `Challenger`, and the player they are challenging – that is, `Opponent`. These two models are contained in an `IssueChallengeRequest` object. The `IssueChallenge` function will need to perform the following actions:

- Validate input
- Create a challenge object
- Send a challenge to the opponent
- Wait for a response from the opponent
- Send the response to the challenger

To create the function and implement these actions, follow these steps:

1. Open the `GameHub.cs` file in the `SticksAndStones.Functions` project.

2. Add the `IssueChallenge` function declaration, as follows:

```
[FunctionName("IssueChallenge")]
public async Task<IssueChallengeResponse> IssueChallenge(
    [HttpTrigger(AuthorizationLevel.Function, "post", Route =
$"Challenge/Issue")] HttpRequest req,
    ILogger log)
{
}
```

The `FunctionName` attribute tells Azure Functions that this is an available function. The `req` parameter is an `HttpRequest` object, and the Azure Functions runtime will provide its instance when the function is called. It is attributed by the `HttpTrigger` attribute, which makes the function available via an `Http` API call. `HttpRequest` must use the `POST` method and not `GET` when making the call and the function's `Route` or `Url` will end with `Challenge/Issue`. The function returns `IssueChallengeResponse` to the caller.

3. The first action the function performs is to validate the inputs. The `IssueChallengeRequest` object is sent as part of the body of the `Http POST` request. To retrieve the instance, use the following code in the `IssueChallenge` function:

```
{
    var result = await JsonSerializer.
DeserializeAsync<IssueChallengeRequest>(req.Body, jsonOptions);

}
```

This is the same way we retrieved the arguments that were passed in the `Connect` function. The main difference is that the type of object being returned by the `DeserializeAsync` method is `IssueChallengeRequest`, not `ConnectRequest`. The `jsonOptions` field is already defined in the `GameHub` class.

4. Now, we need to check whether the challenger and opponent are valid. Valid means that they exist in our database, and neither are currently in a match. We will use **Linq** queries to check our database for each object. Add the following code to the `IssueChallenge` function to verify that the players exist:

```
using var context = contextFactory.CreateDbContext();

Guid challengerId = result.Challenger.Id;
var challenger = (from p in context.Players
                  where p.Id == challengerId
                  select p).FirstOrDefault();

Guid opponentId = result.Opponent.Id;
var opponent = (from p in context.Players
                where p.Id == opponentId
                select p).FirstOrDefault();

if (challenger is null)
    throw new ArgumentException(paramName: nameof(challenger),
message: $"{challenger.GamerTag} is not a valid player.");
if (opponent is null)
    throw new ArgumentException(paramName: nameof(opponent),
message: $"{opponent.GamerTag} is not a valid player.");
```

First, we capture the `Id` value of the player. We use this `Id` to query the `Players` DbSet in the database context for a matching `Id` and return the `Player` object if it exists; otherwise, we return `null`. If the object is `null`, then we exit the function by throwing `ArgumentException` and passing the name of the object as the `paramName` argument and a message detailing the issue. This can be used on the client to display an error message.

5. The following code will check whether the players are currently engaged in a match with another player. Add the following code to the end of the `IssueChallenge` function:

```
var challengerInMatch = (from g in context.Matches
                         where g.PlayerOneId == challengerId ||
g.PlayerTwoId == challengerId
                         select g).Any();

var opponentInMatch = (from g in context.Matches
                       where g.PlayerOneId == opponentId ||
g.PlayerTwoId == opponentId
                       select g).Any();

if (challengerInMatch)
    throw new ArgumentException(paramName: nameof(challenger),
message: $"{challenger.GamerTag} is already in a match!");
```

```
if (opponentInMatch)
    throw new ArgumentException(paramName: nameof(opponent),
message: $"{opponent.GamerTag} is already in a match!");
```

Again, we use **Linq** to query the database, but this time, we are querying the `Matches` `DbSet`. We are not looking for the `Match` instance, just the fact that one does exist, where either `PlayerOneId` or `PlayerTwoId` is the player's `Id` value. We use the `Any` function to return `true` if there are any results and `false` if there are no results. Again, we throw `ArgumentException` if either player is in a match with an appropriate message.

6. At this point, we have validated that both players exist and can join a new game. We will need to capture the game `Id` value if the challenge is accepted, so let's create the variable and log some details before moving on:

```
Guid matchId = Guid.Empty;

log.LogInformation($"{challenger.GamerTag} has challenged
{opponent.GamerTag} to a match!");
```

The next step in the `IssueChallenge` function is to create a `Challenge` object. But because we want to track how long `Challenge` is waiting so that we can time it out, we need a helper class to abstract that detail away from the function.

> **Don't reinvent the wheel**
>
> The implementation of `ChallengeHandler` is heavily based on `AckHandler` from the **Azure SignalR AckableChatRoom** sample. The source for the sample is available at `https://github.com/aspnet/AzureSignalR-samples/tree/main/samples/AckableChatRoom`.

Let's create the `ChallengeHandler` class by following these steps:

1. Create a new folder in the `SticksAndStones.Functions` folder named `Handlers`.

2. Create a new class named `ChallengeHandler` in the `Handlers` folder, and change the access modifier from `internal` to `public`.

3. Add a constructor for the class that has three parameters – `completeAcksOnTimeout` as `bool`, `ackThreshold` as `TimeSpan`, and `ackInterval` as `TimeSpan`. The constructor will create a timer to periodically clear out old challenges and store the `ackThreshold` value in a class field. The class's contents should look like this:

```
private readonly TimeSpan ackThreshold;
private readonly Timer timer;
```

```
public ChallengeHandler(bool completeAcksOnTimeout, TimeSpan
ackThreshold, TimeSpan ackInterval)
{
    if (completeAcksOnTimeout)
    {
        timer = new Timer(_ => CheckAcks(), state: null,
dueTime: ackInterval, period: ackInterval);
    }

    this.ackThreshold = ackThreshold;
}
```

You will need to add a `using` declaration for the `System.Threading` namespace to use the `Timer` type. The `CheckAcks` method will be created later in this section.

4. To provide some reasonable defaults for the constructor, we will create a parameterless constructor and provide the defaults, as shown in the following snippet:

```
private readonly Timer timer;

public ChallengeHandler() : this(
    completeAcksOnTimeout: true,
    ackThreshold: TimeSpan.FromSeconds(30),
    ackInterval: TimeSpan.FromSeconds(1))
    {
    }

public ChallengeHandler(bool completeAcksOnTimeout, TimeSpan
ackThreshold, TimeSpan ackInterval)
```

This will provide the default values to the main constructor.

5. Now, to create a new `Challenge`, the `IssueChallenge` function will call a method named `CreateChallenge`, as shown here:

```
public (Guid id, Task<Challenge> responseTask)
CreateChallenge(Player challenger, Player opponent)
{
    var id = Guid.NewGuid();
    var tcs = new
TaskCompletionSource<Challenge>(TaskCreationOptions.
RunContinuationsAsynchronously);
    handlers.TryAdd(id, new(id, tcs, DateTime.UtcNow, new(id,
challenger, opponent, ChallengeResponse.None)));
    return (id, tcs.Task);
}
```

Add this method to the `ChallengeHandler` class, below the main constructor. The method will use `TaskCompletionSource` to track the `Challenge` object. The `Challenge` and `TaskCompletionSource` instances are held by `ChallengeRecord`, which looks like this:

```
private record struct ChallengeRecord(Guid Id,
TaskCompletionSource<Challenge> ResponseTask, DateTime Created,
Challenge Challenge);
```

Add this declaration to the top of the `ChallengeHandler` class. This `record` has an `Id` value – for uniqueness, `TaskCompletionSource` – a `DateTime` value to track the creation time, and the `Challenge` object itself. We keep a list of `ChallengeRecord` instances in another field called `handlers`. The `handlers` field, which is declared right after the `ChallengeRecord` class, is as follows:

```
private readonly ConcurrentDictionary<Guid, ChallengeRecord>
handlers = new();
```

We use `ConcurrentDictionary` since we may be accessing the field from several different threads at the same time. `ConncurrentDictionary` is designed to prevent data corruption in multithreaded situations, like this one.

Once `TaskCompletionSource`, `Challenge`, and `ChallengeRecord` have been created, the `Challenge Id` value and the `Task` value associated with `TaskCompletionSource` are returned as a `Tuple` value to the `IssueChallenge` function. We will see what happens to that data later in this section when we complete the `IssueChallege` function.

Finally, to resolve the missing namespaces, add the following highlighted namespace declarations to your `ChallengeHandler` class file:

```
using SticksAndStones.Models;
using System;
using System.Collections.Concurrent;
using System.Threading;
using System.Threading.Tasks;
```

6. To acknowledge the challenge, either accepting it or declining it, the `AcknowledgeChallenge` function will call a method named `Respond`. The `Respond` method will remove `ChallengeRecord` from the dictionary, if it exists, and return the associate `Challenge`. If there is no `ChallengeRecord`, then a new empty `Challenge` record is returned, as shown in the following code:

```
public Challenge Respond(Guid id, ChallengeResponse response)
{
    if (handlers.TryRemove(id, out var res))
    {
        var challenge = res.Challenge;
        challenge.Response = response;
        res.ResponseTask.TrySetResult(challenge);
        return challenge;
    }
    return new Challenge();
}
```

7. The `CheckAcks` method, which is called periodically to check for `Challenge` objects that have expired and have not been responded to, looks like this:

```
private void CheckAcks()
{
```

```
        foreach (var pair in handlers)
        {
            var elapsed = DateTime.UtcNow - pair.Value.Created;
            if (elapsed > ackThreshold)
            {
                pair.Value.ResponseTask.TrySetException(new
    TimeoutException("Response time out"));
            }
        }
    }
}
```

This method will iterate over all the pairs in the `handlers` dictionary. For each one, it will determine whether the elapsed time is greater than the threshold provided in the constructor. If it is, then the task fails with a `TimeOutException` error.

8. To wrap up this class, we need to make sure that we clean up any remaining tasks when the service shuts down. We will handle canceling tasks in a `Dispose` method, which is implemented via the `IDisposable` interface. Add the `IDisposable` interface to the `ChallengeHandler` class and add the following `Dispose` method to the end of the class:

    ```
    public void Dispose()
    {
        timer?.Dispose();

        foreach (var pair in handlers)
        {
            pair.Value.ResponseTask.TrySetCanceled();
        }
    }
    ```

The `Dispose` method will dispose of the timer since we don't want that firing any longer. Then, it iterates over the handlers and cancels each of the tasks.

That should complete the `ChallengeHandler` class. We can now resume the implementation of the `IssueChallenge` function:

1. Open the `GameHub.cs` file and locate the constructor, modifying it as highlighted in the following code:

    ```
    private readonly GameDbContext context;
    private readonly ChallengeHandler challengeHandler;

    public GameHub(GameDbContext dbcontext, ChallengeHandler
    handler)
    {
        context = dbcontext;
        challengeHandler = handler;
    }
    ```

Since we will need the `ChallengeHandler` class, and it needs to maintain state, we will use dependency injection and have the Azure Functions runtime supply us with the instance.

2. Open the `Startup.cs` file in the `SticksAndStones.Function` project and add the following line of code at the end of the `Configure` method:

```
builder.Services.AddSingleton<ChallengeHandler>();
```

This will register `ChallengeHandler` with dependency injection to allow the Azure Functions runtime to manage the instance creation and lifetime.

3. Open the `GameHub.cs` file and navigate to the bottom of the `IssueChallenge` function.

4. Add the following lines of code:

```
var challengeInfo = challengeHandler.CreateChallenge(challenger,
opponent);
log.LogInformation($"Challenge [{challengeInfo.id}] has been
created.");

log.LogInformation($"Waiting on response from {opponent.
GamerTag} for challenge[{challengeInfo.id}].");
await Clients.User(opponent.Id.ToString()).SendAsync(Constants.
Events.Challenge, new ChallengeEventArgs(challengeInfo.id,
challenger, opponent));
```

This code will first call `CreateChallenge` using the `ChallengeHandler` instance that we are getting in the constructor. `challengeInfo` is a `Tuple` value of the `Challenge Id` type and `task`.

Next, the opponent is sent a SignalR `Challenge` message with `ChallengeEventArgs`. This message is sent slightly differently since this message will only be sent to the client that matches the opponent's `Id`.

5. Now, we need to wait for the opponent's response, or a timeout from `ChallengeHandler`, by using the following code:

```
ChallengeResponse response;
try
{
    var challenge = await challengeInfo.responseTask.
ConfigureAwait(false);
    log.LogInformation($"Got response from {opponent.GamerTag}
for challenge[{challengeInfo.id}].");
    response = challenge.Response;
}
catch
{
    log.LogInformation($"Never received a response from
{opponent.GamerTag} for challenge[{challengeInfo.id}], it timed
out.");
    response = ChallengeResponse.TimeOut;
}
return new(response);
```

The real trick in this code is the challengeInfo.responseTask await. responseTask is the task that is created as part of TaskCompletionSource in ChallengeHandler. By awaiting it, we do not continue until either the Respond method is called and the task is completed, or the task is failed by setting a TimeoutException error in the CheckAcks method of ChallengeHandler.

Once one of those conditions is true, the method completes and we can get the response from the returned Challenge, or in the case of a timeout, handle the exception and return the response to the client in a new instance of IssueChallengeResponse.

The IssueChallenge function is now complete. The client can call the function and it will send a message to the opponent's client and wait for the response. If the opponent client does not respond in a defined amount of time, which is 30 seconds by default, then the challenge will time out. Now, let's work on accepting or declining a challenge. As with the Connect function, you can try it out using the command line. You just need to connect two players, and then have one challenge the other!

Creating the AcknowledgeChallenge function

The AcknowledgeChallenge function is used by the client to respond to an open challenge from another player. Let's create the function by following these steps:

1. Add a new function to the GameHub class, as follows:

    ```
    [FunctionName("AcknowledgeChallenge")]
    public async Task AcknowledgeChallenge(
        [HttpTrigger(AuthorizationLevel.Function, "post", Route =
    $"Challenge/Ack")] HttpRequest req,
        ILogger log)
    {
    }
    ```

2. In the body of the function, deserialize the arguments using the following line of code:

    ```
    var result = await JsonSerializer.
    DeserializeAsync<AcknowledgeChallengeRequest>(req.Body,
    jsonOptions);
    ```

3. Use challengeHandler to Respond to the challenge:

    ```
    var challenge = challengeHandler.Respond(result.Id, result.
    Response);
    if (challenge.Id == Guid.Empty)
    {
        return;
    }
    ```

4. If the response is Declined, then just log a message:

    ```
    var challenger = challenge.Challenger;
    var opponent = challenge.Opponent;
    ```

```
if (result.Response == ChallengeResponse.Declined)
{
    log.LogInformation($"{opponent.GamerTag} has declined the
challenge from {challenger.GamerTag}!");
}
```

5. If the response is `Accepted`, then create a match and notify the players:

```
if (result.Response == ChallengeResponse.Accepted)
{
    log.LogInformation($"{opponent.GamerTag} has accepted the
challenge from {challenger.GamerTag}!");

    using var context = contextFactory.CreateDbContext();

    var game = Match.New(challenger.Id, opponent.Id);
    context.Matches.Add(game);

    opponent.MatchId = challenger.MatchId = match.Id;

    context.Players.Update(opponent);
    context.Players.Update(challenger);
    context.SaveChanges();

    log.LogInformation($"Created match {match.Id} between
{opponent.GamerTag} and {challenger.GamerTag}!");

    // Create Group for Game
    await UserGroups.AddToGroupAsync(opponent.Id.ToString(),
$"Match[{match.Id}]");
    await UserGroups.AddToGroupAsync(challenger.Id.ToString(),
$"Match[{match.Id}]");
    await Clients.Group($"Match[{match.Id}]").
SendAsync(Constants.Events.MatchStarted, new
MatchStartedEventArgs(match));

    await Clients.All.SendAsync(Constants.Events.PlayerUpdated,
new PlayerUpdatedEventArgs(opponent));
    await Clients.All.SendAsync(Constants.Events.PlayerUpdated,
new PlayerUpdatedEventArgs(challenger));

}
```

So, ignoring all the logging, since that is non-functional, the preceding code starts by creating a new `Match` instance and assigning `PlayerOneId`, `PlayerTwoId`, and `NextPlayerId`.

The `Match` object's `Id` property is then assigned to both of the players, and all the changes are saved to the database.

Next, is the SignalR messages. First, we create a SignalR group with just the two players in it and use the `Match` object's `Id` property in the name. This way we can send messages to the group and both players will receive them.

The first message we will send will indicate the start of a new game and it will send the `match` instance wrapped in `MatchStartedEventArgs`.

Finally, we send a message for each player to all players, indicating a change in their status.

That completes the functionality for one player to challenge another player to a match of Sticks and Stones! It's time to move on to playing a match. But first, we will need a function to return the game to the player.

Getting the match

You may be wondering why we need this functionality since, in the previous function, we sent the `Match` object to both players through a SignalR message. The answer is rather simple – if the user accidentally closes the Sticks and Stones app while in the middle of a game, then when they return to the Sticks and Stones app and log back in, the app will detect that they are still in a match and navigate to the `Match` page. It will use this function to retrieve the `Match` object in this case since it wasn't sent during `Connect`, just `Id`.

So, let's create a function to return a game by its `Id`, starting with the messages.

Creating the messages

This function will only need a response message object. Unlike the previous functions, the `GetMatch` function will use the `Http` GET method, and we will pass the match `Id` value as part of the URL. The response from the `GetGame` function will be the `Match` instance. To create the `GetGameResponse` message, follow these steps:

1. In the `SticksAndStones.Shared` project, create a new file named `GetGameMessages.cs` in the `Messages` folder.

2. Modify the contents of the file so that it's as follows:

    ```
    using SticksAndStones.Models;

    namespace SticksAndStones.Messages;

    public record struct GetMatchResponse(Match Match);
    ```

With the response message class in place, we can create the `GetMatch` function.

Getting a match by its ID

The GetMatch function will accept a single integer named id as a parameter. The parameter is bound to a part of the URL that's used to call the function. Let's look at an example. If we wanted to get Match identified by a Guid type of c39c7490-f4bc-425a-84ab-0a4ad916ea48, then the URL would be http://localhost:7024/api/Game/c39c7490-f4bc-425a-84ab-0a4ad916ea48.

Follow these steps to implement the GetMatch function:

1. Open the GameHub.cs file in the Hubs folder of the SticksAndStones.Functions project.

2. Add the following method declaration after the AcknowledgeChallenge method:

    ```
    [FunctionName("GetMatch")]
    public IActionResult GetMatch(
    [HttpTrigger(AuthorizationLevel.Function, "get", Route =
    "Match/{id}")] HttpRequest req,
        Guid id,
        ILogger log)
    {
    }
    ```

    There are a few differences from the other functions. First, the Http method that's used is "get", not "post". Second, Route is set to "Game/{id}"; {id} in Route tells the Azure Functions runtime that this function has a parameter named id and that the value in that position of the URL should be passed in as an argument. You can see that the third change is that there is an id parameter of the Guid type. This means that whatever is on the URL in the {id} position must be able to be converted into the Guid type; otherwise, the Azure Functions runtime will return an HTTP 500 Internal Server error.

3. To query our database for the Match object that matches the id value, use the following lines of code:

    ```
    using var context = contextFactory.CreateDbContext();

    Match match = (from m in context.Matches where m.Id == id select
    m).FirstOrDefault();
    ```

4. If the method gets this far, then it has been completed successfully, so we can return
    `OkObjectResult`. The object that's returned will be a `GetMatchResponse` instance
    with the `Match` instance that was found, or `null`:

    ```
    return new OkObjectResult(new GetMatchResponse(match));
    ```

Since this function uses the HTTP GET method, you can test it out in your favorite browser:

1. Press *F5* to start the project in debug mode.

2. Wait for txhe service to start, then copy the URL for the `GetMatch` function from the output
    window – for example, `http://localhost:####/api/Game/{id}`, where ### is
    your port number.

3. Open your browser and paste the URL in the address bar.

4. Change `{id}` to anything.

5. Press *Enter*.

    You should get an error page in your browser. Try a valid `Guid` value such as `c39c7490-`
    `f4bc-425a-84ab-0a4ad916ea48`. You should get a response similar to the following:

    ```
    {
        "match": null
    }
    ```

    Since there are no active games, you won't be able to retrieve an actual `Match` instance.

Now that we can retrieve the `Match` object, we will tackle how players make and receive moves and
how to determine the score and winner of the game.

Playing the game

Sticks and Stones is an interactive, fast-paced, turn-based game.

Let's review the following diagram to get a better understanding of what we are building:

Figure 9.11 – Processing turns

Once a match has started, players will take turns choosing a location to place one of their sticks. The client application will then send a message to the GameHub's `ProcessTurn` function. The `ProcessTurn` function will then validate the move, recalculate the score, check for a winner, and finally, send an update to the players.

Creating the ProcessTurn messages and models

The `ProcessTurn` function has three parameters – the `Match Id`, the player making the move, and the position of the move. The function will return an updated `Match` instance. Follow these steps to add the `ProcessTurn` messages:

1. Add a new file named `ProcessTurnMessages.cs` to the `Messages` folder in the `SticksAndStones.Shared` project.

2. Modify the contents of the file so that it looks as follows:

    ```
    using SticksAndStones.Models;
    using System;

    namespace SticksAndStones.Messages;

    public record struct ProcessTurnRequest(Guid MatchId, Player
    Player, int Position);
    public record struct ProcessTurnResponse(Match Match);
    ```

As part of the turn, `ProcessTurn` will send the updated `Match` instance to the players. This will require a new SignalR event. Perform the following steps to add it:

1. `SaveMatchAndSendUpdates` sends a new event via SignalR to the clients, so we need to add that to our constants. Add the highlighted code in the following snippet to the `Constants.cs` file in the `SticksAndStones.Shared` project:

    ```
    public static class Events
    {
        public static readonly string PlayerUpdated =
    nameof(PlayerUpdated);
        public static readonly string Challenge = nameof(Challenge);
        public static readonly string MatchStarted =
    nameof(MatchStarted);
        public static readonly string MatchUpdated =
    nameof(MatchUpdated);
    }
    ```

2. Add the `MatchUpdatedEventArgs` class that we used to send the updated `Match` to the players when the `MatchUpdated` event is set by adding a new file named `MatchUpdatedEventArgs.cs` to the `Messages` folder in the `SticksAndStones.Shared` project.

3.   Modify the contents of the file so that it's as follows:

```
using SticksAndStones.Models;

namespace SticksAndStones.Messages;

public record struct MatchUpdatedEventArgs(Match Match);
```

That concludes the additional models, events, and messages that are used by the `ProcessTurn` function. Next, we can start working on the ProcessTurn (P-Code) function.

Processing turns

The `ProcessTurn` function has a few responsibilities. It will need to do the following:

- Validate the turn
- Make the necessary changes to the `Match` object
- Recalculate the score
- Determine whether there is a winner
- Notify the players

To start the implementation of the `ProcessTurn` function, we will create stubs for each of the methods that we will call when processing a turn. Follow these steps to create the method stubs:

1.   Add a new method declaration to the `GameHub` class for the `ProcessTurn` function:

```
[FunctionName("ProcessTurn")]
public async Task<IActionResult> ProcessTurn(
    [HttpTrigger(AuthorizationLevel.Function, "post", Route =
$"Game/Move")] HttpRequest req,
    ILogger log)
{
}
```

2.   Add a new method declaration for `ValidateProcessTurnRequest`:

```
private Exception ValidateProcessTurnRequest(ProcessTurnRequest
args)
{
    return null;
}
```

The method accepts `ProcessTurnRequest` as the only argument. If there are no errors in the arguments, then it will default to returning `null`. If there is an error, then it will return an `Exception` error.

3. We will use another method to verify the match state before processing the move. Create a new method in the GameHub class named VerifyMatchState, as follows:

```
private Exception VerifyMatchState(Match match,
ProcessTurnRequest args)
{
    return null;
}
```

Just like with ValidateProcessTurnRequest, we will return null for the error if everything is okay; otherwise, we will return an error.

Now that we have the helper method signatures in place, let's implement them, starting with ValidateProcessTurnRequest. Follow these steps to add the implementation to ValidateProcessTurnRequest:

1. In ValidateProcessTurnRequest, add the following at the top of the method to check for a valid position:

```
if (args.Position <= 0 || args.Position > 23)
{
    return new IndexOutOfRangeException("Position is out of
range, must be between 1 and 24");
}
```

2. In ValidateProcessTurnRequest, add the following at the top of the method to check for a valid player:

```
if (args.Player is null)
{
    return new ArgumentException("Invalid Player");
}
```

3. In ValidateProcessTurnRequest, add the following at the top of the method to check for a valid game Id value:

```
if (args.MatchId == Guid.Empty)
{
    return new ArgumentException("Invalid MatchId");
}
```

That completes the ValidateProcessTurnRequest method. Now, we can add the code to VerifyMatchState, as follows:

1. In VerifyMatchState, add the following at the top of the method to check that the position hasn't already been played:

```
if (match.Sticks[args.Position] != 0)
{
    return new ArgumentException($"Position [{args.Position}]
```

```
has already been played");
}
```

2. In `VerifyMatchState`, add the following at the top of the method to check that the correct player is taking their turn:

```
if (args.Player.Id != game.NextPlayerId)
{
    return new ArgumentException($"It is not {args.Player.
GamerTag}'s turn");
}
```

3. In `VerifyMatchState`, add the following at the top of the method to check that the game isn't over already:

```
if (match.WinnerId != Guid.Empty)
{
    return new ArgumentException("Match is complete");
}
```

4. In `VerifyMatchState`, add the following at the top of the method to check that the game object exists:

```
if (match is null)
{
    return new ArgumentException("Invalid MatchId");
}
```

Now that we have created these helper methods, we can implement the `ProcessTurn` method by following these steps:

1. Deserialize the arguments that are passed to the `ProcessTurn` function using `JsonSerializer`, as follows:

```
var args = await JsonSerializer.
DeserializeAsync<ProcessTurnRequest>(req.Body, jsonOptions);
```

2. In the `ProcessTurn` method, we can call `ValidateProcessTurnRequest`. If there is an error, we can handle it, as follows:

```
var error = ValidateProcessTurnRequest(args);
if (error is not null)
{
    log.LogError(error, "Error validating turn request");
    return new BadRequestObjectResult(error);
}
```

3. With the arguments verified, we can query the database for the game, and fail if it doesn't exist:

```
using var context = contextFactory.CreateDbContext();

var game = (from g in context.Matches where m.Id ==
args.MatchId select m).FirstOrDefault() ?? throw new
ArgumentException("Invalid MatchId.");
```

4. Now, we can call `VerifyGameState`. If there is an error, we can handle it, as follows:

```
error = VerifyGameState(game, args);
if (error is not null)
{
    log.LogError(error, "Error validating game state");
    return new BadRequestObjectResult(error);
}
```

5. We must do one final check before making the move and updating the scores – we need to check to see whether the player made their selection before their turn expired using the following code:

```
if (turnHandler.EndTurn(args.GameId) == TurnHandler.TurnStatus.
Forfeit)
{
    error = new ArgumentException($"The turn has expired.");
    log.LogError(error, $"Player did not respond in the time
alloted.");
    return new BadRequestObjectResult(error);
}
```

With the validation of the input and game out of the way, we can now focus on applying the player's move to the current state of the game, updating the score, and determining a winner. To make the code for updating the score simpler, we will need a complex data structure. Let's explore this further.

When a player chooses a location to place one of their sticks, it may complete a square. If it does, then the player who placed the stick gets the square and an additional five points. The trick is how to determine that a stick has completed a square. We are storing the state of all sticks and stones as an array of integers. What we need is a map from a stone index (0–8) to the sticks that make up its sides. But we can simplify the logic a bit more once we know what sticks make up a square since we are only interested in a single stick, and a single stick can complete, at most, two squares. So, we can now have a structure that maps each stick position (0–23) to an array of tuples. Each tuple has an integer that is the index for the stone and another integer array that is the other three stick indexes that make up the square.

Let's use an example to illustrate this. Pretend that we have a game that's in the following state:

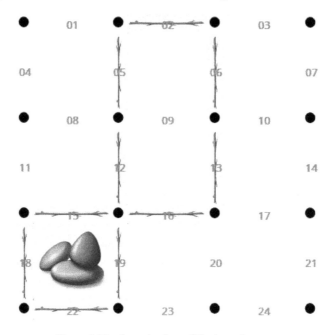

Figure 9.12 – Sample view of the board game

Now, pretend that a player has chosen to place their stick at location 9, highlighted in aqua:

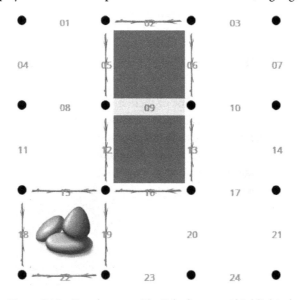

Figure 9.13 – Board game with stick placement highlighted

We will only need to check the two squares highlighted in brown. This means that we need to check whether there are sticks in positions 2, 5, and 6 for the stone in the upper brown box, and 12, 13, and 16 for the stone in the lower brown box.

This means we need two tuples – one for stone 2 and a second for stone 5 – each with an array of integers of the sides – f example, { (2, {2, 5, 6}), (5, {12, 13, 16} }. Using that data, we can check the two possible squares that could be completed by placing a stick at position 9.

Using old-fashioned sticky notes with a pencil and eraser, we can determine that the complete mapping will look like this:

```
(int stone, int[] sticks)[][] stickToStoneMap = new (int, int[])[][] {
/* 1 */ new (int, int[])[] { (1, new int[] { 4, 5, 8}), (0, new int[]
{ 0, 0, 0})},
/* 2 */ new (int, int[])[] { (2, new int[] { 5, 6, 9}), (0, new int[]
{ 0, 0, 0})},
/* 3 */ new (int, int[])[] { (3, new int[] { 6, 7,10}), (0, new int[]
{ 0, 0, 0})},
/* 4 */ new (int, int[])[] { (1, new int[] { 1, 5, 8}), (0, new int[]
{ 0, 0, 0})},
/* 5 */ new (int, int[])[] { (1, new int[] { 1, 4, 8}), (2, new int[]
{ 2, 6, 9})},
/* 6 */ new (int, int[])[] { (2, new int[] { 2, 5, 9}), (3, new int[]
{ 3, 7,10})},
/* 7 */ new (int, int[])[] { (3, new int[] { 3, 6,10}), (0, new int[]
{ 0, 0, 0})},
/* 8 */ new (int, int[])[] { (1, new int[] { 1, 4, 5}), (4, new int[]
{11,12,15})},
/* 9 */ new (int, int[])[] { (2, new int[] { 2, 5, 6}), (5, new int[]
{12,13,16})},
/*10 */ new (int, int[])[] { (3, new int[] { 3, 6, 7}), (6, new int[]
{13,14,17})},
/*11 */ new (int, int[])[] { (4, new int[] { 8,12,15}), (0, new int[]
{ 0, 0, 0})},
/*12 */ new (int, int[])[] { (4, new int[] { 8,11,15}), (5, new int[]
{ 9,13,16})},
/*13 */ new (int, int[])[] { (5, new int[] { 9,12,16}), (6, new int[]
{10,14,17})},
/*14 */ new (int, int[])[] { (6, new int[] {10,13,17}), (0, new int[]
{ 0, 0, 0})},
/*15 */ new (int, int[])[] { (4, new int[] { 8,11,12}), (7, new int[]
{18,19,22})},
/*16 */ new (int, int[])[] { (5, new int[] { 9,12,13}), (8, new int[]
{19,20,23})},
/*17 */ new (int, int[])[] { (6, new int[] {13,14,17}), (9, new int[]
{20,21,24})},
/*18 */ new (int, int[])[] { (7, new int[] {15,19,22}), (0, new int[]
{ 0, 0, 0})},
/*19 */ new (int, int[])[] { (7, new int[] {15,18,22}), (8, new int[]
{16,20,23})},
/*20 */ new (int, int[])[] { (8, new int[] {16,19,23}), (9, new int[]
{17,21,24})},
/*21 */ new (int, int[])[] { (9, new int[] {17,20,24}), (0, new int[]
{ 0, 0, 0})},
```

```
/*22 */ new (int, int[])[] { (7, new int[] {15,18,19}), (0, new int[]
{ 0, 0, 0})},
 /*23 */ new (int, int[])[] { (8, new int[] {16,19,20}), (0, new int[]
{ 0, 0, 0})},
/*24 */ new (int, int[])[] { (9, new int[] {17,20,21}), (0, new int[]
{ 0, 0, 0})},
};
```

Add the preceding code to the GameHub class after the turnHandler field declaration. Now that we have declared the data structure for finding completed boxes, let's continue processing the turn:

1. Return the ProcessTurn method, and add the following code at the end:

    ```
    match.Sticks[args.Position] = args.Player.Id == match.
    PlayerOneId ? 1 : -1;
    ```

    This will assign the position -1 or 1, depending on who the active player is. We will use a value of -1 and 1 later in determining a winner, in the case of three stones in a row.

2. Now that we have placed the stick, we need to adjust the player's score, as follows:

    ```
    if (args.Player.Id == game.PlayerOneId)
    {
        match.PlayerOneScore += 1;
    }
    else
    {
        match.PlayerTwoScore += 1;
    }
    ```

3. The following code will use the data structure from earlier to determine whether placing the stick completed any squares:

    ```
    // Determine if this play creates a square
    foreach (var tuple in stickToStoneMap[args.Position])
    {
        if (tuple.stone == 0) continue;

        var stickCompletesABox =
        (
            Math.Abs(match.Sticks[tuple.sticks[0] - 1]) +
            Math.Abs(match.Sticks[tuple.sticks[1] - 1]) +
            Math.Abs(match.Sticks[tuple.sticks[2] - 1])
        ) == 3;

        if (stickCompletesABox)
        {
            // If so, place stone, and adjust score
            var player = args.Player.Id == match.PlayerOneId ? 1 :
    -1;
    ```

```
        match.Stones[tuple.stone - 1] = player;
        if (player > 0)
        {
            match.PlayerOneScore += 5;
        }
        else
        {
            match.PlayerTwoScore += 5;
        }
    }
}
```

This code will iterate over the tuples declared at the array position of the newly placed stick, adjusting for C# arrays being 0-based. It will then use the array of stick positions from the tuple to index into the array of sticks in the match The value at that location will be either 1 for player one, -1 for player two, or 0 for unclaimed. We can add the absolute value of all three locations and if it is 3, then the newly placed stick completes a box. If so, then assign the stone location from the tuple, adjusting for 0-based arrays again, to the player, and give them five points.

4. To help determine whether the match is over, we are going to use a couple of helper functions to make the code cleaner and easier to read. The first of those returns a `boolean` value if all the sticks have been played in the match. Add the following code to the end of the `GameHub` class:

```
private static bool AllSticksHaveBeenPlayed(Match match)
{
    return !(from s in match.Sticks where s == 0 select
s).Any();
}
```

This function uses a straightforward LINQ query to search the `Sticks` array for any element that has a value of 0, meaning unclaimed. If there are, the function returns.

5. The next function is a little more complex as it is used to determine whether a player has three stones in a row, either horizontally, vertically, or diagonally, and returns the `Id` value of the player that does. Add the following code to the end of the `GameHub` class:

```
private static int HasThreeInARow(List<int> stones)
{
    for (var rc = 0; rc < 3; rc++)
    {
        var rowStart = rc * 3;
        var rowValue = stones[rowStart] + stones[rowStart + 1] +
stones[rowStart + 2];
        if (Math.Abs(rowValue) == 3) // we Have a winner!

        {
            return rowValue;
        }
```

```
                var colStart = rc;
                var colValue = stones[colStart] + stones[colStart + 3] +
    stones[colStart + 6];
                if (Math.Abs(colValue) == 3) // We have a winner!
                {
                    return colValue ;
                }
            }
        var tlbrValue = stones[0] + stones[4] + stones[8];
        var trblValue = stones[2] + stones[4] + stones[6];
        if (Math.Abs(tlbrValue) == 3) { return tlbrValue; }
        if (Math.Abs(trblValue) == 3) { return trblValue; }
        return 0;
    }
```

This method starts by checking all the rows and columns for 3 stones in a row, for the same player. Since there are nine stones arranged in a 3x3 grid, we only need to check three columns and three rows. Using a single iterator, each row or column is checked by adding the values stored at each position in the row or column and if the absolute value of the sum is 3, then a single player has a winning row. If the sum is positive, player one has won; otherwise, player two has won. Since there are only two possible diagonals, those checks use the same logic but are done individually, rather than looping.

6. Now, we can use those two functions to determine whether there is a winner. To do so, we can use the following code at the end of the `ProcessTurn` function:

```
// Does one player have 3 stones in a row?
var winner = Guid.Empty;
var threeInARow = HasThreeInARow(match.Stones);
if (threeInARow != 0)
    winner = threeInARow > 0 ? match.PlayerOneId : match.
PlayerTwoId;

if (winner == Guid.Empty) // No Winner yet
{
    // Have all sticks been played, if yes, use top score.
    if (HaveAllSticksBeenPlayed(match))
    {
        winner = match.PlayerOneScore > match.PlayerTwoScore ?
match.PlayerOneId : match.PlayerTwoId;
    }
}
```

Here, we use the two methods we just created to do the main checks and assign the winner. We capture the winner as the `Guid` type from the `Id` property of the `Player` class, so some translation is needed.

7. Next, we can set the next player's turn, or if there is a winner, complete the match, as follows:

```
if (winner == Guid.Empty)
{
    match.NextPlayerId = args.Player.Id == match.PlayerOneId ?
match.PlayerTwoId : match.PlayerOneId;
}
else
{
    match.NextPlayerId = Guid.Empty;
    match.WinnerId = winner;
    match.Completed = true;
}
```

8. The final steps are to save any changes we have made and send updates to the players. We will use a helper method called `SaveGameAndSendUpdates` to handle that as we will need the same code when a turn expires. Add the following code to the end of the GameHub class:

```
private async Task SaveMatchAndSendUpdates(GameDbContext
context, Match match)
{
    context.Matches.Update(match);
    await context.SaveChangesAsync();
    await Clients.Group($"Match[{match.Id}]").
SendAsync(Constants.Events.MatchUpdated, new
MatchUpdatedEventArgs(match));
    if (match.Completed)
    {
        await UserGroups.RemoveFromGroupAsync(match.PlayerOneId.
ToString(), $"Match[{match.Id}]");
        await UserGroups.RemoveFromGroupAsync(game.PlayerTwoId.
ToString(), $"Match[{match.Id}]");
    }
}
```

This function will save the current match state to the database, then sends a message to the SignalR group for the match indicating that there have been updates to the match. If the match is over, then we remove the players from the group.

1. The following final three lines of code complete the `ProcessTurn` method:

    ```
    await SaveMatchAndSendUpdates(context, match);

    return new OkObjectResult(new ProcessTurnResponse(match));
    ```

    After saving the match changes and notifying the players of the match updates, if the match is not over yet, we notify the next player that it is their turn to play. To wrap things up we return the updated match object back to the player that just made their move.

2. We also need to call `SaveGameAndSendUpdates` when there is an error after calling `VerifyGameState`. Modify that section of code in `ProcessTurn` using the following snippet:

    ```
    error = VerifyMatchState(game, args);
    if (error is not null)
    {
        await SaveMatchAndSendUpdates(game);
        log.LogError(error, "Error validating match state.");
        return new BadRequestObjectResult(error);
    }
    ```

We have now completed all the required functions to connect a player to the service, challenge another player to a match, and then process each player's turn and determine the winner.

Let's take a short look back at what we have accomplished so far in this chapter. We started by creating the Azure services that our game server backend would need, a SignalR service for real-time communication, and finally, the Functions service to host our backend functions. We then implemented the Azure functions that would provide the functionality for our game:

- `Connect`: To register players to the game service

- `IssueChallenge`: To allow one player to request a game with another player

- `AcknowledgeChallenge`: To accept or decline a request

- `ProcessTurn`: To manage the gameplay between two players and determine the winner

Our backend is now complete, and we are ready to publish it to Azure so that we can consume the services from the game app in *Chapter 10*.

Deploying the functions to Azure

The final step in this chapter is to deploy the functions to Azure. You can do that as a part of a **continuous integration/continuous deployment (CI/CD)** pipeline – for example, with Azure DevOps. But the easiest way to deploy the functions, in this case, is to do it directly from Visual Studio. Perform the following steps to deploy the functions:

1. Right-click on the `SticksAndStones.Functions` project and select **Publish**.

2. Select **Azure** as the destination for publishing and click **Next**:

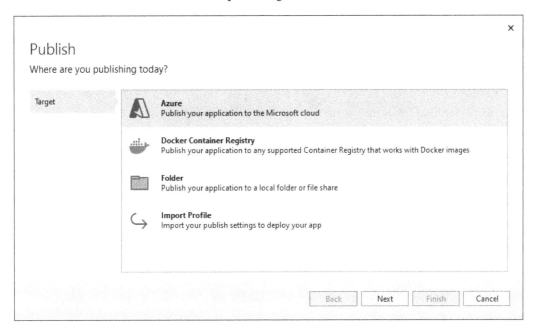

Figure 9.14 – Target selection when publishing

3.   Choose **Azure Function App (Windows)** in the **Specific target** tab, then click **Next**:

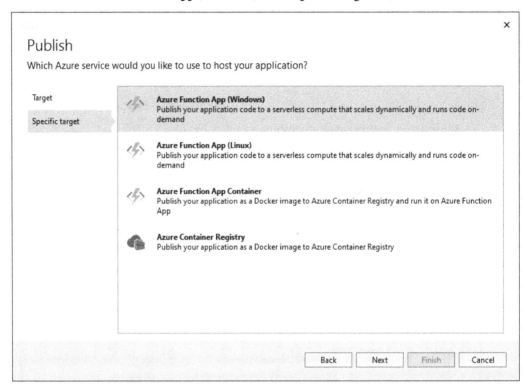

Figure 9.15 – Container selection when publishing

4.   Sign in to the same Microsoft account that we used in the Azure portal when we were creating the **Function App** resource.

5.   Select the subscription that contains the function app. All function apps we have in the subscription will now be loaded.

6.   Select the function app and click **Finish**. If your app isn't showing up, click **Back** and choose the **Azure Function App (Linux)** option as you may not have changed the default when creating the service in the *Creating the Azure service for functions* section.

7.   When the profile is created, click the **Publish** button.

The following screenshot shows the last step. After that, the publishing profile will be created:

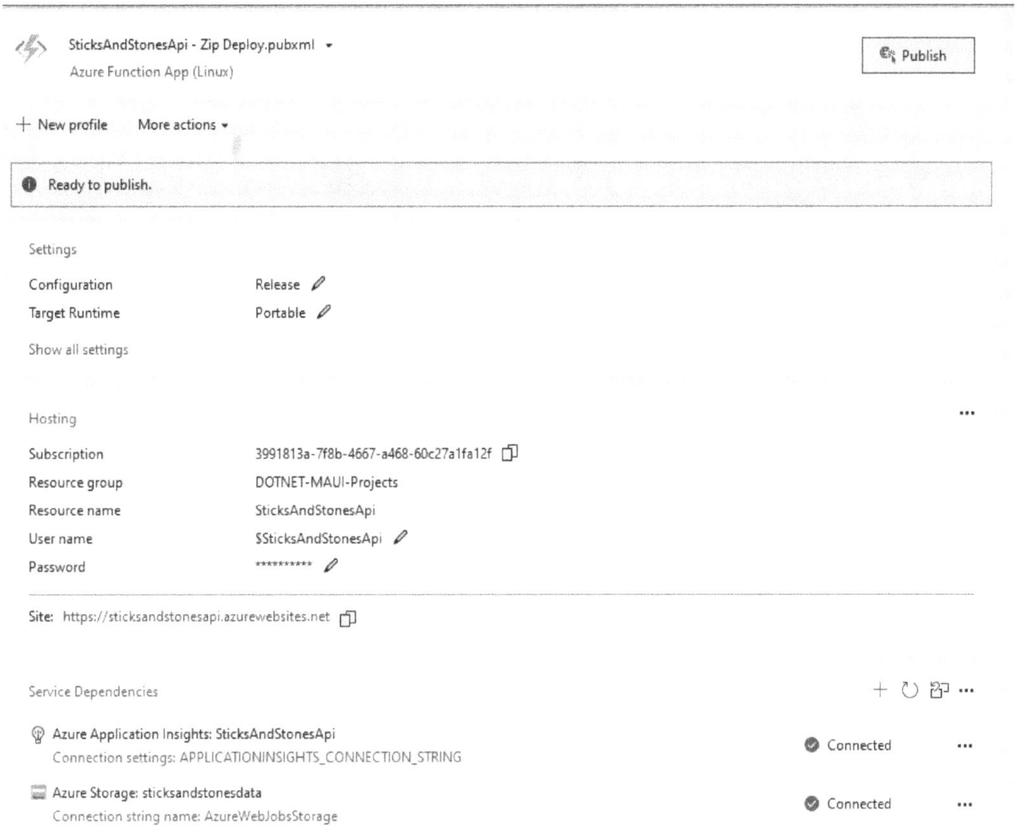

Figure 9.16 – Publishing Azure functions

Summary

In this chapter, we started by learning about a few Azure services, including SignalR, and Functions. Then, we created the services in Azure that our game server backend would need – a SignalR service for real-time communication, and finally, the Functions service to host our backend functions. After this, we implemented the Azure functions that would provide the functionality for our game.

We wrapped up this chapter by publishing our function code to the Azure Functions instance in Azure.

In the next chapter, we will build a game app that will use the backend we have built in this project.

Building a Real-Time Game

In this chapter, we will build a multi-player, head-to-head game app with real-time communication. In the app, you will be able to connect to a game server and view a list of other players that are also connected. You can then select a player to request a game with them and, provided they accept, play a game of *Sticks & Stones*. We will look at how we can use SignalR to implement a real-time connection with the server.

The following topics will be covered in this chapter:

- How to use SignalR in a .NET MAUI app

- How to use control templates

- How to use XAML triggers to update the interface

- How to use XAML styling in a .NET MAUI app

Let's get started.

Technical requirements

Before you start building the app for this project, you need to build the backend that we detailed in *Chapter 9, Setting Up a Backend for a Game Using Azure Services*. You will also need to have Visual Studio for Mac or PC installed, as well as the .NET MAUI components. See *Chapter 1, Introduction to .NET MAUI*, for more details on how to set up your environment. The source code for this chapter is available in this book's GitHub repository: `https://github.com/PacktPublishing/MAUI-Projects-3rd-Edition`.

Project overview

When building a head-to-head game app, it is really important to have real-time communication because the user expects the other players' moves to arrive more or less immediately. To achieve this, we will use SignalR, which is a library for real-time communication. SignalR will use WebSockets if they are available and, if not, it will have several fallback options it can use instead. In the app, we will use SignalR to send updates on player and game status through the Azure Functions that we built in *Chapter 9*.

The build time for this project is about 180 minutes.

Getting started

We can use either Visual Studio on a PC or Mac to complete this project. To build an iOS app using Visual Studio for PC, you have to have a Mac connected. If you don't have access to a Mac at all, you can choose to just build the Android part of the app.

Let's review from *Chapter 9* what the game is all about.

An overview of the game

Sticks & Stones is a turn-based social game based on the concepts of two childhood games mashed into one, Dots and Boxes (https://en.wikipedia.org/wiki/Dots_and_boxes) and Tic-Tac-Toe (https://en.wikipedia.org/wiki/Tic-tac-toe). The game board is laid out in a three-by-three grid. Each player will take a turn placing a stick along the side of a box, between two dots, to earn one point. If a stick completes a box, then the player takes ownership of the box, earning five points. The game is won when a player owns three boxes in a row, horizontally, vertically, or diagonally. If no player can own three boxes in a row, the winner of the game is determined by the player with the highest score.

To keep the app and the service side relatively simple, we will eliminate a lot of state management. When the player opens the app, they will have to connect to the game service. They will have to provide a gamer tag, or username, and an email address. Optionally, they can upload a picture of themselves to use as a profile picture.

Once connected, the player will then see a list of all the other players connected to the same game service; this is called the Lobby. The player's status of either **Ready to play** or **In a match** is displayed along with the player's gamer tag and profile picture. If the player is not in a match, a button is available to challenge the player to a match.

Challenging a player to a match will cause the app to prompt the opponent to respond to the challenge, either accept or decline it. If the opponent accepts the challenge, then both players are navigated to a new game board where the player who received the challenge will have the first turn. Both players' statuses will update to **In a match** in all the other players' lobbies.

Play will alternate between players as they choose a location to place a single stick. Each time a stick is placed by a player, the game board and score will update on both players' devices. When a stick is placed that completes one or more squares, the player then wins that square, and a pile of stones is placed in the center of the square. When all sticks have been placed, or a player owns three stones in a row, the game is over, the players navigate back to the Lobby, and their status is updated to "Ready to play."

If a player leaves the app during a game, then they will have forfeited the match and the remaining opponent will be credited with the win and navigated back to the Lobby.

Now that we understand what we want to build, let's get down to the details.

We recommend that you use the same solution we used in *Chapter 9, Setting Up a Backend for a Game Using Azure Services*, because this will make code sharing easier. If you don't want to go through all of *Chapter 9*, you can get the completed source from *Chapter 9*, at `https://github.com/PacktPublishing/MAUI-Projects-3rd-Edition/tree/chapters/nine/main`.

We will build this app in four sections:

- **Services** – All the classes that are needed to connect and interact with the Azure Functions backend that was built in *Chapter 9, Setting Up a Backend for a Game Using Azure Services*.

- **Connect page** – This will consist of the view and view model needed to allow a user to connect to the game server as a player.

- **Lobby page** – The Lobby is where the player can send and receive challenges with other players. In this section, we will build the view and view model for the lobby.

- **Game page** – This is where players can take turns playing a game of *Sticks and Stones*. In this section, we will build the view and view model needed to make that happen.

Let's start by creating the project for the .NET MAUI app.

Building the game app

It's time to start building the app. Open the `SticksAndStones` solution from the previous chapter and follow these steps to create the project:

1. Open the **Create a new project** wizard by selecting **File**, **Add**, then **New Project…** from the Visual Studio menu:

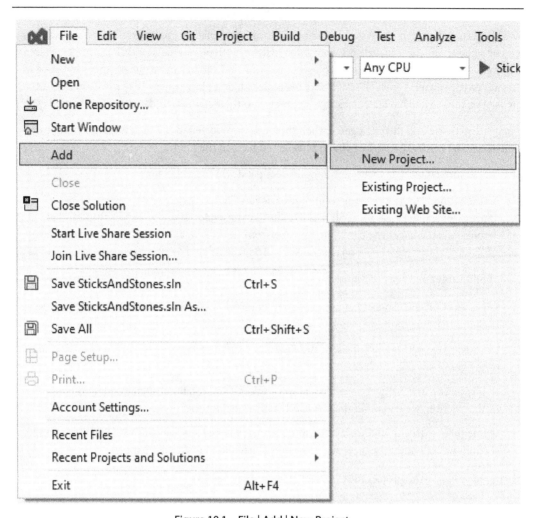

Figure 10.1 – File | Add | New Project…

2. In the search field, type maui and select the **.NET MAUI App** item from the list, or select it from **Recent project templates** if it is listed:

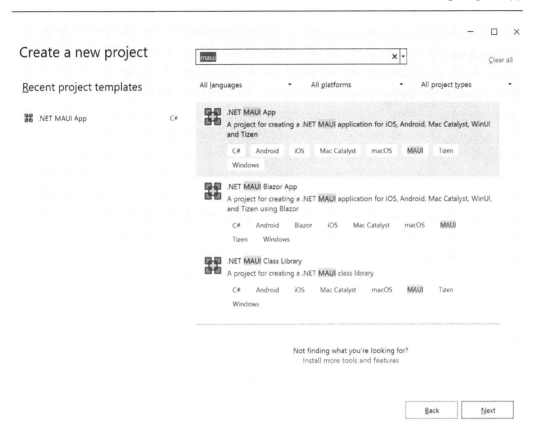

Figure 10.2 – Create a new project

3. Click **Next**.

4. Enter `SticksAndStones.App` as the name of the app and, under **Solution**, select **Add to solution**, as shown in the following screenshot:

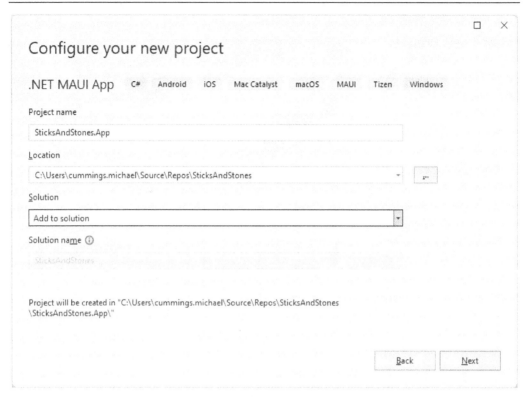

Figure 10.3 – Configure your new project

5.   Click **Next**.

6.   The last step will prompt you for the version of .NET Core to support. At the time of writing, .NET 6 is available as **Long-Term Support** (**LTS**), and .NET 7 is available as **Standard Term Support**. For the purposes of this book, we will assume that you will be using .NET 7:

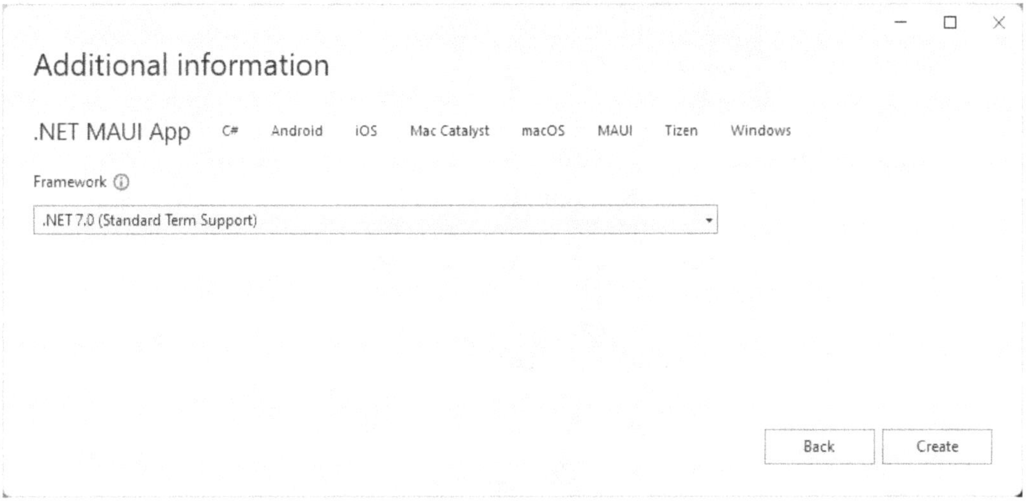

Figure 10.4 – Additional information

7.   Finalize the setup by clicking **Create** and wait for Visual Studio to create the project.

Now we have created the .NET MAUI project for our game screens, let's configure it so that it's ready to add the services and views. We will need to add a project reference to the `SticksAndStones.Shared` project, as well as a few NuGet packages. Follow these steps to complete the setup of the `SticksAndStones.App` project:

1.   Right-click the `SticksAndStones.App` project in **Solution Explorer** and select **Properties**.

2.   In the **Properties** window, use the search box at the top to search for `Default namespace`.

3.   Change the **Default namespace** property value to `$(MSBuildProjectName.Split(".")`
      `[0].Replace(" ", "_"))`.

      This will split the project name on `"."`, using only the first part and replacing any spaces with underscores.

4.   Add a NuGet Package reference to `CommunityToolkit.Mvvm` as, in other chapters, we will be using this package to simplify the implementation of data bindings to properties and commands.

5.   Add a NuGet Package reference to `CommunityToolkit.Maui`. We will be using the `GravatarImageSource` class from this package to render an avatar for the user. For .NET 7, you will need to use version 6.1.0 of the NuGet package. 7.0+ has .NET 8 as a dependency.

6.   Open the `MauiProgram.cs` file and add the highlighted line shown here:

```
using CommunityToolkit.Maui;
using Microsoft.Extensions.Logging;
```

```
namespace SticksAndStones.App
{
    public static class MauiProgram
    {
        public static MauiApp CreateMauiApp()
        {
            var builder = MauiApp.CreateBuilder();
            builder
                .UseMauiApp<App>()
                .ConfigureFonts(fonts =>
                {
                    fonts.AddFont("OpenSans-Regular.ttf",
"OpenSansRegular");
                    fonts.AddFont("OpenSans-Semibold.ttf",
"OpenSansSemibold");
                })
                .UseMauiCommunityToolkit();
#if DEBUG
            builder.Logging.AddDebug();
#endif
            return builder.Build();
        }
    }
}
```

This will configure `CommunityToolkit` for use within the app.

7. Add a NuGet package reference to `Microsoft.Extensions.Logging.Abstractions`. This package is used to log messages from the Azure Functions functions for debugging.

8. Add a NuGet package reference to `Microsoft.Extensions.Logging.Debugging`. This package is used to log messages from the Azure Functions functions for debugging.

9. Add a NuGet package reference to `Microsoft.AspNetCore.SignalR.Client`. This package is required for the app to connect to the SignalR Hub we created in *Chapter 9* and receive messages.

10. Add a Project reference to the `SticksAndStones.Shared` project. This will give us access to the messages and objects we created in *Chapter 9*.

That's it for project creation. Next, we will start with creating classes that interact directly with our service.

Creating the game services

The first thing we will do is create a service that will be used to communicate with the Azure Functions functions service created in *Chapter 9, Setting Up a Backend for a Game Using Azure Services*. The service will be broken down into three main classes:

- `GameService` – Methods and properties for calling the Azure Functions and receiving SignalR messages.

- `ServiceConnection` – Holds the references to `HttpClient` and SignalR Hub instances. Also provides methods for safely making calls using `HttpClient`.

- `Settings` – Stores and retrieves the URL for the server used by `HttpClient`. It also stores the connection details provided by the user.

We will start with the `Settings` class since both `GameService` and `ServiceConnection` will depend on `Settings`.

Creating the Settings service

The `Settings` service is used to store values that are needed between app runs. It will use the .NET MAUI `Preferences` class to store these values in a cross-platform manner. Use the following steps to implement the `Settings` class:

1. In the `SticksAndStones.App` project, create a new folder named `Services`.

2. In the newly created `Services` folder, create a new class named `Settings`.

3. Make the class public.

4. Create a `const string` field named `LastPlayerKey` and initialize it like so:

   ```
   private const string LastPlayerKey = nameof(LastPlayerKey);
   ```

5. Create a `const string` field named `ServerUrlKey` and initialize it like this:

   ```
   private const string ServerUrlKey = nameof(ServerUrlKey);
   ```

   These two fields are used by the .NET MAUI `Preferences` class to store the values for the server URL and the login details for the last time the user logged in.

6. Add a `private const string` field named `ServerUrlDefault` as follows:

   ```
   #if DEBUG && ANDROID
       private const string ServerUrlDefault =
   "http://10.0.2.2:7071/api";
   #else
       private const string ServerUrlDefault = "http://
   localhost:7071/api";
   #endif
   ```

This snippet uses the conditional compilation features of C# to switch the `ServerlUrlDefault` value for Android devices. The `10.0.2.2` IP address is a special value used by the Android emulators to be able to access the host computer's `localhost` address. This is very useful when testing the app using the **Azurite** development environment for Azure Functions.

You may need to adjust the port number, highlighted in the preceding listing, for your specific development environment. Azure Functions will display the server URL when started from Visual Studio, as shown in the following screenshot:

Figure 10.5 – Azure Functions console output

Using the Azure Functions hosted in Azure

If you followed the steps in the *Chapter 9* section called *Deploying the functions to Azure*, then you can use the URL for the Azure Function App created in *Chapter 9*, in the *Creating the Azure service for functions* section. The URL is displayed on the **Overview** tab of the Azure Functions App.

7. Now, add a `public string` property named `ServerUrl` with the following implementation:

```
public string ServerUrl
{
    get => Preferences.ContainsKey(ServerUrlKey) ?
                Preferences.Get(ServerUrlKey, ServerUrlDefault)
    :
                ServerUrlDefault;
    set => Preferences.Set(ServerUrlKey, value);
}
```

This code will get the server URL from the `Preferences` store if it is present; if not, it will use the `serverUrlDefault` value. The property will store the new value in the `Preferences` store.

8. Add the following using declarations at the top of the `Settings.cs` file:

```
using SticksAndStones.Models;
using System.Text.Json;
```

This will enable us to use our model and the `JsonSerializer` classes.

9. Create a new property named `LastPlayer` that is of the `Player` type, as follows:

```
public Player LastPlayer
{
    get
    {
        if (Preferences.ContainsKey(LastPlayerKey))
        {
            var playerJson = Preferences.Get(LastPlayerKey,
string.Empty);
            return JsonSerializer.
Deserialize<Player>(playerJson, new
JsonSerializerOptions(JsonSerializerDefaults.Web)) ?? new();
        }
        return new();
    }
    set => Preferences.Set(LastPlayerKey,
JsonSerializer.Serialize(value, new
JsonSerializerOptions(JsonSerializerDefaults.Web)));
}
```

Here, the property `set` method will convert the `Player` object to a `Json` string before storing it in `Preferences` and, when getting the property, if it exists in the `Preferences` store, convert the stored `Json` to a `Player` object. If there is no value in the `Preferences` store, then the `get` method will return an empty `Player` object.

10. The final step for the `Settings` class is to register it with the dependency injection container. Open the `MauiProgram.cs` file in the `SticksAndStones.App` project, then add the following highlighted code to the `CreateMauiApp` method:

```
public static MauiApp CreateMauiApp()
{
    var builder = MauiApp.CreateBuilder();
    builder
        .UseMauiApp<App>()
        .ConfigureFonts(fonts =>
        {
```

```
                fonts.AddFont("OpenSans-Regular.ttf",
        "OpenSansRegular");
                fonts.AddFont("OpenSans-Semibold.ttf",
        "OpenSansSemibold");
            });

    #if DEBUG
            builder.Logging.AddDebug();
    #endif
        builder.Services.AddSingleton<Services.Settings>();


            return builder.Build();
        }
```

With the Settings class complete, we can now focus on the ServiceConnection class.

Creating the ServiceConnection class

The ServiceConnection class encapsulates the functionality needed to communicate with the Azure Functions service. It has methods to call the function methods and return the results, with appropriate error handling. It is also responsible for initializing the SignalR Hub instance that is used for real-time communication. The ServiceConnection class has a couple of dependencies that we need, so let's put them together first.

The first thing to add is logging. Having logging during debugging can be very helpful in figuring out problems, especially when dealing with asynchronous processes. Communicating with Azure Functions will have a lot of asynchronous operations. To enable logging while debugging, add the highlighted code to the CreateMauiApp method in the MauiProgram class in the SticksAndStones. App project:

```
#if DEBUG
        builder.Logging.AddDebug();
        builder.Services.AddLogging(configure =>
        {
            configure.AddDebug();
        });
    #endif
        builder.Services.AddSingleton<Services.Settings>();


        return builder.Build();
```

This will add an instance of ILoggingProvider to the services container. The ILoggerProvider instance will provide instances of ILogger<T>. This will enable the use of ILogger<T> as a dependency in the ServiceConnection class constructor.

> **More on logging providers**
>
> Read more about how logging providers work, and logging in general, at `https://learn.microsoft.com/en-us/dotnet/core/extensions/logging-providers`.

Now, when making requests to APIs using HTTP, it is a common and good practice to use asynchronous calls so that you do not block the main or UI thread. All UI updates, such as animation, button clicks, taps on the screen, or text changes, happen on the UI thread. HTTP calls can take a non-trivial amount of time to complete, which can cause the app to become unresponsive to a user.

Error handling in asynchronous programming can be difficult. To help with errors when making API calls, we are going to use a couple of classes to encapsulate the exceptions; these classes are `AsyncError` and `AsyncExceptionError`. We need `AsyncError` and `AsyncExceptionError` because it is a bad practice to serialize and deserialize any class instances that derive from `System.Exception`. Not all classes derived from `System.Exception` are serializable, and even if they are, you may not be able to deserialize them due to a missing type – for example, the type is available on the server but not on the client. Create a new file named `AsyncError.cs` in the `SticksAndStones.App` project and replace the contents with the following code:

```
using System.Text.Json.Serialization;

namespace SticksAndStones;

public record AsyncError
{
    [JsonPropertyName("message")]
    public string Message { get; set; }
}

public record AsyncExceptionError : AsyncError
{
    [JsonPropertyName("innerException")]
    public string InnerException { get; set; }
}
```

The `AsyncError` class has a single property, `Message`. The `Message` property is decorated with the `JsonPropertyName` attribute so that it can be serialized if needed, using a lowercase version of the property name. `AsyncExceptionError` inherits from `AsyncError` and adds an additional property, `InnerException`. The `InnerException` property is also attributed with `JsonPropertyName`.

The last class we will need is `AsyncLazy<T>`. You may have already used `Lazy<T>` in other applications you have written. It's very handy when you want to delay the creation of a class until right before you need it. If you never need it, it doesn't get created. But `Lazy<T>` does not work great with asynchronous programming, so if you wanted to lazy instantiate a class that is created asynchronously, that becomes tedious. Luckily for us, Stephen Toub, who works for Microsoft on the

.NET team, created `AsyncLazy<T>`. To add it to the `SticksAndStones.App` project, create a new file named `AsyncLazy~1.cs` and replace the contents with the following:

```
using System.Runtime.CompilerServices;

namespace SticksAndStones;

// AsyncLazy<T>, Microsoft, Stephen Toub, .NET Parallel Programming
Blog, https://devblogs.microsoft.com/pfxteam/asynclazyt/

public class AsyncLazy<T> : Lazy<Task<T>>
{
    public AsyncLazy(Func<T> valueFactory) :
        base(() => Task.Factory.StartNew(valueFactory))
    { }

    public AsyncLazy(Func<Task<T>> taskFactory) :
        base(() => Task.Factory.StartNew(() => taskFactory()).
Unwrap())
    { }

    public TaskAwaiter<T> GetAwaiter() { return Value.GetAwaiter(); }
}
```

> **Learn more about AsyncLazy<T>**
>
> Visit the .NET blog to learn more about how Stephen Toub created the `AsyncLazy<T>` class: `https://devblogs.microsoft.com/pfxteam/asynclazyt/`.

That completes the changes needed to start implementing the `ServiceConnection` class. To create the class, follow these steps:

1. Create a new class named `ServiceConnection` in the `SticksAndStones.App` project in the `Services` folder.

    Change the class to `public sealed` and inherit from `IDisposable`:
    ```
    public sealed class ServiceConnection : IDisposable
    ```

2. Modify the namespace declarations at the top of the file to the following:
    ```
    using Microsoft.AspNetCore.Http.Connections.Client;
    using Microsoft.AspNetCore.SignalR.Client;
    using Microsoft.Extensions.Logging;
    using SticksAndStones.Models;
    using System.Net;
    using System.Net.Http.Json;
    using System.Text.Json;
    ```

These are needed to reference the types needed in the following steps.

3. Add the following `private` fields to the class:

    ```
    private readonly ILogger log;
    private readonly HttpClient httpClient;
    private readonly JsonSerializerOptions serializerOptions;
    ```

    The `serializerOptions` is used to make sure the JSON that is sent and received from the Azure Functions functions can be serialized and deserialized properly.

4. Now, add a `public` property named Hub. The type for Hub is `AsyncLazy<HubConnection>`. `HubConnection` is the type from the SignalR client library that is used to receive messages from the SignalR service. The property should look like the following:

    ```
    public AsyncLazy<HubConnection> Hub { get; private set; }
    ```

    `HubConnection` is initialized in the `ConnectHub` method. But first, let's add the constructor.

5. The constructor for the `ServiceConnection` class has two parameters: `ILogger<ServiceConnection>` and a `Settings` parameter. In the body of the constructor, the `private` fields created in *step 3* are initialized as follows:

    ```
    public ServiceConnection(ILogger<ServiceConnection> logger,
    Settings settings)
    {
        httpClient = new()
        {
            BaseAddress = new Uri(settings.ServerUrl)
        };
        httpClient.DefaultRequestHeaders.Accept.
    Add(new("application/json"));

        serializerOptions = new
    JsonSerializerOptions(JsonSerializerDefaults.Web);

        log = logger;
    }
    ```

    The `logger` and `settings` parameters are provided by the .NET MAUI dependency injection service. The `httpClient` field is initialized and it's `BaseAddress` is assigned the settings `ServerUrl` property as a URI. Then, `DefaultHeaders` is modified to indicate to the server that the results are expected to be in JSON format. The `serializerOptions` instance is initialized to the defaults for `Web`, which is consistent with the formatting used by Azure Functions. Finally, the `log` field is initialized with the `logger` parameter value.

6. Now, let's implement the `Dispose` method. It will clean up any values that will potentially hold on to any native resources, such as networks, file handles, and so on. The two values that

this class has references to that need to be disposed of are `httpClient` and Hub. Note that we will not have to call `Dispose` ourselves as the .NET MAUI dependency injection system will do that. Add the following code to the `ServiceConnection` class:

```
public void Dispose()
{
    httpClient?.Dispose();
    Hub?.Value?.Dispose();
    GC.SuppressFinalize(this);
}
```

7. Now, add the class to dependency injection by adding the following highlighted line of code to the `MauiProgram.cs` file:

```
builder.Services.AddSingleton<Services.Settings>();
builder.Services.AddSingleton<Services.ServiceConnection>();

return builder.Build();
```

8. Initializing the Hub property will happen in the `ConnectHub` method. The configuration for the SignalR SignalR Hub connection is returned to the app in the `Connect` function result. Since we haven't and won't make that call before this class is constructed, we can't create the Hub in the constructor. The configuration is needed before you can initialize the Hub instance. The `ConnectHub` method has a single parameter of `ConnectionInfo`. Add the method using the following code snippet:

```
public void ConnectHub(ConnectionInfo config)
{
    Hub = new(async () =>
    {
        var connectionBuilder = new HubConnectionBuilder();

        connectionBuilder.WithUrl(config.Url,
(HttpConnectionOptions obj) =>
        {
            obj.AccessTokenProvider = async () => await Task.
FromResult(config.AccessToken);
        });
        connectionBuilder.WithAutomaticReconnect();
        var hub = connectionBuilder.Build();
        await hub.StartAsync();
        return hub;
    });
}
```

This method initializes the Hub property to a new `AsyncLazy<HubConnection>` instance. The constructor for `AsyncLazy<T>` takes `Func<T>`, which is provided using the anonymous method syntax. The anonymous method is also decorated as an `async` method, meaning that it will contain an awaited method call. The anonymous method takes no parameters and, in the body, starts by creating a new `HubConnectionBuilder`. Then, the `WithUrl` extension method is called on `HubConnectionBuilder` to set the URL for the SignalR service and provide the `AccessToken` value needed to make the connection. `AccessTokenProvider` is a `Task<string>` so the `config.AccessToken` is provided through another `async` anonymous function. The `WithAutomaticReconnect` method sets `HubConnection` instance to automatically try reconnecting the SignalR service if the connection is lost. If `WithAutomaticReconnect` isn't called, then the app is responsible for reconnecting if the connection is lost. The `HubConnection` instance is created by calling `HubConnectionBuilder.Build`. The Hub instance is then started with `StartAsync`, which is awaited, and then the Hub is returned. The thing to remember here is that when `ConnectHub` is called, the anonymous function isn't executed. The method won't be called until the first time a property or method of the Hub property is accessed.

The `ServiceConnection` class contains two helper functions that are used from the `GameService` class to make HTTP requests to the Azure Functions service. The first, `GetAsync<T>`, takes two parameters: a URL and a dictionary of query parameters to pass along with the URL. It returns an instance, `T`, and `AsyncError` as a `Tuple`. The `GetAsync` method will use the GET HTTP method when making the HTTP request. The other helper method, `PostAsync<T>`, uses the POST HTTP method and accepts two parameters: a URL, and an object to send in the body of the request formatted as JSON. It will return an instance of `T` from the response.

The `GetAsync<T>` and `PostAsync<T>` use a couple of helper methods; use the following code snippet to add them to the `ServiceConnection` class:

```
UriBuilder GetUriBuilder(Uri uri, Dictionary<string, string>
parameters)
=> new(uri)
{
    Query = string.Join("&",
    parameters.Select(kvp =>
            $"{kvp.Key}={kvp.Value}"))
};

async ValueTask<AsyncError?> GetError(HttpResponseMessage
responseMessage, Stream content)
{
    AsyncError? error;
    if (responseMessage.StatusCode == HttpStatusCode.Unauthorized)
    {
```

```
        log.LogError("Unauthorized request {@Uri}", responseMessage.
RequestMessage?.RequestUri);
        return new()
        {
            Message = "Unauthorized request."
        };
    }

    try
    {
        error = await JsonSerializer.
DeserializeAsync<AsyncError>(content, serializerOptions);
    }
    catch (Exception e)
    {
        error = new AsyncExceptionError()
        {
            Message = e.Message,
            InnerException = e.InnerException?.Message,
        };
    }

    log.LogError("{@Error} {@Message} for {@Uri}", responseMessage.
StatusCode, error?.Message, responseMessage?.RequestMessage?.
RequestUri);
    return error;
}
```

The `GetUriBuilder` method will return a new `UriBuilder` from the provided URL and `Dictionary` of key-value pairs to use in the query string. The `GetError` method will return either an `AsyncError` object or `AsyncExceptionError` object based on the status code or the contents of the response from the HTTP method call.

Now, we can add the `GetAsync<T>` method to the `ServiceConnection` class using the following code:

```
public async Task<(T Result, AsyncError Exception)> GetAsync<T>(Uri
uri, Dictionary<string, string> parameters)
{
    var builder = GetUriBuilder(uri, parameters);
    var fullUri = builder.ToString();
    log.LogDebug("{@ObjectType} Get REST call @{RestUrl}", typeof(T).
Name, fullUri);
    try
    {
```

```
        var responseMessage = await httpClient.GetAsync(fullUri);
        log.LogDebug("Response {@ResponseCode} for {@RestUrl}",
responseMessage.StatusCode, fullUri);
        if (responseMessage.IsSuccessStatusCode)
        {
            try
            {
                var content = await responseMessage.Content.
ReadFromJsonAsync<T>();

                log.LogDebug("Object of type {@ObjectType} parsed for
{@RestUrl}", typeof(T).Name, fullUri);
                return (content, null);
            }
            catch (Exception e)
            {
                log.LogError("Error {@ErrorMessage} for when parsing
${ObjectType} for {@RestUrl}", e.Message, typeof(T).Name, fullUri);
                return (default, new AsyncExceptionError()
                {
                    InnerException = e.InnerException?.Message,
                    Message = e.Message
                });
            }
        }
        log.LogDebug("Returning error for @{RestUrl}", fullUri);
        return (default, await GetError(responseMessage, await
responseMessage.Content.ReadAsStreamAsync()));
    }
    catch (Exception e)
    {
        log.LogError("Error {@ErrorMessage} for REST call ${ResUrl}",
e.Message, fullUri);
        // The service might not be happy with us, we might have
connection issues etc..
        return (default, new AsyncExceptionError()
        {
            InnerException = e.InnerException?.Message,
            Message = e.Message
        });
    }
}
```

While this method is a little long, what it is doing is not all that complicated. First, it uses the
GetUriBuilder method to create the UriBuilder instance and build the fullUri string value
from that. Then, it makes an HTTP GET call using the HttpClient instance to the URL. If there

is any failure, the exception handler will catch it and return an `AsynExceptionError`. If there are no errors in making the request and the response code indicates success, then the result is processed and returned. Otherwise, the result content is read for an error, and if it is found, it is returned. When the `GetAsync<T>` method returns, it will always return two items: the response of the `T` type and `AsyncError`. If either one of them isn't present, then their default value is returned or `null`.

Review and add the following code snippet to the `ServiceConnection` class to implement the `PostAsync<T>` method:

```
public async Task<(T Result, AsyncError Exception)> PostAsync<T>(Uri
uri, object parameter)
{
    log.LogDebug("{@ObjectType} Post REST call @{RestUrl}", typeof(T).
Name, uri);
    try
    {
        var responseMessage = await httpClient.PostAsJsonAsync(uri,
parameter, serializerOptions);
        log.LogDebug("Response {@ResponseCode} for {@RestUrl}",
responseMessage.StatusCode, uri);
        await using var content = await responseMessage.Content.
ReadAsStreamAsync();
        if (responseMessage.IsSuccessStatusCode)
        {
            if(string.IsNullOrEmpty(await.responseMessage.Content.
ReadAsStringAsync()))
                return (default, null);

            try
            {
                log.LogDebug("Parse {@ObjectType} SUCCESS for {@
RestUrl}", typeof(T).Name, uri);
                var result = await responseMessage.Content.
ReadFromJsonAsync<T>();
                log.LogDebug("Object of type {@ObjectType} parsed for
{@RestUrl}", typeof(T).Name, uri);
                return (result, null);
            }
            catch (Exception e)
            {
                log.LogError("Error {@ErrorMessage} for when parsing
${ObjectType} for {@RestUrl}", e.Message, typeof(T).Name, uri);
                return (default, new AsyncExceptionError()
                {
                    InnerException = e.InnerException?.Message,
```

```
                    Message = e.Message
                });
            }
        }
        log.LogDebug("Returning error for @{RestUrl}", uri);
        return (default, await GetError(responseMessage, content));
    }
    catch (Exception e)
    {
        log.LogError("Error {@ErrorMessage} for REST call ${ResUrl}",
e.Message, uri);
        // The service might not be happy with us, we might have
connection issues etc..
        return (default, new AsyncExceptionError()
        {
            InnerException = e.InnerException?.Message,
            Message = e.Message
        });
    }
}
```

This method is mostly the same as GetAsync<T> with a couple of minor changes. First, it does not need to call GetUriBuilder to add the parameters to the Uri QueryString, as the parameters are sent as part of the request body. Second, it uses the HTTP POST method instead of GET. With those changes, much of the method is error handling to make sure that we return the right data.

And that completes the ServiceConnection class. ServiceConnection and the Settings service classes will be used in the next section, where we create the GameService class.

Creating the GameService class

The GameService class is a layer between the UI and the network. It uses the ServiceConnection class, which handles the specific network calls to create the logic we need to interact with the Azure Functions functions. For each of the functions that we created in *Chapter 9*, there is a corresponding method in the GameService class to make the call to the function and return the result, if any.

Follow these steps to create and initialize the class:

1. Create a new class named GameService in the SticksAndStones.App project under the Services folder.

2. Change the class definition to public sealed and inherit from the IDisposable interface:

    ```
    public sealed class GameService : IDisposable
    ```

3. Add the following namespace declarations to the top of the file:

```
using System.Collections.ObjectModel;
using CommunityToolkit.Mvvm.Messaging;
using Microsoft.AspNetCore.SignalR.Client;
using SticksAndStones.Messages;
using SticksAndStones.Models;
```

4. The `GameService` class will depend on both the `Settings` service and the `ServiceConnection` service, so we need to add them to the constructor and store the references in class fields, as shown here:

```
private readonly ServiceConnection service;
private readonly Settings settings;

public GameService(Settings settings, ServiceConnection service)
{
    this.service = service;
    this.settings = settings;
}
```

Add the preceding code snippet to the `GameService` class. .NET MAUI will provide the `Settings` and `ServiceConnection` instances through dependency injection.

5. Implement the `IDisposable` interface by adding the following method to the GameService class:

```
public void Dispose()
{
    service.Dispose();
    GC.SuppressFinalize(this);
}
```

6. Now, add the class to dependency injection by adding the following highlighted line of code to the `MauiProgram.cs` file:

```
#if DEBUG
            builder.Logging.AddDebug();
#endif
            builder.Services.AddSingleton<Services.Settings>();
            builder.Services.AddSingleton<Services.
ServiceConnection>();

            builder.Services.AddSingleton<Services.
GameService>();

            return builder.Build();
```

We will start with the `Connect` method. `Connect` will accept a `Player` object to connect as and return an updated `Player` object. Additionally, if the connection is successful, it will configure the SignalR Hub. To create the `Connect` function, follow these steps:

1. Create a `private` field of `SemaphoreSlim` called `semaphoreSlim` and initialize the field with a new instance with an initial and maximum count of 1:

    ```
    public sealed class GameService : IDisposable
    {
        private readonly SemaphoreSlim semaphoreSlim = new(1, 1);

        private readonly ServiceConnection service;
    ```

    The `SemaphoreSlim` class is a great way to limit the number of threads performing an operation at a time. In our case, we only want one thread making the network calls at a time. It will be used in all the methods that make network calls from the `GameService` class.

2. `GameService` will track the current player in a `public` property called `CurrentPlayer`; add the property to the class using the following code:

    ```
    private readonly Settings settings;

    public Player CurrentPlayer { get; private set; } = new Player()
    { Id = Guid.Empty, GameId = Guid.Empty };
    ```

    The property is initialized to an empty `Player` object.

3. Once the user has connected as a player, we will also need somewhere to store the list of online players. To do that, add the following property to the `GameService` class:

    ```
    public ObservableCollection<Player> Players { get; } = new();
    ```

4. The `GameService` class also tracks the current status of the connection in a property called `IsConnected`; add the property using the following code snippet to the `GameService` class:

    ```
    public bool IsConnected { get; private set; }
    ```

5. Add a `public async` method named `Connect` to the `GameService` class. It should return `Task<Player>` and take a single `Player` as a parameter, as shown:

    ```
    public async Task<Player> Connect(Player player)
    {
    }
    ```

6. Within the `Connect` method, the first step is to make sure there is only one thread operating in the method at a time:

    ```
    await semaphoreSlim.WaitAsync();
    ```

This uses the `async/await` structures in C# to create a lock that only releases when there are enough open slots in `SemaphoreSlim`. Since the `SemaphoreSlim` instance was only initialized with a single slot, only one thread can process the `Connect` method at a time.

7. To make sure the `SemaphoreSlim` instance releases the slot, we need to add exception handling around the rest of the method and call `Release` at the end. Add the following code snippet to the `Connect` method:

```
try
{

}
finally
{
    semaphoreSlim.Release();
}
return CurrentPlayer;
```

The `try/finally` block ensures that we will always call `Release` at the conclusion of the method, which will prevent the `SemaphoreSlim` instance from being starved, preventing any additional thread from entering the method. Lastly, we return the value of `CurrentPlayer`, which we will set next within the `try` block.

There is another way to handle SemaphoreSlim

Using a `try/catch/finally` block works, but it is a little clunky if you handle all your exceptions properly, or don't have any. Tom Dupont has published a helper class on his blog that allows you to use a `using` statement to manage the lifetime of the `SemaphoreSlim` instance. You can read his post at `http://www.tomdupont.net/2016/03/how-to-release-semaphore-with-using.html`. Here is an example of using his extension:

```
using var handle = semaphoreSlim.UseWaitAsync();
```

---

8. Within the `try` block, add the following lines of code:

```
CurrentPlayer = player;

var (response, error) = await service.
PostAsync<ConnectResponse>(new($"{settings.ServerUrl}/Connect"),
new ConnectRequest(player));
if (error is null)
{
    service.ConnectHub(response.ConnectionInfo);

    response.Players.ForEach(Players.Add);
    CurrentPlayer = response.Player;
```

```
        IsConnected = true;
    }
    else
    {
        WeakReferenceMessenger.Default.
Send<ServiceError>(new(error));
    }
```

This block of code handles the call to the Connect function in the Azure Functions service. We start by setting the passed-in player details as the CurrentPlayer property. Then, the player instance is packaged into a ConnectRequest object and we pass that into a call to PostAsync<T> on the ServiceConnection instance. The URL is created from the ServerUrl property stored in the Settings service concatenated with /Connect. The response is expected to be of the ConnectResponse type and we store that in response. If we do not get any error, then we can call ConnectHub on the ServiceConnection instance, populate our Players collection, and set the CurrentPlayer property to the returned Player instance, which will have additional details from the server. If anything goes awry, then the error object will be populated, and we will send a message that contains that error to the UI.

ServiceError is the first message that we need to send to ViewModel instances from the GameService class. It is used to send errors from the ServiceConnection instance to ViewModel instances. We will add the ServiceError class in the next steps.

1. In the SticksAndStones.App project, create a new folder named Messages.

2. Create a new class named ServiceError in the Messages folder of the SticksAndStones. App project.

3. The ServiceError message is a simple wrapper around the AyncError object that can be used to send a message back to a view model. Replace the contents of the ServiceError. cs file with the following:

```
using CommunityToolkit.Mvvm.Messaging.Messages;

namespace SticksAndStones.Messages;

internal class ServiceError : ValueChangedMessage<AsyncError>
{
    public ServiceError(AsyncError error) : base(error)
    {
    }
}
```

4. Finally, since we are using `SemaphoreSlim` and it can hold onto native resources, we should make sure that those are released properly. Add the following highlighted code to the `Dispose` method to clean up the `semaphoreSlim` field:

```
public void Dispose()
{
    semaphoreSlim.Release();
    semaphoreSlim.Dispose();
    service.Dispose();
    GC.SuppressFinalize(this);
}
```

The concludes the `Connect` method for now.

The next three methods are called from the `Lobby` page. The first method is used to refresh the list of players. It is called when the user pulls down the list causing a refresh, or if the SignalR Hub is reconnected. To implement the `RefreshPlayerList` method, follow these steps:

1. the `RefreshPlayerList` method takes no arguments and returns `Task`; add the method to the `GameService` class as follows:

```
public async Task RefreshPlayerList()
{
    await semaphoreSlim.WaitAsync();
    try
    {

        var getAllPlayers = service.
GetAsync<GetAllPlayersResponse>(new($"{settings.ServerUrl}/
Players/GetAll"), new Dictionary<string, string> { { "id",
$"{CurrentPlayer.Id}" } });
        var (response, error) = await getAllPlayers;
        if (error is null)
        {
            Players.Clear();
            response.Players.ForEach(Players.Add);
        }
        else
        {       WeakReferenceMessenger.Default.
Send<ServiceError>(new(error));
        }
    }
    finally
    {
        semaphoreSlim.Release();
```

```
            }
        }
```

2. To refresh the list of players when the SignalR Hub reconnects, add the following highlighted code to the Connect method:

```
if (error is null)
{
    service.ConnectHub(response.ConnectionInfo);
    response.Players.ForEach(Players.Add);
    CurrentPlayer = response.Player;
    (await service.Hub).Reconnected += (s) => { return
RefreshPlayerList(); };

}
```

This line of code is interesting. First, we await service.Hub, then set the Reconnected event to an anonymous function that calls RefreshPlayerList. If you recall, the Hub property in the ServiceConnection class is AsyncLazy<T>. The first time we reference the Hub property, it will initialize itself, asynchronously, hence the await call.

The next method that is used from the Lobby page is IssueChallenge. The IssueChallenge method is called from the Lobby page when a player wishes to play a match against another player. The IssueChallenge method does not return any value since the actual response to the challenge will come back through the SignalR Hub. The method will send the request to the server and handle any errors, as shown here:

```
public async Task IssueChallenge(Player opponent)
{
    await semaphoreSlim.WaitAsync();
    try
    {
        var (response, error) = await service.
PostAsync<IssueChallengeResponse>(new($"{settings.ServerUrl}/
Challenge/Issue"), new IssueChallengeRequest(CurrentPlayer,
opponent));
        if (error is not null)
        {           WeakReferenceMessenger.Default.
Send<ServiceError>(new(error));
        }
    }
    finally
    {
        semaphoreSlim.Release();
    }
}
```

Add the preceding code to the GameService class. The SendChallengeResponse method, which is called when the opposing player responds to a challenge, is very similar to the IssueChallenge method, as shown:

```
public async Task SendChallengeResponse(Guid challengeId, Models.
ChallengeResponse challengeResponse)
{
    await semaphoreSlim.WaitAsync();
    try
    {
        var (response, error) = await service.
PostAsync<string>(new($"{settings.ServerUrl}/Challenge/Ack"), new
AcknowledgeChallengeRequest(challengeId, challengeResponse));
        if (error is not null)
        {        WeakReferenceMessenger.Default.
Send<ServiceError>(new(error));
        }
    }
    finally
    {
        semaphoreSlim.Release();
    }
}
```

Add the SendChallengeResponse method to the GameService class. That completes the methods needed to support the Lobby page. The final page in our app is the Game page. There are three more methods that are needed by the Game page. Follow these steps to add them:

1. Add the EndTurn method, which will send the player's move to the Game server, using the following code snippet:

```
public async Task<(Game?, string?)> EndTurn(Guid gameId, int
position)
{
    await semaphoreSlim.WaitAsync();
    try
    {
        var (response, error) = await service.
PostAsync<ProcessTurnResponse>(new($"{settings.ServerUrl}/
Game/Move"), new ProcessTurnRequest(gameId, CurrentPlayer,
position));
            if (error is not null)
            {
                return (null, error.Message);
            }
            else return (response.Game, null);
```

```
        }
        finally
        {
            semaphoreSlim.Release();
        }
    }
```

EndTurn is very similar to the IssueChallenge and SendChallengeResponse methods, with a minor exception: it returns the updated Game object and an error message if present.

2. the GetPlayerId method is a small helper function to search the Players list and return the Player instance that matches the ID passed in. Use the following code snippet to add the GetPlayerById method:

```
public Player? GetPlayerById(Guid playerId)
{
    if (playerId == CurrentPlayer.Id)
        return CurrentPlayer;
    return (from p in Players where p.Id == playerId select
p).FirstOrDefault();
}
```

3. the GetMatchById method is the last method that will make a call to the backend. In this case, it will retrieve a Match object given an ID. Using the following code snippet, add GetMatchById to the GameService class:

```
public async Task<Match> GetMatchById(Guid matchId)
{
    await semaphoreSlim.WaitAsync();
    try
    {
        var (response, error) = await service.
GetAsync<GetMatchResponse>(new($"{settings.ServerUrl}/Match/
{matchId}"), new());
        if (error != null) { }
        if (response.Match != null)
            return response.Match;
        return new Match();
    }
    finally
    {
        semaphoreSlim.Release();
    }
}
```

The final piece to the GameService class is the handling of the events received through the SignalR Hub. To refresh our memory from *Chapter 9*, the backend functions will send the following events to the clients via SignalR:

- PlayerUpdatedEventArgs
- ChallengeEventArgs
- GameStartedEventArgs
- GameUpdatedEventArgs

We will handle each of these events in the GameService method. To implement the handlers for these events, follow these steps:

1. When the Hub receives PlayerUpdatedEventArgs, we will need to update Player in the Players collection with the new values. We will create a helper function to handle that work, as follows:

    ```
    private void PlayerStatusChangedHandler(PlayerUpdatedEventArgs
    args)
    {
        var changedPlayer = (from player in Players
                             where player.Id == args.Player.Id
                             select player).FirstOrDefault();
        if (changedPlayer is not null)
        {
            changedPlayer.MatchId = args.Player.MatchId;
        }
        else if (args.Player.Id != CurrentPlayer.Id)
        {
            Players.Add(args.Player);
        }
    }
    ```

    The PlayerStatusChangedHandler method will locate the changed player in the Players collection and update the relevant fields of the instance or add it if it doesn't exist.

2. To call the PlayerStatusUpdateHandler class when the PlayerUpdated event is received, add the following highlighted code to the Connect method:

    ```
    if (error is null)
    {
        service.ConnectHub(response.ConnectionInfo);

        response.Players.ForEach(Players.Add);
        CurrentPlayer = response.Player;
    ```

```
    IsConnected = true;

    (await service.Hub).On<PlayerUpdatedEventArgs>(Constants.
Events.PlayerUpdated, PlayerStatusChangedHandler);

    (await service.Hub).Reconnected += (s) => { return
RefreshPlayerList(); };

}
```

The other three events will send messages to the ViewModel instance using WeakReferenceManager. First, we need to add the message types, using the following steps:

1. Add a new class named ChallengeReceived to the Messages folder in the SticksAndStones.App project.

2. Replace the contents of the ChallengeReceived.cs file with the following:

```
using CommunityToolkit.Mvvm.Messaging.Messages;
using SticksAndStones.Models;

namespace SticksAndStones.Messages;

public class ChallengeRecieved : ValueChangedMessage<Player>
{
    public Guid Id { get; init; }
    public ChallengeRecieved(Guid id, Player challenger) :
base(challenger)
    {
        Id = id;
    }
}
```

3. Add a new class named MatchStarted in the Messages folder of the SticksAndStones. App project.

4. Replace the contents of the MatchStarted.cs file with the following code:

```
using CommunityToolkit.Mvvm.Messaging.Messages;
using SticksAndStones.Models;

namespace SticksAndStones.Messages;

public class MatchStarted : ValueChangedMessage<Match>
{
    public MatchStarted(Match match) : base(match)
    {
```

```
        }
    }
```

5. Add a new class named `MatchUpdated` to the `SticksAndStones.App` project in the `Messages` folder.

6. Replace the contents of the `MatchUpdated.cs` file with the following code:

```csharp
using CommunityToolkit.Mvvm.Messaging.Messages;
using SticksAndStones.Models;

namespace SticksAndStones.Messages;

class MatchUpdated : ValueChangedMessage<Match>
{
    public MatchUpdated(Match match) : base(match)
    {
    }
}
```

7. To send a message when the event is received, add the following highlighted code to the `Connect` method in the `GameService` class:

```csharp
service.ConnectHub(response.ConnectionInfo);

response.Players.ForEach(Players.Add);
CurrentPlayer = response.Player;
IsConnected = true;

(await service.Hub).On<PlayerUpdatedEventArgs>(Constants.Events.
PlayerUpdated, PlayerStatusChangedHandler);
(await service.Hub).On<ChallengeEventArgs>(Constants.Events.
Challenge, (args) => WeakReferenceMessenger.Default.Send(new
ChallengeRecieved(args.Id, args.Challenger)));
(await service.Hub).On<MatchStartedEventArgs>(Constants.Events.
MatchStarted, (args) => WeakReferenceMessenger.Default.Send(new
MatchStarted(args.Game)));
(await service.Hub).On<MatchUpdatedEventArgs>(Constants.Events.
MatchUpdated, (args) => WeakReferenceMessenger.Default.Send(new
MatchUpdated(args.Game)));

(await service.Hub).Reconnected += (s) => { return
RefreshPlayerList(); };
```

That concludes the `GameService` class. We have all the methods needed to interact with the backend functions and we are handling the events that are being sent to the clients. The next portion of the chapter will add the pages needed to present the screens to the user, starting with the **Connect** page.

Creating the Connect page

The **Connect** page, as shown in *Figure 10.6*, is the first screen a user is presented with after the app loads. The page contains four main elements: an entry box for the player's gamer tag, an entry box for the player's email address, an image control for the player's avatar, and the **Connect** button.

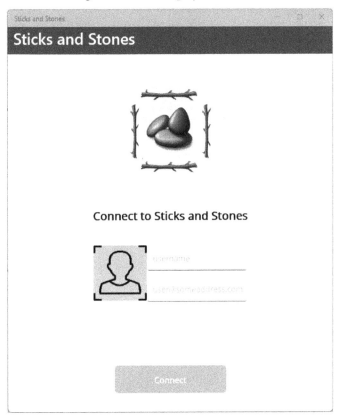

Figure 10.6 – The Connect page

The **Connect** page will consist of several parts:

- A ViewModel file called ConnectViewModel.cs

- A XAML file called ConnectView.xaml, which contains the layout

- A code-behind file called ConnectView.xaml.cs, which will carry out the data-binding process

- A XAML file containing the layout for a custom button control, called ActivityButton. xaml

- The code-behind for ActivityButton in ActivityButton.xaml.cs

We will begin the implementation of the **Connect** page by implementing `ConnectViewModel` first.

Adding ConnectViewModel

`ConnectViewModel` – along with `LobbyViewModel` and `GameView` model – will inherit from a single base class called `ViewModelBase`. The `ViewModelBase` class provides the necessary functionality to implement refreshing the page. Not all pages will use this feature, but it will be available. To add `ViewModelBase`, follow these steps:

1. Create a new folder named `ViewModels` in the `SticksAndStones.App` project.

2. Add a new class named `ViewModelBase` in the `ViewModels` folder.

3. Add the following namespace declarations at the top of the `ViewModelBase.cs` file:

    ```
    using CommunityToolkit.Mvvm.ComponentModel;
    using CommunityToolkit.Mvvm.Input;
    ```

4. Change the class declaration to `public abstract partial` and derive the class from `ObservableRecipient`:

    ```
    public abstract partial class ViewModelBase :
    ObservableRecipient
    ```

    `ObservableRecipient` comes from `CommunityToolkit`. If you have worked through the other chapters in this book, you will have seen view models that derive from `ObservableObject`, which implements `INotifyPropertyChanged`. `ObservableRecipient` extends `ObservableObject` and adds built-in support for working with implementations of the .NET MAUI `IMessage` interface. To learn more about `ObservableRecipient`, visit https://learn.microsoft.com/en-us/dotnet/communitytoolkit/mvvm/observablerecipient.

5. Add a `private bool` field named `canRefresh` with the `ObservableProperty` attribute:

    ```
    [ObservableProperty]
    private bool canRefresh;
    ```

6. Add a `private bool` field named `isRefreshing` with the `ObservableProperty` attribute:

    ```
    [ObservableProperty]
    private bool isRefreshing;
    ```

7. Add a `private` method named `CanExecuteRefresh` that takes no parameters and returns a `bool`. The method signature and implementation are in the following code snippet:

    ```
    private bool CanExecuteRefresh() => CanRefresh && !IsRefreshing;
    ```

8. Add a new `protected virtual` method named `RefreshInternal` that returns a `Task` and its implementation returns `Task.CompletedTask`, as shown here:

```
protected virtual Task RefreshInternal() => Task.CompletedTask;
```

9. Add the `Refresh` method as shown here:

```
[RelayCommand(CanExecute = nameof(CanExecuteRefresh))]
public async Task Refresh()
{
    IsRefreshing = true;
    await RefreshInternal();
    IsRefreshing = false;
    return;
}
```

The `Refresh` method is a `Command`, meaning that it can be bound to XAML elements as `RefreshCommand`. The `CanExecuteRefresh` method is used to determine the enabled/disabled state for the command. The command itself flips the `IsRefreshing` Boolean and calls the `RefreshInternal` method where classes derived from `ViewModelBase` would put the specific implementation.

Now that `ViewModelBase` has been implemented, we can implement `ConnectViewModel`. The `ConnectViewModel` class has bindable properties for the player's gamer tag and email address and various states for commands. Finally, there is a command to establish a connection to the game services. Let's implement the `ConnectViewModel` class by following these steps:

1. Create a new class named `ConnectViewModel` in the `ViewModels` folder.

2. Alter the class definition to `public partial` and derive from `ViewModelBase`, as shown here:

```
public partial class ConnectViewModel : ViewModelBase
{
}
```

3. `ConnectViewModel` depends on `GameService` and the `Settings` service so let's add a constructor to acquire them through dependency injection and `private` fields to store their values, as shown here:

```
public partial class ConnectViewModel : ViewModelBase
{
    private readonly GameService gameService;
    private readonly Settings settings;

    public ConnectViewModel(GameService gameService, Settings settings)
```

```
    {
            this.gameService = gameService;
            this.settings = settings;
    }
}
```

4. To use the `GameService` and `Settings` classes, you'll need to add a namespace declaration to the top of the file:

    ```
    using SticksAndStones.Services;
    ```

5. Add a `private string` field named `gamerTag` attributed with `ObservableProperty` to make it bindable, as shown in the following code snippet:

    ```
    [ObservableProperty]
    private string gamerTag;
    ```

6. Add a `private string` field named `emailAddress` attributed with `ObservableProperty` to make it bindable, as shown in this code snippet:

    ```
    [ObservableProperty]
    private string emailAddress;
    ```

7. In the constructor for `ConnectViewModel`, initialize the bindable properties from the last time the user connected, as shown in the following snippet:

    ```
    {
            this.gameService = gameService;
            this.settings = settings;

            // Load Player settings
            var player = settings.LastPlayer;
            Username = player.GamerTag;
            EmailAddress = player.EmailAddress;
    }
    ```

8. The **Connect** page does not need to refresh the view, so disable that functionality by adding the following line of code to the beginning of the constructor:

    ```
    CanRefresh = false;
    ```

9. To implement the `Connect` command, we will need four things: a `string` indicating the status of the command, a `bool` to indicate the current state of the command, a method to return if the command is enabled, and the method for the command itself. To add the status as a string, add the following code above the constructor in the `ConnectViewModel` class:

    ```
    [ObservableProperty]
    private string connectStatus;
    ```

We mark this field with `ObservableProperty` so that it is bindable into the view.

10. To add the `isConnecting` field to track the state of the command, add the following code under the `connectStatus` field:

```
[ObservableProperty]
private bool isConnecting;
```

11. The `CanExecuteConnect` method will return `true` if the command is enabled, and `false` if not. Add the method using the following code snippet under the `isConnecting` field:

```
private bool CanExecuteConnect() => !string.
IsNullOrEmpty(GamerTag) && !string.IsNullOrEmpty(EmailAddress)
&& !IsConnecting;
```

12. The `Connect` command will call a `Connect` method to establish the connection with the game server. This is mostly just to keep the methods small and manageable. Add the following private `Connect` method to the `ConnectViewModel` class:

```
private async Task<Player> Connect(Player player)
{
    // Get SignalR Connection
    var playerUpdate = await gameService.Connect(player);

    if (gameService.IsConnected)
    {
        // If the player has an in progress match, take them to
it.
        if (gameService.CurrentPlayer?.MatchId != Guid.Empty)
        {
            await Shell.Current.GoToAsync($"///Match", new
Dictionary<string, object>() { { "MatchId", gameService.
CurrentPlayer.MatchId } });
        }
        else
        {
            await Shell.Current.GoToAsync($"///Lobby");
        }
    }
    return playerUpdate;
}
```

This method will call the `Connect` method on the `GameService` class passing in the player details. If the connection is successful, then the user is navigated to the Lobby page, unless they are currently in a game, in which case, they are navigated to the Game page.

Navigation in .NET MAUI Shell

In .NET MAUI, navigation is performed by calling `GotoAsync` from the `Shell` object. The `Shell` object can be obtained by either casting `App.Current.MainPage` to a `Shell` object, or by using the `Shell.Current` property. The route passed to `GotoAsync` can be either relative to the current location or absolute. The valid forms of relative and absolute routes are as follows:

- `route` – The route will be searched for upward from the current position and, if found, pushed onto the navigation stack

- `/route` – The route will be searched for downward from the current position and, if found, pushed onto the navigation stack

- `//route` – The route will be searched for upward from the current position and, if found, will replace the navigation stack

- `///route` – The route will be searched for downward from the current position and, if found, will replace the navigation stack

To learn more about routes and navigation, visit `https://learn.microsoft.com/en-us/dotnet/maui/fundamentals/shell/navigation`.

13. Add the method implementing the `Connect` command to the bottom of the `Connect ViewModel` class using the following code snippet:

```
[RelayCommand(CanExecute = nameof(CanExecuteConnect))]
public async Task Connect()
{
    IsConnecting = true;
    ConnectStatus = "Connecting...";

    var player = settings.LastPlayer;

    player.GamerTag = GamerTag;
    player.EmailAddress = EmailAddress;

    player.Id = (await Connect(player)).Id;

    settings.LastPlayer = player;

    ConnectStatus = "Connect";
    IsConnecting = false;
}
```

The command is very straightforward. It sets the `IsConnecting` and `ConnectStatus` properties, then updates the `Player` values from the view. Then, it calls `Connect`, passing in the current `Player` instance. The ID of the returned player is captured and set back on

LastPlayer in Settings. Finally, the ConnectStatus and IsConnecting properties are set back to their defaults.

14. To wrap things up, we need to add a couple of attributes to make sure that properties are updated appropriately as values change. For instance, when the IsConnecting property is changed, we need to ensure that the CanExecuteConnect method is evaluated again. To do this, we add the NotifyCanExecuteChangeFor attribute to the IsConnecting field, as shown here:

```
[ObservableProperty]
[NotifyCanExecuteChangedFor(nameof(ConnectCommand))]
private bool isConnecting;
```

Since the gamerTag field and emailAddress field are also referenced in the CanExecuteConnect method, we should add the attribute to those fields as well, as shown here:

```
[ObservableProperty]
[NotifyCanExecuteChangedFor(nameof(ConnectCommand))]
private string gamerTag;

[ObservableProperty]
[NotifyCanExecuteChangedFor(nameof(ConnectCommand))]
private string emailAddress;
```

We have nearly completed ConnectViewModel. The final feature to implement is the handling of messages that we may receive from GameService. The ObservableObject implementation from CommunityToolkit provides a feature to make subscribing and unsubscribing to these messages clean. To implement the message handlers, follow these steps:

1. Add a new private method named OnServiceError to the ConnectViewModel class, using the following code snippet:

```
private void OnServiceError(AsyncError error)
{
    MainThread.BeginInvokeOnMainThread(async () =>
    {
        await Shell.Current.CurrentPage.DisplayAlert("There is a
problem...", error.Message, "Ok");
    });
}
```

2. We will subscribe to the ServiceError messages from the OnActivated event method of the ObservableObject class, Add the following method to the ConnectViewModel class to subscribe to the ServiceError message:

```
protected override void OnActivated() => Messenger.
Register<ServiceError>(this, (r, m) => OnServiceError(m.Value));
```

3. To unsubscribe from the `ServiceError` messages, add the following method to the `ConnectViewModel` class:

```
protected override void OnDeactivated() => Messenger.
Unregister<ServiceError>(this);
```

4. In the constructor for `ConnectViewModel`, we need to enable the `OnActivated` and `OnDeactivated` events that are raised by `ObservableObject`. These events are the recommended places to subscribe and unsubscribe to messages. Add the following line of code to the end of the constructor to enable the events:

```
IsActive = true;
```

Setting `IsActive` to `true` will cause the `OnActivated` event to fire. Setting it to `false` will cause the `OnDeactivated` event to fire.

5. To have the `OnDeactivated` event fire, we need to set `IsActive` to `false`. In the `Connect` method, add the highlighted line of code:

```
private async Task<Player> Connect(Player player)
{
    // Get SignalR Connection
    var playerUpdate = await gameService.Connect(player);

    if (gameService.IsConnected)
    {

        IsActive = false;

        // If the player has an in progress match, take them to
it.
        if (gameService.CurrentPlayer?.MatchId != Guid.Empty)
        {
            await Shell.Current.GoToAsync($"///Match", new
Dictionary<string, object>() { { "MatchId", gameService.
CurrentPlayer.MatchId } });
        }
        else
        {
            await Shell.Current.GoToAsync($"///Lobby");
        }
    }
    return playerUpdate;
}
```

6. Open the `MauiProgram.cs` file in the `SticksAndStones.App` project and add the following highlighted line to register `ConnectViewModel` with dependency injection:

```
builder.Services.AddSingleton<Services.GameService>();

builder.Services.AddTransient<ViewModels.ConnectViewModel>();
return builder.Build();
```

This concludes the implementation of `ConnectViewModel`. The `ConnectViewModel` class controls the entry of the user's gamer tag and email. It connects the user to the game server using their player details.

Adding the Connect view

The `Connect` view looks fairly simple, but there is a lot to it. We will break down the creation of the view into three sections:

- Creating the `ActivityButton` control:

  The `ActivityButton` control is the button used to initiate a connection to the backend services. While a simple button might do the trick, what if we had an animation that indicated the `connect` operation was in progress and the text of the button updated as well? That is what `ActivityButton` will do, in a reusable control.

- Creating the images:

  There are a few images used on this page. All of them were generated using AI. We'll explore how that was done.

- Building the view:

  This is where we bring `ActivityButton` together with our custom images and the built-in controls of .NET MAUI to make `ConnectView` appear as it does in the figures.

Let's start by building the `ActivityButton` control.

Creating the ActivityButton control

So, what is this `ActivityButton` control? It's basically a button with an `ActivityIndicator` that will only show up while the task behind the button is doing its work. The complexity of this control comes from the fact that we are making a general-purpose control instead of a specialized control. So, for all intents and purposes, it needs to act like a normal `Button` and `ActivityIndicator`. We are only going to implement the feature that we need for this application, but even then, it's still a reusable control.

From `Button`, we want to have the following XAML attributes:

- `Text`, `FontFamily`, and `FontSize`
- `Command` and `CommandParameter`

From `ActivityIndicator`, we will have `IsRunning`.

Each of these XAML elements will be bindable, like their original properties. An example of what the XAML might look like for declaring this control as an element would be as follows:

```
<controls:ActivityButton IsRunning="{Binding IsConnecting}"
                         Text="{Binding ConnectStatus}"
                         BackgroundColor="#e8bc65"
                         Command="{Binding ConnectCommand}"
                         HorizontalOptions="Center"
                         WidthRequest="200"
                         HeightRequest="48"/>
```

This listing comes from the actual XAML we will be creating for `ConnectView.xaml` later in this section.

The attributes that are copied from the two underlying controls need to be able to bind to the view model. This requires that they are implemented as bound properties. To create a bound property, you need two things: a property and a `BindableProperty` that references the property. The `BindableProperty` provides the functionality to keep the view model property, which implements `INotifyPropertyChanged`, with the property of the control. Let's create the Command bindable property as an example:

1. Create a new folder named `Controls` in the `SticksAndStones.App` project.

2. Add a new .NET MAUI `ContentView` (XAML) called `ActivityButton`.

3. Open the `ActivityButton.xaml.cs` file.

4. Create a new `public` `ICommand` property named Command, as shown in the following listing:

   ```
   public ICommand Command
   {
       get => (ICommand)GetValue(CommandProperty);
       set { SetValue(CommandProperty, value); }
   }
   ```

`BindableProperty` properties have a circular reference with the properties they are bound to, so you will get red squigglies until we complete the next step. This looks like almost every other property we have created, except that the `get` and `set` methods are just delegating to the `GetValue` and `SetValue` methods, respectively. `GetValue` and `SetValue` are provided by the `BindableObject` class, which `ContentView` ultimately inherits from. `GetValue` and `SetValue` are the `BindableObject` equivalents to `INotifyPropertyChanged` for view models. Calling them not only stores the values but also sends notifications that the value has changed.

5. Now, add the `BindableProperty` property for the `Command` property, using the following code snippet:

```
public static readonly BindableProperty CommandProperty =
BindableProperty.Create(
     propertyName: nameof(Command),
     returnType: typeof(ICommand),
     declaringType: typeof(ActivityButton),
     defaultBindingMode: BindingMode.TwoWay);
```

`CommandProperty` is of the `BindableProperty` type and is created by using the `Create` factory method of the `BindableProperty` class. We pass in the name of the property we are binding to (`Command`), the type that property returns (`Icommand`), the declaring type (which is `ActivityButton` in this case), and then what mode we want the binding to have. There are four options for `BindingMode`:

- `OneWay` – The default, propagates changes from the source (the view model) to the target (the control)

- `OneWayToSource` – This is the reverse of `OneWay`, propagating changes from the target (the control) to the source (the view model)

- `TwoWay` – This propagates changes in both directions

- `OneTime` – This propagates the changes only when `BindingContext` changes and all `INotifyPropertyChanged` events are ignored

These two pieces – the normal property that you would use in most of your C# classes, and `BindableProperty` – provide the complete functionality we need to create the custom control.

Now that we understand how to implement `BindableProperty` on a XAML control, we can complete the implementation of `ActivityButton`.

Let's start by updating the XAML and then we will follow that with the remaining `BindableProperty` implementations. The following steps will walk you through creating the control:

1. The template we chose to create the XAML and `.cs` files is not quite what we need for `ActivityButton`. We will need to alter the underlying root control from `ContentView` to `Frame`. We use `Frame` to wrap our layout with a border. Open the `ActivityButton.cs` file and update the class definition to inherit from `Frame` instead of `ContentView`, as follows:

```
public partial class ActivityButton : Frame
```

2. Now, open the `ActivityButton.xaml` file and modify it to look like the following:

```
<Frame  xmlns="http://schemas.microsoft.com/dotnet/2021/maui"
             xmlns:x="http://schemas.microsoft.com/winfx/2009/
xaml"
             x:Class="SticksAndStones.Controls.ActivityButton">
</Frame>
```

In addition to changing `ContentView` to `Frame`, also remove the contents of `Frame` as we won't be reusing it.

3. Let's name our control to make it easier to reference it later. Typically, in C#, if you want to reference the instance of the class, you will use the `this` keyword. That doesn't exist by default in XAML so add the `x:Name` attribute with the value of `this` to mimic C#.

4. Update the `Frame` element and add the `BackgroundColor` attribute with a value of `{x:StaticResource Primary}`. `Primary` is defined in the `Resources/Styles/Colors.xaml` file and we can reference it using the `StaticResource` extension method.

5. Update the `Frame` element and add the `CornerRadius` attribute with a value of 5. This will give our button rounded corners.

6. Add the `Padding` attribute with a value of 12 to the `Frame` element. This will ensure that there is plenty of whitespace around the control. The `Frame` element should now look like the following:

```
<?xml version="1.0" encoding="utf-8" ?>
<Frame xmlns="http://schemas.microsoft.com/dotnet/2021/maui"
             xmlns:x="http://schemas.microsoft.com/winfx/2009/
xaml"
             x:Class="SticksAndStones.Controls.ActivityButton"
        x:Name="this"
        BackgroundColor="{x:StaticResource Primary}"
        CornerRadius="5"
        Padding="12">
</Frame>
```

7. To get `ActivityIndicator` and `Label` centered side by side within `Frame`, we will use `HorizontalStackLayout` contained in `VerticalStackLayout`. `StackLayout` controls ignore the alignment options for the direction of the control, so, for example, `VerticalStackLayout` ignores the `VerticalOptions` property of its children and `HorizontalStackLayout` ignores the `HorizontalOptions` property of its children. This is because, by its nature, `HorizontalStackLayout` is in control of laying out its children in the horizontal plane, and the same is true for `VerticalStackLayout`, except in the vertical plane. Add the following highlighted code to the XAML:

```
<Frame xmlns="http://schemas.microsoft.com/dotnet/2021/maui"
             xmlns:x="http://schemas.microsoft.com/winfx/2009/
```

```
xaml"
                x:Class="SticksAndStones.Controls.ActivityButton"
        BackgroundColor="{x:StaticResource Primary}"
        CornerRadius="5"
        Padding="12">
    <VerticalStackLayout>
        <HorizontalStackLayout
HorizontalOptions="CenterAndExpand" Spacing="10">
        </HorizontalStackLayout>
    </VerticalStackLayout>
</Frame>
```

8. Within the `HorizontalStackLayout` element, add the following XAML:

```
<ActivityIndicator HeightRequest="15" WidthRequest="15"
                   Color="{x:StaticResource White}"
                   IsRunning="{Binding Source={x:Reference
this},Path=IsRunning}"
                   IsVisible="{Binding Source={x:Reference
this},Path=IsRunning}"
                   VerticalOptions="CenterAndExpand"/>
```

`ActivityIndicator` will have a `Height` and `Width` value of 15 and a `Color` value of `White`. The `IsRunning` and `IsVisible` properties are bound to the control's `IsRunning` property. We haven't created the `IsRunning` property yet, so this won't work until we do. The `x:Reference` markup extension allows us to bind the property to the parent control, which we named `this` in *step 3*.

9. Now, we can add `Label` within `HorizontalStackLayout` using the following XAML:

```
<Label x:Name="buttonLabel" TextColor="{x:StaticResource White}"
       Text="{Binding Source={x:Reference this},Path=Text}"
       FontSize="15"
       VerticalOptions="CenterAndExpand"
       VerticalTextAlignment="Center"
       HorizontalTextAlignment="Start" />
```

10. When the user taps or clicks anywhere in `Frame`, `Command` should be run. To configure that, we will use `GestureRecognizer`. `GestureRecognizer` is XAML's way of providing event handlers. There are several different kinds of `GestureRecognizer`:

   - `DragGestureRecognizer` and `DropGestureRecognizer`

   - `PanGestureRecognizer`

   - `PinchGestureRecognizer`

   - `PointerGestureRecognizer`

- SwipeGestureRecognizer

- TapGestureRecognizer

For `ActivityButton`, we are interested in `TapGestureRecognizer`. Since the action to take is undefined until this control is used on a view, `TapGestureRecognizer` will invoke a command when `Frame` is tapped. Add the following XAML to the `Frame` element to create `TapGestureRecognizer`:

```
<Frame.GestureRecognizers>
    <TapGestureRecognizer Command="{Binding Source={x:Reference
this},Path=Command}" CommandParameter="{Binding
Source={x:Reference this},Path=CommandParameter}" />
</Frame.GestureRecognizers>
```

The `Command` attribute and the `CommandParameter` attribute of `TapGestureRecognizer` are set to bind to the parent controls' `Command` and `CommandParameter` properties.

If the `IsRunning` property is `true`, then `Frame` should be disabled, and the reverse is true as well. `DataTrigger` is a XAML way of setting properties of one control based on changes in another control's properties. To add the triggers for `Frame`, add the highlighted XAML to the control:

```
<Frame.Triggers>
    <DataTrigger TargetType="Frame" Binding="{Binding
Source={x:Reference this},Path=IsBusy}" Value="True">
        <Setter Property="IsEnabled" Value="False" />
    </DataTrigger>
    <DataTrigger TargetType="Frame" Binding="{Binding
Source={x:Reference this},Path=IsBusy}" Value="False">
        <Setter Property="IsEnabled" Value="True" />
    </DataTrigger>
</Frame.Triggers>
```

11. That concludes the XAML portion of the control. Open the `ActivityButton.xaml.cs` file and we can add the missing properties, starting with `CommandParameter`:

```
public static readonly BindableProperty CommandParameterProperty
= BindableProperty.Create(
    propertyName: nameof(CommandParameter),
    returnType: typeof(object),
    declaringType: typeof(ActivityButton),
    defaultBindingMode: BindingMode.TwoWay);

public object CommandParameter
{
    get => GetValue(CommandParameterProperty);
```

```
    set { SetValue(CommandParameterProperty, value); }
}
```

Add the previous code listing to the `ActivityButton` class. Other than the name, there isn't anything significant about this property from the `Command` property. `CommandParameter` allows you to specify parameters to pass to `Command`, but using XAML.

12. The `Label` control is populated from the `Text` property. To add the `Text` property, add the following code to the `ActivityButton` class:

```
public static readonly BindableProperty TextProperty =
BindableProperty.Create(
    propertyName: nameof(Text),
    returnType: typeof(string),
    declaringType: typeof(ActivityButton),
    defaultValue: string.Empty,
    defaultBindingMode: BindingMode.TwoWay);

public string Text
{
    get => (string)GetValue(TextProperty);
    set { SetValue(TextProperty, value); }
}
```

In the case of the `Text` property, `returnType` has changed to `string`, but otherwise, it's similar to `Command` and `CommandParameter`.

13. The next property we need to implement is the `IsRunning` property, as follows:

```
public static readonly BindableProperty IsRunningProperty =
BindableProperty.Create(
    propertyName: nameof(IsRunning),
    returnType: typeof(bool),
    declaringType: typeof(ActivityButton),
    defaultValue: false);

public bool IsRunning
{
    get => (bool)GetValue(IsRunningProperty);
    set { SetValue(IsRunningProperty, value); }
}
```

14. To allow the size and font of the text to be changed, we implement the `FontSize` and `FontFamily` properties:

```
public static readonly BindableProperty FontFamilyProperty =
BindableProperty.Create(
```

```
        propertyName: nameof(FontFamily),
        returnType: typeof(string),
        declaringType: typeof(ActivityButton),
        defaultValue: string.Empty,
        defaultBindingMode: BindingMode.TwoWay);

    public string FontFamily
    {
        get => (string)GetValue(Label.FontFamilyProperty);
        set { SetValue(Label.FontFamilyProperty, value); }
    }

    public static readonly BindableProperty FontSizeProperty =
    BindableProperty.Create(
        nameof(FontSize),
        typeof(double),
        typeof(ActivityButton),
        Device.GetNamedSize(NamedSize.Small, typeof(Label)),
        BindingMode.TwoWay);

    public double FontSize
    {
        set { SetValue(FontSizeProperty, value); }
        get { return (double)GetValue(FontSizeProperty); }
    }
```

That completed `ActivityButton`. We will use `ActivityButton` in the *Creating the Connect view* section right after we create the images we need for the game.

Creating images using Bing Image Creator

There are a few images that are used in the game. They are as follows:

- A horizontal stick
- A vertical stick
- A pile of stones

Creating these images can be quite time-consuming and, based on your artistic abilities, not quite what you expected. It is quite possible that for your app you may opt to hire a graphics designer or artist to create your digital assets for you. Recently, a new option has become available, and that is to use AI to generate images. In this section, we will look at how to use **Bing Image Creator** to create the images that are needed for the game.

Bing Image Creator uses an English description of the scene that you would like to see and attempts to create it. There are a few keywords that you can use to direct the Image Creator in the artistic style of the image to create, such as *game art*, *digital art*, or *photorealistic*.

Let's get started by creating the stick image:

1. Open `https://bing.com/create` in Microsoft Edge or your favorite web browser.

2. If asked, log in with your Microsoft account. This can be the same account that you used in *Chapter 9* to log in to the Azure portal.

3. In the prompt box, type in the following prompt, then press **Create**:

    ```
    A single wood stick, positioned horizontally, with five stubs
    where branches would be and no leaves, no background, game art
    ```

    Image Creator will generate four different images based on your description. If you aren't satisfied with the result, adjust the description slightly and try again. The more descriptive you are, the better your result. Try to get a stick that is nearly vertical or horizontal since it will be easier to rotate and crop the image. It will also look much better if it is on a bright white background.

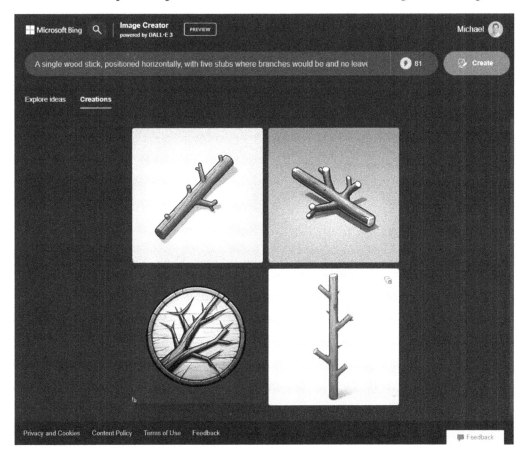

Figure 10.7 – An Image Creator sample set of images

4. Once you have an image you are satisfied with, click on the image to open it up.

5. Click the **Download** button to save the image to your local computer.

6. Now, open the downloaded file in your favorite image editor. The images created by Image Creator are roughly 1024 x 1024, and ideally, the image should be a 3:9 ratio, or around 300 x 900 pixels. Using your image editor tools, crop the image so that it is roughly 300 x 900 pixels.

7. Save the image as either `hstick.jpeg` if the stick is orientated horizontally, or `vstick.jpeg` if vertically, in the `Resources/Images` folder of the `SticksAndStones.App` project.

8. Using the same image editing tools, rotate the image 90 degrees so that it is the opposite orientation and save the image in the `Resources/Images` folder as `hstick.jpeg` if the stick is now orientated horizontally, or `vstick.jpeg` if vertically.

We have nearly half of the images that we need to create. Let's work on creating the stones next:

1. Open `https://bing.com/create` in Microsoft Edge or your favorite web browser.

2. If asked, log in with your Microsoft account. This can be the same account that you used in *Chapter 9* to log in to the Azure portal.

3. In the prompt box, type in the following prompt, then press **Create**:

```
3 grey stones, arranged closely together, no background, game
art
```

4. Work the prompt to get three stones nicely piled together, preferably on a white background, as shown here:

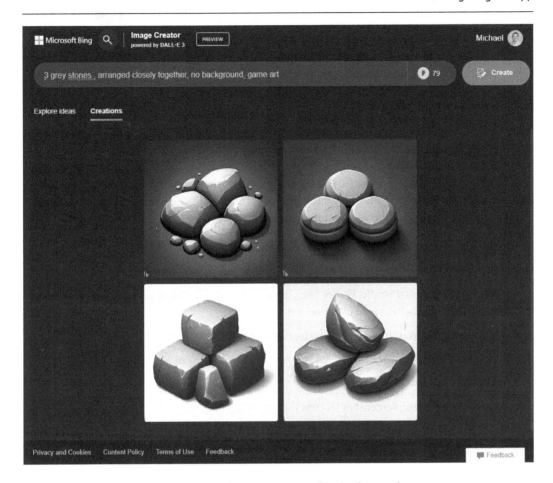

Figure 10.8 – Three stones on a white background

5. When you are satisfied with the generated image, click the image to open it.

6. Click the **Download** button to save the image to your local computer.

7. Now, open the downloaded file in your favorite image editor. Since the stones are supposed to be a square image, we can just save the file into the `Resources/Images` folder as `stones.jpeg`.

> **Don't have an image editor?**
>
> Don't have a favorite image editor? If you are on Windows, Paint does the job nicely, or you can use Visual Studio to edit images. On macOS, you can use Preview.

Excellent, we now have the sticks and the stones needed to play the game, and that concludes the use of Image Creator to generate our game's images. You can always go back to the site and review previous results, which is a nice feature. Now, we can move forward with creating the `Connect` view.

Creating the Connect view

The `Connect` view is the first UI other than the splash screen the user is going to see in the app. *Figure 10.6* provides a representation of what the final view will look like. The images may be different if you decide to generate your own of course. We will break this section into three parts. First, we will create the top portion of the view containing the static content, then move on to creating the middle portion of the view containing the entry controls, and then, finally, the **Connect** button. Let's get started with the top section of the view by following these steps:

1. In the `SticksAndStones.App` project, create a folder named `Views`.

2. Right-click on the `Views` folder, select **Add**, and then click **New Item....**

    If you are using Visual Studio 17.7 or later, click the **Show All Templates** button in the dialog that pops up; otherwise, move to the next step.

3. Under the **C# Items** node on the left, select **.NET MAUI**.

4. Select **.NET MAUI ContentPage (XAML)** and name it `ConnectView.xaml`.

5. Click **Add** to create the page.

    Refer to the following screenshot to view the preceding information:

Figure 10.9 – Adding a new .NET MAUI ContentPage (XAML)

6. Change the title of the view to `Sticks &amp; Stones`. Since XAML is a dialect of XML, the ampersand (`&`) must be escaped as `&amp;` in the string.

    Add the following highlighted namespaces to the `ContentView` element. They will provide us access to the classes in the `ViewModels`, `Controls`, and `Toolkit` namespaces:

    ```
    <ContentPage xmlns="http://schemas.microsoft.com/dotnet/2021/
    maui"
                    xmlns:x="http://schemas.microsoft.com/winfx/2009/
    xaml"
            xmlns:viewModels="clr-namespace:SticksAndStones.
    ViewModels"
            xmlns:controls="clr-namespace:SticksAndStones.Controls"
                    xmlns:toolkit="http://schemas.microsoft.com/
    dotnet/2022/maui/toolkit"
                    x:Class="SticksAndStones.Views.ConnectView"
                    Title="Sticks and Stones">
    ```

7. To make IntelliSense happy with the bindings we will be adding, define the view model that the view is using by adding the `x:DataType` attribute to the `ContentView` element, as shown:

    ```
    <ContentPage  xmlns="http://schemas.microsoft.com/dotnet/2021/
    maui"
                    xmlns:x="http://schemas.microsoft.com/winfx/2009/
    xaml"
            xmlns:viewModels="clr-namespace:SticksAndStones.
    ViewModels"
            xmlns:controls="clr-namespace:SticksAndStones.Controls"
                    xmlns:toolkit="http://schemas.microsoft.com/
    dotnet/2022/maui/toolkit"
                    x:Class="SticksAndStones.Views.ConnectView"
            x:DataType="viewModels:ConnectViewModel"
                    Title="Sticks and Stones">
    ```

8. We don't want the user to use any navigation, such as the `Shell`-provided `Back` button, other than what is provided on this page, so disable it using the highlighted code in the following listing:

    ```
    <ContentPage xmlns="http://schemas.microsoft.com/dotnet/2021/
    maui"
                    xmlns:x="http://schemas.microsoft.com/winfx/2009/
    xaml"
            xmlns:viewModels="clr-namespace:SticksAndStones.
    ViewModels"
            xmlns:controls="clr-namespace:SticksAndStones.Controls"
                    xmlns:toolkit="http://schemas.microsoft.com/
    dotnet/2022/maui/toolkit"
                    x:Class="SticksAndStones.Views.ConnectView"
            x:DataType="viewModels:ConnectViewModel"
    ```

```
Title="Sticks and Stones"
NavigationPage.HasNavigationBar="False">
```

9. Finally, set `BackgroundColor` of the entire view to `White`, which will make the images blend better, by adding the following highlighted code:

```
<ContentPage xmlns="http://schemas.microsoft.com/dotnet/2021/
maui"
                xmlns:x="http://schemas.microsoft.com/winfx/2009/
xaml"
        xmlns:viewModels="clr-namespace:SticksAndStones.
ViewModels"
        xmlns:controls="clr-namespace:SticksAndStones.Controls"
                xmlns:toolkit="http://schemas.microsoft.com/
dotnet/2022/maui/toolkit"
                x:Class="SticksAndStones.Views.ConnectView"
        x:DataType="viewModels:ConnectViewModel"
        Title="Sticks &amp; Stones"
        NavigationPage.HasNavigationBar="False"
        BackgroundColor="White">
```

10. To lay out the view, use a `Grid` control that has four rows defined; add the following code within the `ContentPage` element:

```
<Grid Margin="40">
    <Grid.RowDefinitions>
        <RowDefinition Height="8*"/>
        <RowDefinition Height="2*"/>
        <RowDefinition Height="8*"/>
        <RowDefinition Height="1*"/>
    </Grid.RowDefinitions>
</Grid>
```

`Grid` uses a `Margin` value of `40` to provide plenty of whitespace around the images and controls. The first row at 8 units will contain the logo for the app. The second row will contain the text `Sticks & Stones`. The third row will have the avatar image, email, and gamer tag entry controls. The final row will contain the `Connect` button.

Recall that the `Height` units are relative, so row 0, the first row, will be four times higher than row 1 and eight times higher than row 3, the final row. The `*` symbol in the `Height` value indicates that the row can expand if needed.

11. To arrange our generated images into a box-like layout, another `Grid` control is used. Add the following listing to the view between the `</Grid.RowDefinitions>` and `</Grid>` tags:

```
<Grid Grid.Row="0" WidthRequest="150" HeightRequest="150">
    <Grid.ColumnDefinitions>
```

```
        <ColumnDefinition Width="1*" />
        <ColumnDefinition Width="5*" />
        <ColumnDefinition Width="1*" />
    </Grid.ColumnDefinitions>
    <Grid.RowDefinitions>
        <RowDefinition Height="1*" />
        <RowDefinition Height="4*" />
        <RowDefinition Height="1*" />
    </Grid.RowDefinitions>

    <Image Grid.Row="0" Grid.Column="1" Source="hstick.jpeg"
Aspect="Fill"/>
    <Image Grid.Row="1" Grid.Column="0" Source="vstick.jpeg"
Aspect="Fill"/>
    <Image Grid.Row="1" Grid.Column="1" Source="stones.jpeg"
Aspect="AspectFit"/>
    <Image Grid.Row="1" Grid.Column="2" Source="vstick.jpeg"
Aspect="Fill"/>
    <Image Grid.Row="2" Grid.Column="1" Source="hstick.jpeg"
Aspect="Fill"/>
</Grid>
```

This Grid control defines three rows and three columns. The content of Grid is entirely made up of Image controls. Grid is positioned in row 0 of its parent Grid. The children of the grid, the Image controls, are positioned by setting the Grid.Row and Grid.Column attributes on the Image control. The stick images use the Aspect attribute set to Fill. Fill allows the image to scale to completely fill the content area; to do so, it may not scale uniformly along both the *x* and *y* axis. The stones use an Aspect value of AspectFit. This will uniformly scale the image till at least one side fits, which may cause letterboxing. There are two more options for the Aspect property: Center, which does no scaling, and AspectFill, which will scale until both axes have filled the view, which may cause clipping.

12. A Label element containing the text Connect to Sticks & Stones is added to the outer Grid control and it's placed in row 1, which is the second row of Grid. Add the following code to the outer Grid control after the inner Grid control added in *step 11*:

```
<Label Grid.Row="1" Text="Connect to Sticks &amp; Stones"
FontSize="20" TextColor="Black" FontAttributes="Bold"
Margin="0,0,0,20" HorizontalOptions="Center"/>
```

13. The next section of the page contains the avatar image, gamer tag entry, and email entry controls. HorizontalStackLayout and VerticalStackLayout controls are used to arrange the controls. Add the following snippet to the outer Grid control after the Label control added in *step 12*:

```
<HorizontalStackLayout Grid.Row="2" HorizontalOptions="Center">
    <VerticalStackLayout Spacing="10" >
```

```
        <Image HeightRequest="96" WidthRequest="96"
BackgroundColor="LightGrey">
            <Image.Source>
                <toolkit:GravatarImageSource
                    Email="{Binding EmailAddress}"
                    Image="MysteryPerson" />
            </Image.Source>
        </Image>
    </VerticalStackLayout>
    <VerticalStackLayout Spacing="10" >
        <Entry Placeholder="username" Keyboard="Email"
Text="{Binding Username}" HorizontalTextAlignment="Start"
HorizontalOptions="FillAndExpand"/>
        <Entry Placeholder="user@someaddress.
com" Keyboard="Email" Text="{Binding
EmailAddress}" HorizontalTextAlignment="Start"
HorizontalOptions="FillAndExpand"/>
    </VerticalStackLayout>
</HorizontalStackLayout>
```

The HorizontalStackLayout control is assigned to the Grid row 2, the third row. It is also centered horizontally within the row. The first VerticalStackLayout arranges the controls that make up the avatar. It contains an Image element, whose source is set to an instance of GravatarImageSource. GravatarImageSource uses the **Gravatar** service (https:// gravatar.com) to provide avatars when given an email address. The EmailAddress property of ConnectViewModel is bound to the Email property of GravatarImageSource. The image will automatically update on changes to EmailAddress. The Image property uses the MysteryPerson value to provide a plain profile when there isn't a Gravatar available for the email address. The second VerticalStackLayout contains two Entry controls: the first, for the gamer tag, is bound to the Username property of ConnectViewModel, and the second is bound to the EmailAddress property of ConnectViewModel. The Keyboard attribute determines which virtual keyboard is displayed when the focus is on the control. See https://learn.microsoft.com/en-us/dotnet/maui/user-interface/controls/entry#customize-the-keyboard for more information on customizing the keyboard.

14. The final control to add to ConnectView is the Connect button. Use the following snippet to add the button to the view:

```
<controls:ActivityButton Grid.Row="3"
                         IsRunning="{Binding IsConnecting}"
                         Text="{Binding ConnectStatus}"
                         BackgroundColor="#e8bc65"
                         Command="{Binding ConnectCommand}"
                         HorizontalOptions="Center"
```

```
                         WidthRequest="200"
                         HeightRequest="48"/>
```

The `Connect` button uses the `ActivityButton` control created in the *Creating the ActivityButton control*. The control is positioned in row 3, the fourth row, and the `IsRunning` attribute is bound to the `ConnectViewModel.IsConnecting` method. The `Text` attribute of the button is bound to the `ConnectViewModel.ConnectStatus` property, and finally, `Command` is bound to the `ConnectViewModel.Connect` method.

15. Open the `MauiProgram.cs` file in the `SticksAndStones.App` project and add the following highlighted line to register `ConnectView` with dependency injection:

```
builder.Services.AddSingleton<Services.GameService>();

builder.Services.AddTransient<ViewModels.ConnectViewModel>();
builder.Services.AddTransient<Views.ConnectView>();

return builder.Build();
```

16. Now, we need to consume the `ConnectViewModel` instance through dependency injection and set it as the binding object. Open the `ConnectView.Xaml.cs` file and modify it as follows:

```
using SticksAndStones.ViewModels;

namespace SticksAndStones.Views;

public partial class ConnectView : ContentPage
{
    public ConnectView(ConnectViewModel viewModel)
    {
        this.BindingContext = viewModel;
        InitializeComponent();
    }
}
```

17. Finally, we need to set `ConnectView` as the first view displayed. Open the `AppShell.xaml` file in the `SticksAndStones.App` project and update the contents of the `Shell` element as shown:

```
<Shell
    x:Class="SticksAndStones.App.AppShell"
    xmlns="http://schemas.microsoft.com/dotnet/2021/maui"
    xmlns:x="http://schemas.microsoft.com/winfx/2009/xaml"
    xmlns:local="clr-namespace:SticksAndStones.App"
    xmlns:views="clr-namespace:SticksAndStones.Views"
    Shell.FlyoutBehavior="Disabled">
```

```
<ShellItem Route="Connect">
    <ShellContent ContentTemplate="{DataTemplate
views:ConnectView}" />
    </ShellItem>

</Shell>
```

The first of the three views in the app is complete. To test it out, follow these steps:

1. In Visual Studio, right-click the `SticksAndStones.Functions` project in **Solution Explorer**, then select **Debug | Start Without Debugging**.

2. In **Solution Explorer**, right-click the `SticksAndStones.App` project and select **Set as Startup Project**.

3. Press *F5* to start the `SticksAndStones.App` project using the debugger.

The **Connect** page is the first page our users will see when they launch the app. We have added image controls and entry controls, used data binding to bind to our view model, and extended XAML by creating a custom control and a converter. In the next section, we will add the `Lobby` page, which will allow us to start games with other players.

Creating the Lobby page

The `Lobby` page displays the list of connected players and allows a player to challenge another to a match. As additional players connect to the server, they are added to the list of available players. *Figure 10.10* shows the two views for the page, one with connected players, and the empty view when there are no additional players connected.

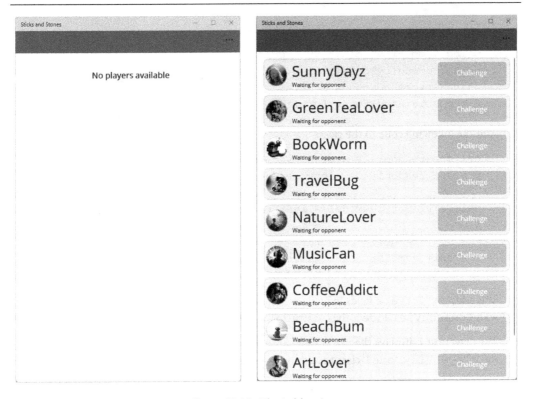

Figure 10.10 – The Lobby views

Each player is displayed on a card with their avatar image, gamer tag, status, and a button to allow the player to challenge the other to a match.

This page is comprised of two `ViewModel` classes, not one. As you might expect, there is the `LobbyViewModel` class, the `LobbyViewModel` which class has a collection of `PlayerViewModel` instances. In addition to the `ViewModel` classes, there is the `LobbyView` class. Let's get started by creating the `PlayerViewModel` class.

Adding PlayerViewModel

`PlayerViewModel` is very much like all our other `ViewModel` classes but with one slight difference: it isn't bound directly to a view in the same way. Otherwise, it has the same purpose: abstract the model, in this case, `Player`, away from the UI that displays it. `PlayerViewModel` provides all the needed binding properties to display each individual player card in `LobbyView`. To add `PlayerViewModel`, follow these steps:

1. In the `SticksAndStones.App` project, under the `ViewModels` folder, create a new class named `PlayerViewModel`.

2. Add the following namespaces to the top of the file:

```
using CommunityToolkit.Mvvm.ComponentModel;
using CommunityToolkit.Mvvm.Input;
using SticksAndStones.Models;
using SticksAndStones.Services;
```

3. Add the following code to the class:

```
private readonly Player playerModel;
private readonly GameService gameService;

public PlayerViewModel(Player player, GameService gameService)
{
    playerModel = player;
    this.gameService = gameService;
}
```

This adds two `private` fields to hold the values of the arguments passed to the constructor. As with `ConnectViewModel`, the constructor arguments are provided by dependency injection.

4. The player card displays the player's gamer tag, avatar, and status. Add the following code to the `PlayerViewModel` class to add the `Id` and `GamerTag` properties:

```
public Guid Id => playermodel.Id;

public string GamerTag => playerModel.GamerTag;
```

For `PlayerViewModel`, some of the properties that we bind to are not implemented with `ObservablePropertyAttribute`. That is because we are providing their values from the `Player` model directly. So, the `get` method of the property just returns the corresponding property of the model object. There is no defined `set` method, so this property is essentially a one-way data binding.

5. The `Status` property is a little different since it does not exist on our `Player` model. The `Status` property is a textual indication of whether the player is in a match or not. The `Player` model does have a `MatchId` property, so if the `Player` model has a valid `MatchId` (i.e., not `Guid.Empty`), then the status would be `"In a match"`; otherwise, that status would be `"Waiting for opponent"`. Add the following code to `PlayerViewModel` to implement the `Status` property:

```
public bool IsInMatch => !(playerModel.MatchId == Guid.Empty);

public string Status => IsInMatch switch
{
    true => "In a match",
```

```
        false => "Waiting for opponent"
    };
```

The IsInMatch property is used to simplify the Status property implementation. It will
also be used later in the class. The Status property is a simple switch on IsInMatch and
returns the proper string value.

6. To add a command to handle the Challenge button, add the following code to the
    PlayerViewModel class:

```
[ObservableProperty]
[NotifyPropertyChangedFor(nameof(ChallengeStatus))]
private bool isChallenging = false;

public string ChallengeStatus => IsChallenging switch
{
    true => "Challenging...",
    false => "Challenge"
};

public bool CanChallenge => !IsInMatch && !IsChallenging;

[RelayCommand(CanExecute = nameof(CanChallenge))]
public void Challenge(PlayerViewModel opponent)
{

    MainThread.BeginInvokeOnMainThread(async () =>
    {
        IsChallenging = true;
        bool answer = await Shell.Current.CurrentPage.
DisplayAlert("Issue Challenge!", $" You are about to challenge
{GamerTag} to a match!\nAre you sure?", "Yes", "No");
        if (answer)
        {
            await gameService.IssueChallenge(opponent.Player);
        }
        IsChallenging = false;
    });
    return;
}
```

The command is prevented from executing while it is currently waiting for a challenge response,
which makes sense – no need to nag the other player. The IsChallenging property is set
to true while challenging and false when it is complete. The CanChallenge property is
a combination of IsInMatch and IsChallenging, meaning that you can't challenge the

same player while you have an existing challenge in progress, and you can't challenge a player who is already in a match with another player. ChallengeStatus, which is used as the text for the button, is bound to the IsChallenging value and updates when that property is updated. You may have noticed that our command takes a single parameter. This is used to operate on the correct player.

That completes PlayerViewModel. Next, LobbyViewModel is used to encapsulate the collection of PlayerViewModel objects.

Adding LobbyViewModel

LobbyViewModel is a fairly straightforward implementation. It has a collection of PlayerViewModel objects that are exposed to the UI, it allows the user to pull to refresh the view, and it handles the messages of ChallengeReceived, MatchStarted, and ServiceError. Follow these steps to implement LobbyViewModel:

1. In the SticksAndStones.App project, inside the ViewModels folder, create a new class named LobbyViewModel.

2. Add the following namespaces to the top of the file:

```
using System.Collections.ObjectModel;
using System.Collections.Specialized;
using CommunityToolkit.Mvvm.Messaging;
using SticksAndStones.Messages;
using SticksAndStones.Models;
using SticksAndStones.Services;
```

3. Modify the class declaration to a public partial class that inherits from ViewModelBase, as shown:

```
public partial class LobbyViewModel : ViewModelBase
```

4. Add the following code to the LobbyViewModel class:

```
private readonly GameService gameService;

public ObservableCollection<PlayerViewModel> Players { get;
init; }

public LobbyViewModel(GameService gameService)
{
    this.gameService = gameService;
    Players = new(from p in gameService.Players
                  where p.Id != gameService.CurrentPlayer.Id
                  select new PlayerViewModel(p, gameService));
    CanRefresh = true;
```

```
        IsActive = true;
    }
```

LobbyViewModel receives an instance of GameService via dependency injection. The gameService instance is used to initialize the Players list. The Players property from the GameService class is a collection of the Player model, whereas Players in LobbyViewModel is an ObservableCollection instance of PlayerViewModel. We use ObservableCollection because it provides support for INotifyPropertyChanged and INotifyCollectionChanged when it is bound automatically. A LINQ query is used to get all the current players and add them to the Players ObservableCollection. CanRefresh from ViewModelBase is set to true, which enables RefreshCommand. Finally, IsActive is set to true, which enables the OnActivated and OnDeactivated events.

5.   The GameService.Players list will be updated as players connect to the server. However, these changes do not get propagated to the LobbyViewModel.Players collection automatically. By implementing a handler for the CollectionChanged event of the GameService.Players property, we can then update the LobbyViewModel.Players collection appropriately. Add the following method to the LobbyViewModel class:

```
private void OnPlayersCollectionChanged(object? sender,
NotifyCollectionChangedEventArgs e)
{
    if (e.Action == NotifyCollectionChangedAction.Add)
    {
        foreach (var player in e.NewItems.Cast<Player>())
        {
            Players.Add(new PlayerViewModel(player,
gameService));
        }
    }
    else if (e.Action == NotifyCollectionChangedAction.Remove)
    {
        foreach (var player in e.OldItems.Cast<Player>())
        {
            var toRemove = Players.FirstOrDefault(p => p.Id ==
player.Id);
            Players.Remove(toRemove);
        }
    }
    else if (e.Action == NotifyCollectionChangedAction.Replace)
    {
    }
    else if (e.Action == NotifyCollectionChangedAction.Reset)
    {
        Players.Clear();
```

```
            }
    }
```

The `OnPlayersCollectionChanged` method is an implementation of `Notify CollectionChangedEventHandler`. It is called by the `Observable Collection.CollectionChanged` event. The event is called whenever an item in the collection is added, removed, or updated. It is also called when the entire collection is cleared. This method handled the `NotifyCollectionChangedAction` values of `Add`, `Remove`, and `Reset`.

6. The `Players.CollectionChanged` event is assigned to the `OnPlayers CollectionChanged` method in the `OnActivated` event handler. Add the `OnActivated` and `OnDeactivated` methods using the following listing:

```
protected override void OnActivated()
{
    gameService.Players.CollectionChanged +=
OnPlayersCollectionChanged;

        // If the player has an in progress match, take them to it.
        if (gameService.CurrentPlayer?.MatchId != Guid.Empty)
        {
            MainThread.InvokeOnMainThreadAsync(async () =>
            {
                IsActive = false;
                await Shell.Current.GoToAsync(Constants.
ArgumentNames.MatchId, new Dictionary<string, object>() { {
"MatchId", gameService.CurrentPlayer.MatchId } });
            });
        }
}

protected override void OnDeactivated()
{
    gameService.Players.CollectionChanged -=
OnPlayersCollectionChanged;

}
```

In the `OnActivated` method, the `CollectionChanged` event is assigned to the `OnPlayersCollectionChanged` method and unassigned in the `OnDeactivated` method. In `OnActivated`, there is also a check to see whether the player is already in a match. If they are, then the app navigates to the `Match` view immediately. When navigating to the `Match` view, we send an argument for `Match`. This will be either the `MatchId` or the `Match` model.

7. Open the `Constants.cs` file in the `SticksAndStones.Shared` project to add the following code snippet to the `Constants` class:

```
public class ArgumentNames
{
    public static readonly string Match = nameof(Match);
    public static readonly string MatchId = nameof(MatchId);
}
```

8. While in the Lobby, there are three messages that need to be handled: `ChallengeReceived`, `MatchStarted`, and `ServerError`. Add the code in the following listing to add the handlers for each of these messages:

```
private void OnChallengeReceived(Guid challengeId, Player
opponent)
{
    MainThread.BeginInvokeOnMainThread(async () =>
    {
        bool answer = await Shell.Current.CurrentPage.
DisplayAlert("You have been challenged!", $"{opponent.GamerTag}
has challenged you to a match of Sticks & Stones, do you
accept?", "Yes", "No");
        await gameService.SendChallengeResponse(challengeId,
answer ? Models.ChallengeResponse.Accepted : Models.
ChallengeResponse.Declined);
    });
}

private void OnMatchStarted(Match match)
{
    MainThread.BeginInvokeOnMainThread(async () =>
    {
        IsActive = false;
        await Shell.Current.GoToAsync($"///Match", new
Dictionary<string, object>() { { Constants.ArgumentNames.Match,
match } });
    });
}

private void OnServiceError(AsyncError error)
{
    MainThread.BeginInvokeOnMainThread(async () =>
    {
        IsActive = false;
        await Shell.Current.CurrentPage.DisplayAlert("There is a
problem...",error.Message, "Ok");
```

```
      });
    }
```

In `OnChallengeReceived`, the user is prompted to accept or decline the challenge. Their response is then sent to the challenger via the `SendChallengeResponse` method of the `GameService` class. `OnMatchStarted` will navigate the user to the `Match` view. Finally, `OnServiceError` will display the error to the user.

9. Add the following snippet to the top of the `OnActivated` method to register to receive the messages:

```
Messenger.Register<ChallengeRecieved>(this, (r, m) =>
OnChallengeReceived(m.Id, m.Value));
Messenger.Register<MatchStarted>(this, (r, m) =>
OnMatchStarted(m.Value));
Messenger.Register<ServiceError>(this, (r, m) =>
OnServiceError(m.Value));
```

10. Add the following snippet to the end of the `OnDecactived` method to stop receiving messages:

```
Messenger.Unregister<ChallengeRecieved>(this);
Messenger.Unregister<MatchStarted>(this);
Messenger.Unregister<ServiceError>(this);
```

11. To refresh the `Players` list when the user pulls down on the list in the UI, add the following method to the `LobbyViewModel` class:

```
protected override async Task RefreshInternal()
{
    await gameService.RefreshPlayerList();
    return;
}
```

12. `LobbyViewModel` needs to be registered with dependency injection, so open the `MauiProgram.cs` file and add the following highlighted line of code:

```
builder.Services.AddTransient<ViewModels.ConnectViewModel>();
builder.Services.AddTransient<ViewModels.LobbyViewModel>();

builder.Services.AddTransient<Views.ConnectView>();
```

`LobbyViewModel` is now complete, and it is time to create the view!

Adding the Lobby view

The `Lobby` view simply displays a list of connected players with their avatar, gamertag and current status. To build the `LobbyView` follow these steps:

1. Right-click on the `Views` folder of the `SticksAndStone.App` project, select **Add**, and then click **New Item…**.

If you are using Visual Studio 17.7 or later, click the **Show all Templates** button in the dialog that pops up; otherwise, move to the next step.

2. Under the **C# Items** node on the left, select **.NET MAUI**.

3. Select **.NET MAUI ContentPage (XAML)** and name it `LobbyView.xaml`.

4. Click **Add** to create the page.

Refer to the following screenshot to view the preceding information:

Figure 10.11 – Adding a new .NET MAUI ContentPage (XAML)

5. Open the `LobbyView.xaml.cs` file and add the following `using` declaration:

```
using SticksAndStones.ViewModels;
```

6. Make the following highlighted changes to the constructor:

```
public LobbyView(LobbyViewModel viewModel)
{
    this.BindingContext = viewModel;
    InitializeComponent();
}
```

These changes allow for dependency injection to supply the `LobbyViewModel` instance to the view, which is then assigned to `BindingContext`.

7. Open the `AppShell.xaml` file and add the following code snippet to the `ContentPage` element:

```
<ShellItem Route="Lobby">
    <ShellContent ContentTemplate="{DataTemplate
views:LobbyView}" />
</ShellItem>
```

This registers the `"Lobby"` route and directs it to `LobbyView`.

8. Open the `MauiProgram.cs` file and add the following highlighted line of code:

```
builder.Services.AddTransient<Views.ConnectView>();
builder.Services.AddTransient<Views.LobbyView>();

return builder.Build();
```

This will register `LobbyView` with dependency injection so that `DataTemplate` can locate it.

9. Open the `LobbyView.xaml` file and change the `Title` attribute of the `ContentPage` element to `"Lobby"`.

10. Add the following highlighted namespaces to the `LobbyView` element; they will provide us access to the classes in the `ViewModels`, `Controls`, and `Toolkit` namespaces:

```
<ContentPage xmlns="http://schemas.microsoft.com/dotnet/2021/
maui"
             xmlns:x="http://schemas.microsoft.com/winfx/2009/
xaml"
       xmlns:viewModels="clr-namespace:SticksAndStones.
ViewModels"
       xmlns:controls="clr-namespace:SticksAndStones.Controls"
       xmlns:toolkit=" http://schemas.microsoft.com/
dotnet/2022/maui/toolkit"
             x:Class="SticksAndStones.Views.LobbyView">
```

11. To make IntelliSense happy with the bindings we will be adding, define the view model that the view is using by adding the `x:DataType` attribute to the `LobbyView` element, as shown:

```
<ContentPage xmlns="http://schemas.microsoft.com/dotnet/2021/
maui"
             xmlns:x="http://schemas.microsoft.com/winfx/2009/
xaml"
       xmlns:viewModels="clr-namespace:SticksAndStones.
ViewModels"
       xmlns:controls="clr-namespace:SticksAndStones.Controls"
       xmlns:toolkit=" http://schemas.microsoft.com/
dotnet/2022/maui/toolkit"
       x:DataType="viewModels:LobbyViewModel"
             x:Class="SticksAndStones.Views.LobbyView">
```

12. We don't want the user to use any navigation, such as the Shell-provided Back button, other than what is provided on this page, so disable it using the highlighted code in the following listing:

```
<ContentPage xmlns="http://schemas.microsoft.com/dotnet/2021/
maui"
             xmlns:x="http://schemas.microsoft.com/winfx/2009/
xaml"
        xmlns:viewModels="clr-namespace:SticksAndStones.
ViewModels"
        xmlns:controls="clr-namespace:SticksAndStones.Controls"
        xmlns:toolkit=" http://schemas.microsoft.com/
dotnet/2022/maui/toolkit"
        x:Class="SticksAndStones.Views.LobbyView"
        x:DataType="viewModels:LobbyViewModel"
        NavigationPage.HasNavigationBar="False">
```

13. Finally, set the BackgroundColor value of the entire view to White, which will make the images blend better, by adding the following highlighted code:

```
<ContentPage xmlns="http://schemas.microsoft.com/dotnet/2021/
maui"
                xmlns:x="http://schemas.microsoft.com/winfx/2009/
xaml"
        xmlns:viewModels="clr-namespace:SticksAndStones.
ViewModels"
        xmlns:controls="clr-namespace:SticksAndStones.Controls"
        xmlns:toolkit=" http://schemas.microsoft.com/
dotnet/2022/maui/toolkit"
            x:Class="SticksAndStones.Views.ConnectView"
        x:DataType="viewModels:ConnectViewModel"
        Title="ConnectView"
        NavigationPage.HasNavigationBar="False"
        BackgroundColor="White">
```

14. Replace the contents of ContentPage with the following code snippet:

```
<RefreshView IsRefreshing="{Binding IsRefreshing}"
Command="{Binding RefreshCommand}">
    <ScrollView Padding="5">
        <CollectionView ItemsSource="{Binding Players}"
Margin="5,5,5,0 SelectionMode="None">
        </CollectionView>
    </ScrollView>
</RefreshView>
```

For LobbyView, there is a vertical scrolling list of players. The root element is RefreshView. Its IsRefreshing attribute is bound to the IsRefreshing property of LobbyViewModel. The Command attribute for RefreshView is bound to RefreshCommand, which will end up executing the RefreshInternal method of LobbyViewModel. IsRefreshing and RefreshCommand are implemented in the BaseViewModel class. Inside RefreshView is ScrollView, which provides scrolling capability to have a large list. Inside ScrollView is CollectionView, which will display each Player instance as an individual item, so ItemsSource is bound to the Players property of LobbyViewModel. As there is no real need to select individual Player items, SelectionMode is set to none.

15. When the list is empty, it is nice to display something to the user so that they aren't left wondering what happened. CollectionView has an EmptyView property that is used to configure what is displayed when there are no items. Add the following code snippet immediately following the ContentPage start opening tag:

```
<ContentPage.Resources>
    <ContentView x:Key="BasicEmptyView">
        <StackLayout>
            <Label Text="No players available"
                   Margin="10,25,10,10"
                   FontAttributes="Bold"
                   FontSize="18"
                   HorizontalOptions="Fill"
                   HorizontalTextAlignment="Center" />
        </StackLayout>
    </ContentView>
</ContentPage.Resources>
```

This defines a page resource containing ContentView with a Key value of "BasicEmptyView". The view contains StackLayout, which has Label as a child with the text "No players available". Appropriate styling is applied to make sure it's large enough and has enough surrounding whitespace.

16. Add the following attribute to the CollectionView element:

```
EmptyView="{StaticResource BasicEmptyView}"
```

This binds BasicEmptyView to the EmptyView property of CollectionView. *Figure 10.12* shows the result if you run the app and log in:

Figure 10.12 – Lobby with no players

17. The player card will also use a static resource, which just makes the file a little easier to read, and less indenting. Add the following code snippet to the `ContentView.Resources` element, under the `BasicEmptyView` element:

```
<DataTemplate x:Key="PlayerCardViewTemplate"
x:DataType="viewModels:PlayerViewModel">
    <ContentView>
        <Border StrokeShape="RoundRectangle 10,10,10,10"
BackgroundColor="AntiqueWhite" Padding="3,3,3,3"
Margin="5,5,5,5">
            <Grid>
                <Grid.ColumnDefinitions>
                    <ColumnDefinition Width="50" />
                    <ColumnDefinition Width="4*" />
                    <ColumnDefinition Width="2*" />
                </Grid.ColumnDefinitions>
                <toolkit:AvatarView Grid.Column="0"
Margin="0" BackgroundColor="LightGrey" HeightRequest="48"
WidthRequest="48" CornerRadius="25" VerticalOptions="Center"
```

```
HorizontalOptions="Center">
                <toolkit:AvatarView.ImageSource>
                    <toolkit:GravatarImageSource
                        Email="{Binding EmailAddress}"
                        Image="MysteryPerson" />
                </toolkit:AvatarView.ImageSource>
            </toolkit:AvatarView>
            <VerticalStackLayout Grid.Column="1"
Margin="10,0,0,0">
                <Label Text="{Binding GamerTag}"
HorizontalTextAlignment="Start" FontSize="Large"
BackgroundColor="AntiqueWhite" />
                <Label Text="{Binding Status}"
HorizontalTextAlignment="Start" FontSize="Caption"
BackgroundColor="AntiqueWhite"/>
            </VerticalStackLayout>
            <controls:ActivityButton Grid.Column="2"
IsRunning="{Binding IsChallenging}" Text="{Binding
ChallengeStatus}" BackgroundColor="#e8bc65" Command="{Binding
ChallengeCommand}" CommandParameter="{Binding .}"
IsVisible="{Binding CanChallenge}" Margin="5"/>
        </Grid>
      </Border>
    </ContentView>
</DataTemplate>
```

`DataTemplate` will display the player's avatar. To do so, it will use an `Image` control and `GravatarImageSource` the same way as was done in the *Creating the Connect view* section. A `DataTemplate` element is needed since this is used for `ItemTemplate`, then the obligatory `ContentView`. Then, `Border` is defined. It uses a special `Stroke` shape to round out the edges of the rectangle instead of having square corners, and a color of `AntiqueWhite` is applied as the `BackgroundColor` value. Additional shapes to use for `Border` include `Ellipse`, `Line`, `Path`, `Polygon`, `Polyline`, and `Rectangle`. See the `Border` documentation at `https://learn.microsoft.com/en-us/dotnet/maui/user-interface/controls/border` for more details. Inside `Border`, there is a `Grid` control that defines three columns. The first column contains the avatar image and has a width of 50, the next column contains the gamer tag and the status of the player, stacked vertically, and the third column contains the `Challenge` button.

For the avatar, the `AvatarView` control from `CommunityToolkit` is used. It provides a round version of the image.

The `Challenge` button uses the `ActivityButton` control, and is bound to the `IsChallenging`, `CanChallenge`, `ChallengeStatus`, and `ChallengeCommand` properties of `PlayerViewModel`.

18. To use `PlayerCardViewTemplate` as `ItemTemplate` of `CollectionView`, add the following attribute to the `CollectionView` element:

```
ItemTemplate="{StaticResource PlayerCardViewTemplate}"
```

That completes the `Lobby` page. At this point, you should be able to launch `SticksAndStone.Functions` and connect with the `SticksAndStones.App` project to see the different layouts provided by the `Lobby` view. There is only one more page to create to complete the game, and that is the `Match` page.

Creating the Match page

The `Match` page displays the game board with the players and score. It also manages the gameplay, allowing each player to take their turn placing a stick. As each player takes their turn, the board updates to show the current state of the match. Let's get started with creating the `ViewModel` classes.

Creating the ViewModel classes

There are two different `ViewModel` classes used in the `Match` page, just as there were in the `Lobby` page, one for the game and a second for the player details.

Adding MatchPlayerViewModel

`MatchPlayerViewModel` is the abstraction between the `Player` model and `MatchView`. `MatchPlayerViewModel` needs to expose the `Id`, `GamerTag`, and `EmailAddress` values from the `Player` model to `MatchView`. In addition, since each player has a score, the player's score from the `Match` model is exposed to `MatchView`. There are a couple of additional properties that are needed as well:

- `IsPlayersTurn`:

  This is used to determine whether `MatchPlayerViewModel` is the current player.

- `PlayerToken`:

  This is used to map each player to a token to track which player has placed which stick. A token, either `-1` or `1`, is used because it makes determining a winner easier than if it was using the `Id` property, which is a `Guid`. Review the *Processing turns section of Chapter 9*, for a refresher on how the winner is determined.

To create `MatchPlayerViewModel`, follow these steps:

1. Create a new class named `MatchPlayerViewModel` in the `ViewModels` folder of the `SticksAndStones.App` project.

2. Modify the `using` declarations to the following at the top of the file:

```
using CommunityToolkit.Mvvm.ComponentModel;
using SticksAndStones.Models;
```

3. Add the `public` and `partial` modifiers to the class and make it inherit from `ObservableObject`, as shown:

```
public partial class MatchPlayerViewModel: ObservableObject
{

}
```

4. `MatchPlayerViewModel` is an abstraction of both the `Player` and `Match` models, which will be passed in through the constructor. Create the fields and the constructor, as shown in the following listing:

```
private readonly Player playerModel;
private readonly Match matchModel;

public MatchPlayerViewModel(Player player, Match match)
{
    this.playerModel = player;
    this.matchModel = match;
}
```

5. The `PlayerToken` property is 1 if the `Player` model is `PlayerOne` in the `Match` model; otherwise, it is `-1`. Add the `PlayerToken` property using as shown here:

```
public int PlayerToken => playerModel.Id == matchModel.
PlayerOneId ? 1 : -1;
```

6. The `IsPlayersTurn` property will return `true` if the `Player` model is the `NextPlayer` of the `Match` model, as shown here:

```
public bool IsPlayersTurn => playerModel.Id == matchModel.
NextPlayerId;
```

7. The `Id`, `GamerTag`, and `EmailAddress` properties all just map directly to the corresponding property from the `Player` model. This is the same implementation that was used in `PlayerViewModel` for the Lobby page. Use the following listing to add the properties to `MatchPlayerViewModel`:

```
public Guid Id => playerModel.Id;

public string GamerTag => playerModel.GamerTag;

public string EmailAddress => playerModel.EmailAddress;
```

8. The final property that is needed for `MatchPlayerViewModel` is the `Score` property. The `Score` property is mapped to either the `PlayerOneScore` or `PlayerTwoScore` property from the `Match` model, depending on which player the `Player` model is. Use the following listing to add the `Score` property to `MatchPlayerViewModel`:

```
public int Score => playerModel.Id == matchModel.PlayerOneId ?
matchModel.PlayerOneScore : matchModel.PlayerTwoScore;
```

That is all there is to `MatchPlayerViewModel`. The next section will walk you through creating `MatchViewModel`.

Adding MatchViewModel

`MatchViewModel` needs to provide the functionality for all the gameplay. It provides the two `MatchPlayerViewModel` objects for display in the header of the page, and the board to display where sticks have been played and which stones have been captured. It also provides the needed functionality for players to take their turn and forfeit the game if they choose. To implement `MatchViewModel`, follow these steps:

1. Create a new class in the `ViewModels` folder of the `SticksAndStones.App` project named `MatchViewModel`.

2. Modify the `using` declarations section at the top of the page to match the following listing:

```
using CommunityToolkit.Mvvm.ComponentModel;
using CommunityToolkit.Mvvm.Input;
using CommunityToolkit.Mvvm.Messaging;
using SticksAndStones.Models;
using SticksAndStones.Services;
```

3. Add the `public` and `partial` class modifiers to the class, inherit from `ViewModelBase`, and implement `IQueryAttributable`, as shown:

```
public partial class MatchViewModel : ViewModelBase,
IQueryAttributable
```

Recall that in `ConnectViewModel` and `LobbyViewModel`, when they navigate to `Match`, they pass an argument – either `MatchId` or the `Match` instance itself. `IQueryAttributable` is how that argument is passed to `MatchViewModel`. The implementation for `IQueryAttributable` is provided in a later step.

4. `MatchViewModel` only has a single dependency, `GameService`, so add a field to store the instance and a constructor to accept the instance as a parameter, as shown in the following listing:

```
private readonly GameService gameService;

public MatchViewModel(GameService gameService)
{
```

```
        this.gameService = gameService;
    }
```

5. When `MatchViewModel` is loaded, it will need to process the arguments, either the `Match` instance or a `MatchId` value. Either argument will end up with a `Match` instance that is used for displaying the board in the view, and from that, create two instances of `MatchPlayerViewModel` for player one and player two. Add `match`, `playerOne`, and `playerTwo` fields to the `MatchViewModel` class to hold those instances, as shown here:

```
[ObservableProperty]
private Match match;

[ObservableProperty]
private MatchPlayerViewModel playerOne;

[ObservableProperty]
private MatchPlayerViewModel playerTwo;
```

6. `IQueryAttributable` is used to process the arguments passed to the view model. Well, it is one way of doing it. The `IQueryAttributable` interface has only one method defined, `ApplyQueryAttributes`. The .NET MAUI routing system will automatically call the `ApplyQueryAttributes` method if the view model implements the `IQueryAttributable` interface. To add the implementation for `IQueryAttributable`, use the following listing:

```
public async Task ApplyQueryAttributes(IDictionary<string,
object> query)
{
    Match match = null;
    if (query.ContainsKey(Constants.ArgumentNames.Match))
    {
        match = query[Constants.ArgumentNames.Match] as Match;
    }
    if (query.ContainsKey(Constants.ArgumentNames.MatchId))
    {
        var matchId = new Guid($"{query[Constants.ArgumentNames.
MatchId]}");
        if (matchId != Guid.Empty)
        {
            match = await gameService.GetMatchById(matchId);
        }
    }

        LoadMatch(match);
    });
}
```

```
private void LoadMatch(Match match)
{
    if (match is null) return;

    PlayerOne = new MatchPlayerViewModel(gameService.
GetPlayerById(match.PlayerOneId), match);
    PlayerTwo = new MatchPlayerViewModel(gameService.
GetPlayerById(match.PlayerTwoId), match);

    this.Match = match;
}
```

`ApplyQueryAttributes` has a single parameter query, which is a dictionary of key-value pairs with the key as a string and the value as an object. The key ID is the name of the parameter as was passed in – for example, `"Match"` or `"MatchId"`. The method will check for the existence of the `"Match"` key being present and get the value as `Match` if it is. If the `"MatchId"` key is present, then `GameService` is used to get the `Match` model from `Id`. If there is no value for `match`, then the method returns; otherwise; initialize the two instances of `GamePlayerViewModel` and store them and `Match` in the `ViewModel` properties. The `LoadMatch` method is called from `ApplyQueryAttributes` since we will need the same functionality when the `UpdateMatch` event is received.

7. Before we can allow a player to choose a location to place a stick, it must be their turn. Create a property named `IsCurrentPlayersTurn` using the following listing:

```
public bool IsCurrentPlayersTurn => gameService.CurrentPlayer.Id
== (Match?.NextPlayerId ?? Guid.Empty);
```

8. Anytime the `Match` object is updated, `IsCurrentPlayersTurn` needs to be updated as well, since it depends on values in the `Match` property. To have this happen automatically, use the `NotifyPropertyChangedFor` attribute from `CommunityToolkit`. Add the highlighted line in the following code listing:

```
[ObservableProperty]
[NotifyPropertyChangedFor(nameof(IsCurrentPlayersTurn))]
private Match match;
```

Now, whenever the `Match` property is changed, the `NotfiyPropertyChanged` method will also be called for `IsCurrentPlayersTurn`. See *the Defining a ViewModel base class* section in *Chapter 2*, for a refresher on implementing the `INotifyPropertyChanged` interface.

9. The game allows the player to try out different positions of the stick before committing. If this is the current player's turn, the one connected and using the app, then the `SelectStick` method will place a stick at the location chosen by the user. The choice is not sent to the server until the user clicks the **Send** button in the view; this allows the player a little time

to think and try out different plays before committing. The last choice made by the player is captured in the `lastSelectedStick` field. Add the following listing to implement the `SelectStick` method:

```
int lastSelectedStick = -1;

[RelayCommand(CanExecute = nameof(IsCurrentPlayersTurn))]
private void SelectStick(string arg)
{
    if (gameService.CurrentPlayer is null) return;
    if (Match is null) return;

    if (int.TryParse(arg, out var pos))
    {
        pos--; // adjust for 0 based indexes
        if (lastSelectedStick != -1 && lastSelectedStick != pos)
            Match.Sticks[lastSelectedStick] = 0;

        if (Match.Sticks[pos] != 0)
            return;

        Match.Sticks[pos] = gameService.CurrentPlayer.Id ==
    PlayerOne.Id ? PlayerOne.PlayerToken : PlayerTwo.PlayerToken;
        lastSelectedStick = pos;
        OnPropertyChanged(nameof(Match));
    }
}
```

The value of `-1` for `lastSelectedStick` is used to mean no stick. The `SelectStick` method is exposed as a `Command` instance via `RelayCommandAttribute`. The `IsCurrentPlayersTurn` property is used to determine whether the command can execute. Recall from *Chapter 9* that the `Sticks` elements will have one of three values: `-1` for player one, `0` for empty, and `1` for player two. After determining whether the stick position is valid, the method raises the `OnPropertyChanged` event for the `Match` property, which causes the bindings to update.

10. After deliberating on which position to place their next stick, the player has three options: send their move to the server and end their turn, be indecisive and undo their move, or give up and exit the match. Add the `Play` method to `MatchViewModel` using the following code snippet:

```
[RelayCommand]
private async Task Play()
{
    if (lastSelectedStick == -1)
    {
```

```
        await Shell.Current.CurrentPage.DisplayAlert("Make a
move", "You must make a move before you play.", "Ok");
        return;
    }
    if (await Shell.Current.CurrentPage.DisplayAlert("Make a
move", "Are you sure this is the move you want, this can't be
undone.", "Yes", "No"))
    {
        var (newMatch, error) = await gameService.EndTurn(Match.
Id, lastSelectedStick);
        if (error is not null)
        {
            await Shell.Current.CurrentPage.DisplayAlert("Error
in move", error, "Ok");
            return;
        }
        lastSelectedStick = -1;
    }
}
```

The `Play` method is exposed as a `Command` so that it can be bound to by UI elements.

11. The Undo method is called when the player taps the **Undo** button. This method resets the value of the `lastSelectedStick` position and the value of `lastSelectedStick`. Add the Undo method, as shown in the following code listing:

```
[RelayCommand]
private async Task Undo()
{
    if (lastSelectedStick != -1)
    {
        if (await Shell.Current.CurrentPage.DisplayAlert("Undo
your move", "Are you sure you don't want to play this move?",
"Yes", "No"))
        {
            OnPropertyChanging(nameof(Match));
            Match.Sticks[lastSelectedStick] = 0;
            OnPropertyChanged(nameof(Match));
            lastSelectedStick = -1;
            return;
        }
    }
}
```

Again, the `RelayCommand` attribute is applied to the method to allow it to be bound to by UI elements.

The Forfeit method is called when the player uses the **Back** button to go back to the Lobby. If the match is still in progress, then the player is warned that leaving the match forfeits to the opposing player. Use the following code listing to add the Forfeit method to the MatchViewModel class:

```
[RelayCommand]
private async Task Forfeit()
{
    var returnToLobby = true;
    if (!Match.Completed)
    {
        returnToLobby = await Shell.Current.CurrentPage.
DisplayAlert("W A I T", "Returning to the Lobby will forfeit
your match, are you sure you want to do that?", "Yes", "No"))
        if (returnToLobby)
        {
            await Shell.Current.GoToAsync("///Lobby");
        }
    }
}
```

12. When the opposing player sends their move to the server, it is received in the app as a MatchUpdated event from the SignalR service. Add the handler for the MatchUpdated event using the following listing:

```
void OnMatchUpdated(object r, Messages.MatchUpdated m)
{
    LoadMatch(m.Value);
    if (Match.WinnerId != Guid.Empty && Match.Completed == true)
    {
        MainThread.InvokeOnMainThreadAsync(async () =>
        {
            if (Match.WinnerId == gameService.CurrentPlayer.Id)
            {
                await Shell.Current.CurrentPage.
DisplayAlert("Congratulations!", $"You are victorious!\nPress
the back button to return to the lobby.", "Ok");
            }
            else
            {
                await Shell.Current.CurrentPage.
DisplayAlert("Bummer!", $"You were defeated, better luck next
time!\nPress the back button to return to the lobby.", "Ok");
            }
        });
        return;
```

```
        }
    }
```

13. To register the `MatchUpdated` event handler, the `Register` method is called from
`OnActivated`, and `UnRegister` is called from `OnDeactivated`, as shown in the
following listing:

```
protected override void OnActivated()
{
    Messenger.Register(this, (MessageHandler<object, Messages.
MatchUpdated>)OnMatchUpdated);
}

protected override void OnDeactivated()
{
    Messenger.Unregister<Messages.MatchUpdated>(this);
}
```

14. Register `MatchViewModel` with dependency injection by adding the following highlighted
line of code to the `CreateMauiApp` method in the `MauiProgram.cs` file:

```
builder.Services.AddTransient<ViewModels.ConnectViewModel>();
builder.Services.AddTransient<ViewModels.LobbyViewModel>();
builder.Services.AddTransient<ViewModels.MatchViewModel>();

builder.Services.AddTransient<Views.ConnectView>();
builder.Services.AddTransient<Views.LobbyView>();
```

Why is it called IQueryAttributable? That feels awkward

The reason behind the name of the interface is that naming things is hard. The system for
passing arguments to a view model can be either declarative or not. The declarative way uses
`QueryPropertyAttribute` to map the query parameter to a property on the view model. If
you choose not to use the attributes, but instead handle the mapping yourself manually, you declare
your class `IQueryAttributable`, as in I could have used `QueryPropertyAttribute`
but I choose not to. For more information, visit `https://learn.microsoft.com/en-
us/dotnet/maui/fundamentals/shell/navigation#pass-data`.

Adding the Match view

This page is complex, so we are going to break it down into smaller, more manageable chunks. First,
the basic page layout is defined including the commands that are available to the player: `Play`, `Undo`,
and `Forfeit`. Next, the scoreboard area is defined with the player's gamer tag, Gravatar, and scores.
Finally, the game board is defined and laid out in a three-by-three grid. Let's get started by creating
the view and the layout.

Create the view

The `Match` view is not unlike any of the other views that have been created except that it has many more elements than preview views. Let's get started by creating the view and some basic elements by following these steps:

1. Right-click on the `Views` folder of the `SticksAndStone.App` project, select **Add**, and then click **New Item…**.

   If you are using Visual Studio 17.7 or later, click the **Show all Templates** button in the dialog that pops up; otherwise, move to the next step.

2. Under the **C# Items** node on the left, select **.NET MAUI**.

3. Select **.NET MAUI ContentPage (XAML)** and name it `MatchView`.

4. Click **Add** to create the page.

   Refer to the following screenshot to view the preceding information:

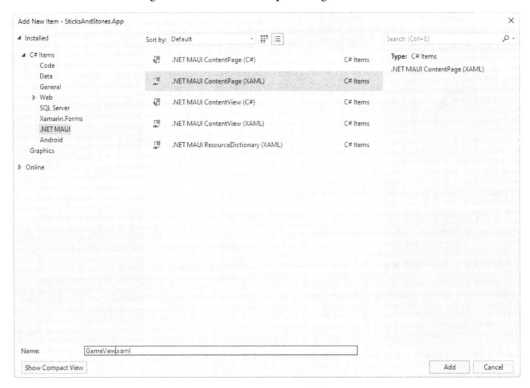

Figure 10.13 – Adding a new .NET MAUI ContentPage (XAML)

5. Open the `MatchView.xaml.cs` file and add the following `using` declaration:

```
using SticksAndStones.ViewModels;
```

6. Make the following highlighted changes to the constructor:

```
public MatchView(MatchViewModel viewModel)
{
    this.BindingContext = viewModel;
    InitializeComponent();
}
```

These changes allow for dependency injection to supply the MatchViewModel instance to the view, which is then assigned to BindingContext.

7. Open the AppShell.xaml file and add the following code snippet to the ContentPage element:

```
<ShellItem Route="Match">
    <ShellContent ContentTemplate="{DataTemplate
views:MatchView}" />
</ShellItem>
```

This registers the "Match" route and directs it to MatchView.

8. Open the MauiProgram.cs file and add the following highlighted line of code:

```
builder.Services.AddTransient<Views.ConnectView>();
builder.Services.AddTransient<Views.LobbyView>();
builder.Services.AddTransient<Views.MatchView>();

return builder.Build();
```

This will register MatchView with dependency injection so that DataTemplate can locate it.

9. Open the MatchView.xaml file and remove the Title attribute of the ContentPage element.

10. Add the following highlighted namespaces to the MatchView element. They will provide us access to the classes in the ViewModels, Converters, and Controls namespaces:

```
<ContentPage xmlns="http://schemas.microsoft.com/dotnet/2021/
maui"
                xmlns:x="http://schemas.microsoft.com/winfx/2009/
xaml"
        xmlns:viewModels="clr-namespace:SticksAndStones.
ViewModels"
        xmlns:controls="clr-namespace:SticksAndStones.Controls"
        xmlns:toolkit=" http://schemas.microsoft.com/
dotnet/2022/maui/toolkit"
        x:Class="SticksAndStones.Views.GameView">
```

11. To make IntelliSense happy with the bindings we will be adding, define the view model that the view is using by adding the `x:DataType` attribute to the `MatchView` element, as shown:

```
<ContentPage xmlns="http://schemas.microsoft.com/dotnet/2021/
maui"
              xmlns:x="http://schemas.microsoft.com/winfx/2009/
xaml"
      xmlns:viewModels="clr-namespace:SticksAndStones.
ViewModels"
      xmlns:controls="clr-namespace:SticksAndStones.Controls"
      xmlns:toolkit=" http://schemas.microsoft.com/
dotnet/2022/maui/toolkit"
      x:DataType="viewModels:GameViewModel"
      x:Class="SticksAndStones.Views.GameView">
```

`MatchView` uses a few icons from the Font Awesome font library, so we will need to download and install the library so that it is available in the app.

Downloading and configuring Font Awesome

Font Awesome is a free collection of images packaged into a font. .NET MAUI has excellent support for using Font Awesome in toolbars, navigation bars, and all over the place. It's not strictly needed to make this app, but we think that it's worth the extra round trip since you are most likely going to need something like this in your new killer app.

Downloading the font is straightforward. Please note the renaming of the file – it is not really needed but it's easier to edit configuration files and such if they have a simpler name. Follow these steps to acquire and copy the font to each project:

1. Browse to `https://fontawesome.com/download`.

2. Click the **Free for Desktop** button to download Font Awesome.

3. Unzip the downloaded file, then locate the `otfs` folder.

4. Rename the `Font Awesome 5 Free-Solid-900.otf` file to `FontAwesome.otf` (you can keep the original name, but it's just less to type if you rename it). Your filename may be different since Font Awesome is continually updating but it should be similar.

5. Copy `FontAwesome.otf` to the `Resources/Fonts` folder in the `SticksAndStones.App` project.

It would be nice if all that was needed was to copy the font file into the project folders. A lot does happen with just that action. The default .NET MAUI template includes all the fonts in the `Resources/Fonts` folder with the following item definition in the `News.csproj` file:

```
<!-- Custom Fonts -->
<MauiFont Include="Resources\Fonts\*" />
```

This ensures that the font files are processed and included in the app package automatically. What is left is to register the font with the .NET MAUI runtime so it is available to our XAML resources. To do that, add the following highlighted line to the `MauiProgram.cs` file:

```
.ConfigureFonts(fonts =>
{
    fonts.AddFont("OpenSans-Regular.ttf", "OpenSansRegular");
    fonts.AddFont("OpenSans-Semibold.ttf", "OpenSansSemibold");
    fonts.AddFont("FontAwesome.otf", "FontAwesome");
})
```

This line adds an alias that we can use in the next section to create static resources. The first parameter is the filename for the font file, and the second is the alias for the font that you can use in the `FontFamily` attribute.

Defining the layout

Now that Font Awesome is installed and configured in .NET MAUI, `TitleView` can use it. Add the custom title area and the main layout by following these steps:

1. First, override `TitleView` of the `Shell` element and provide a new container to hold the buttons:

    ```
    <Shell.TitleView>
        <Grid>
            <HorizontalStackLayout HorizontalOptions="Start">
            </HorizontalStackLayout>
            <HorizontalStackLayout HorizontalOptions="End">
            </HorizontalStackLayout>
        </Grid>
    </Shell.TitleView>
    ```

    The buttons are arranged in two segments, one aligned to the left or start of the window and the other aligned to the right or end of the window.

2. The player can at any point decide they no longer wish to continue playing. To exit the match, the player can use the **Back** button to return to the Lobby. When the button is tapped, the player is prompted to ensure that they no longer wish to continue playing. To add the **Back** button to the `Start` section of `TitleView` and bind `ForfeitCommand` in `MatchViewModel`, add the highlighted code from the following snippet :

    ```
    <HorizontalStackLayout HorizontalOptions="Start">
        <ImageButton Command="{Binding ForfeitCommand}"
    ToolTipProperties.Text="Return to the lobby.">
            <ImageButton.Source>
    ```

```
            <FontImageSource Glyph="&#xf0a8;"
FontFamily="FontAwesome" Color="White" Size="28" />
        </ImageButton.Source>
    </ImageButton>
</HorizontalStackLayout>
```

3. When it is the player's turn, they have two buttons that are enabled, `Play` and `Undo`. The `Play` and `Undo` buttons are placed in the `TitleView` area of the .NET MAUI page. Add the following highlighted code to add the `Play` and `Undo` buttons to `TitleView`:

```
<HorizontalStackLayout HorizontalOptions="End">
    <ImageButton Command="{Binding UndoCommand}"
IsVisible="{Binding IsCurrentPlayersTurn}" ToolTipProperties.
Text="Undo the last stick placement.">
        <ImageButton.Source>
            <FontImageSource Glyph="&#xf0e2;"
FontFamily="FontAwesome" Color="White" Size="28" />
        </ImageButton.Source>
    </ImageButton>
    <ImageButton Command="{Binding PlayCommand}"
IsVisible="{Binding IsCurrentPlayersTurn}" ToolTipProperties.
Text="Send the stick placement, and end my turn.">
        <ImageButton.Source>
            <FontImageSource Glyph="&#xf1d8;"
FontFamily="FontAwesome" Color="White" Size="28" />
        </ImageButton.Source>
    </ImageButton>
</HorizontalStackLayout>
```

4. Remove the default `VerticalStackLayout` element in `ContentView` and add the following code:

```
<ContentView>
    <Grid>
        <Grid.RowDefinitions>
            <RowDefinition Height="4*" />
            <RowDefinition Height="2*" />
            <RowDefinition Height="6*" />
            <RowDefinition Height="2*" />
        </Grid.RowDefinitions>

    </Grid>
</ContentView>
```

This adds a `Grid` control with four rows. The first and third rows will contain the scoreboard and the game board, respectively, while the second and fourth rows are padding.

The main layout is ready. Next, the scoreboard is added to the first row of the layout.

Creating the scoreboard

The scoreboard contains each player's avatar, gamer tag, and score. The elements are bound to the respective fields of `MatchPlayerViewModel` for the player. The `Match` object has two properties, `PlayerOne` and `PlayerTwo`, each of which is a `MatchPlayerViewModel`. To add the scoreboard, follow these steps:

1. Each player is identified by a different color. To add each color as a resource, open the `Colors.xaml` file in the `Resources/Styles` folder of the `SticksAndStones.App` project and add the following lines to the `ResourceDictionary` element:

   ```
   <Color x:Key="PlayerOne">#6495ED</Color>
   <Color x:Key="PlayerTwo">#CD5C5C</Color>
   ```

2. The scoreboard uses `HorizontalStackLayout` for the outer container. Add the following code to the `Grid` element:

   ```
   <HorizontalStackLayout Grid.Row="0"
   HorizontalOptions="CenterAndExpand" Margin="10" BindableLayout.
   ItemsSource="{Binding Players}">
   </HorizontalStackLayout>
   ```

   `HorizontalStackLayout` is assigned to row 0 of the `Grid`, and its contents are bound to the `Players` property of the view model, using `BindableLayout.ItemsSource`. `BindableLayout` is the underlying interface that supports all layout controls, such as `AbsoluteLayout` and `FlexLayout`.

3. Each player will have their own card within `HorizontalStackLayout`. Since the control is bound to the `Players` property, which is an array of `MatchPlayerViewModels`, the `BindableLayout.ItemTemplate` property provides the view that each item in `Players` is displayed with. The cards are laid out using a `Border` element and nested `VerticalStackLayout` elements. Add the following highlighted code to `HorizontalStackLayout`:

   ```
   <HorizontalStackLayout Grid.Row="0"
   HorizontalOptions="CenterAndExpand" Margin="10">
       <BindableLayout.ItemTemplate>
           <DataTemplate>
               <Border x:DataType="viewModels:MatchPlayerViewModel"
   Padding="0" Margin="2" StrokeShape="RoundRectangle 10,10,10,10"
   HeightRequest="175">
                   <VerticalStackLayout Padding="2"
   HorizontalOptions="Center">
                   </VerticalStackLayout>
               </Border>
   ```

```
    </DataTemplate>
  </BindableLayout.ItemTemplate>
</HorizontalStackLayout>
```

4. The `Border` element is the outermost container for the player card. To set the border color and background color of the `Border` element based on `PlayerToken`, triggers are used (https://learn.microsoft.com/en-us/dotnet/maui/fundamentals/triggers) – specifically, `DataTrigger` is used to set attribute values based on some other value. Add the following code to the `Border` element:

```
<Border.Triggers>
    <DataTrigger TargetType="Border" Binding="{Binding
PlayerToken}" Value="1" >
        <Setter Property="Stroke" Value="{StaticResource
PlayerOne}" />
        <Setter Property="BackgroundColor"
Value="{StaticResource PlayerOne}" />
    </DataTrigger>
    <DataTrigger TargetType="Border" Binding="{Binding
PlayerToken}" Value="-1" >
        <Setter Property="Stroke" Value="{StaticResource
PlayerTwo}" />
        <Setter Property="BackgroundColor"
Value="{StaticResource PlayerTwo}" />
    </DataTrigger>
</Border.Triggers>
```

The `DataTrigger` binding attribute is compared to the `Value` attribute. If they are equal, then the `Setter` elements of `DataTrigger` are executed. In this case, if the `PlayerToken` property is `-1`, then set the `Stroke` and `BackgroundColor` attributes of `Border` to the `PlayerOne` color that was defined in *step 1*. Otherwise, if the `PlayerToken` property is equal to `-1`, then set the `Stroke` and `BackgroundColor` attributes to the `PlayerTwo` color.

5. `VerticalStackLayout` contains another `VerticalStackLayout` and `Border` element, as shown in the following highlighted code:

```
<VerticalStackLayout BackgroundColor="{Binding PlayerToken,
Converter={StaticResource PlayerToColor}}" Padding="2"
HorizontalOptions="Center">
    <VerticalStackLayout>
    </VerticalStackLayout>
    <Border Padding="0" WidthRequest="96"
StrokeShape="RoundRectangle 10,10,10,10" StrokeThickness="0">
        <Image IsVisible="{Binding IsPlayersTurn}"
Source="hstick.jpeg" Aspect="AspectFit"
MaximumHeightRequest="36"/>
    </Border>
</VerticalStackLayout>
```

`VerticalStackLayout` will be used to hold `GamerTag`, `AvatarImage`, and the player's score, which is added in the next step. `Border` contains a horizontal stick image whose `IsVisible` attribute is bound to the `IsPlayersTurn` property. The stick is used as a visual indicator of which player's turn it is. If it is not the player's turn, the image is not displayed.

6. Within the second `VerticalStackLayout` are a `Label` and a `FlexLayout`. Add the following highlighted code:

```
<VerticalStackLayout>
    <Label Text="{Binding GamerTag}"
HorizontalOptions="FillAndExpand"
HorizontalTextAlignment="Center" FontSize="18"
FontFamily="OpenSansSemibold"/>
    <FlexLayout Margin="3">
    </FlexLayout>
</VerticalStackLayout>
```

7. `FlexLayout` contains the visual elements to display `AvatarImage` and `Score`. Add the following highlighted code to `FlexLayout`:

```
<FlexLayout Margin="3">
    <toolkit:AvatarView FlexLayout.Order="0" Margin="0"
BackgroundColor="LightGrey" HeightRequest="85"
WidthRequest="85" CornerRadius="50" VerticalOptions="Center"
HorizontalOptions="Center">
        <toolkit:AvatarView.ImageSource>
            <toolkit:GravatarImageSource
                Email="{Binding EmailAddress}"
                Image="MysteryPerson" />
        </toolkit:AvatarView.ImageSource>
        <toolkit:AvatarView.Triggers>
            <DataTrigger TargetType="toolkit:AvatarView"
Binding="{Binding Path=PlayerToken}" Value="-1">
                <Setter Property="FlexLayout.Order" Value="1" />
            </DataTrigger>
        </toolkit:AvatarView.Triggers>
    </toolkit:AvatarView>
    <Label FlexLayout.Order="1" Text="{Binding Score}"
FontSize="48" Padding="5" MinimumWidthRequest="65"
HorizontalTextAlignment="Center">
        <Label.Triggers>
            <DataTrigger TargetType="Label" Binding="{Binding
Path=PlayerToken}" Value="-1">
                <Setter Property="FlexLayout.Order" Value="0" />
            </DataTrigger>
        </Label.Triggers>
    </Label>
</FlexLayout>
```

By using the `FlexLayout` control, the order in which the `FlexLayout` children are displayed is governed by the `FlexLayout.Order` attribute. Similar to `Grid` with its `Grid.Row` and `Grid.Column` properties on its children, the `Order` attribute is set on the child. The order of the children in `FlexLayout` is changed through the use of `Trigger`. On `AvatarView`, `DataTrigger` will set the `FlexLayout.Order` attribute to `"1"` if the `PlayerToken` property is equal to `-1`, which is `PlayerTwo`. On the `Label`, `DataTrigger` sets the `FlexLayout.Order` attribute to `"0"`, effectively swapping the two elements.

And that concludes the scoreboard. The final part of `MatchView` is the largest: the board. Read on to learn how to create the board visuals.

Creating the game board

The game board is composed of three different elements. These elements are dots in the corners of each square, sticks (both horizontal and vertical), and stones. These elements are laid out as shown here:

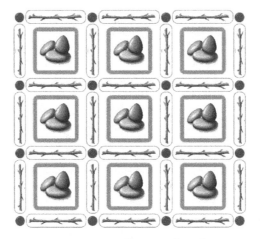

Figure 10.14 – The game board

The board uses a `Grid` control to provide the basic layout. Using 7 columns and 7 rows will provide cells for each of the elements: 16 dots, 9 stones, and 24 sticks. Add the following code to provide the basic layout of the game board to the top-level `Grid` element:

```
<Grid Grid.Row="2" BackgroundColor="White" Margin="10,40,10,0"
MaximumHeightRequest="410" MaximumWidthRequest="400" >
    <Grid.ColumnDefinitions>
        <ColumnDefinition Width="1*" />
        <ColumnDefinition Width="5*" />
        <ColumnDefinition Width="1*" />
        <ColumnDefinition Width="5*" />
        <ColumnDefinition Width="1*" />
        <ColumnDefinition Width="5*" />
```

```
        <ColumnDefinition Width="1*" />
    </Grid.ColumnDefinitions>

    <Grid.RowDefinitions>
        <RowDefinition Height="1*" />
        <RowDefinition Height="4*" />
        <RowDefinition Height="1*" />
        <RowDefinition Height="4*" />
        <RowDefinition Height="1*" />
        <RowDefinition Height="4*" />
        <RowDefinition Height="1*" />
    </Grid.RowDefinitions>
</Grid>
```

Let's start by adding the corners to the grid, since they are the simplest. To define a corner, use a `Label` with the text `"&#x26AB;"`, which is the hexadecimal character code for a dot. To center the dot horizontally and vertically, set `HorizontalOptions` and `VerticalOptions` to `"Center"`. Your basic element would look like the following:

```
<Label Text="&#x26AB" HorizontalOptions="Center"
VerticalOptions="Center" />
```

Without the `Grid.Row` and `Grid.Column` attributes, the `Label` will be put into row 0 and column 0. There are 16 corners in the grid, and they occupy all the even-numbered cells, so `(0,0)`, `(0,2)`, `(0,4)`, `(0,6)`, `(2,0)`, `(2,2)`, and so on. Fully defined labels for the first row would look like the following:

```
            <Label Grid.Row="0" Grid.Column="0" Text="&#x26AB"
HorizontalOptions="Center" VerticalOptions="Center" />
            <Label Grid.Row="0" Grid.Column="2" Text="&#x26AB"
HorizontalOptions="Center" VerticalOptions="Center" />
            <Label Grid.Row="0" Grid.Column="4" Text="&#x26AB"
HorizontalOptions="Center" VerticalOptions="Center" />
            <Label Grid.Row="0" Grid.Column="6" Text="&#x26AB"
HorizontalOptions="Center" VerticalOptions="Center" />
```

When you work this out for all 16 rows, that's a lot of duplication of the `Text`, `HorizontalOptions`, and `VerticalOptions` attributes. By using a `Style` element, that duplication can be eliminated. A `Style` element contains `Setter` elements such as the `DataTrigger` elements. When `Style` is applied to the element, the `Setter` elements are used to update the target element's attributes. Use the following steps to add the corner elements to the `Grid` control using `Style`:

1. Add the following `Style` element to the `ContentPage.Resources` element:

```
<Style x:Key="dotLabel"
        TargetType="Label">
```

```
        <Setter Property="Text" Value="&#x26AB;" />
        <Setter Property="HorizontalOptions" Value="Center" />
        <Setter Property="VerticalOptions" Value="Center" />
    </Style>
```

This `Style` element is identified by the `x:Key` attribute.

2. Add a `Label` to the `Grid` control created at the start of this section.

3. Set the `Grid.Row` attribute of the `Label` to 0.

4. Set the `Grid.Column` attribute of the `Label` to 0.

5. Set the `Style` attribute to the `{StaticResource dotLabel}` value. The `Style` attribute is used to specify which style should be applied to the element. Since `Style` is defined in the `ContentView.Resources` element, it is a `StaticResource`.

6. The completed `Label` should look like the following:

```
<Label Grid.Row="0" Grid.Column="0" Style="{StaticResource
dotLabel}" />
```

7. Now, copy the `Label` just created and increase the `Grid.Column` value by two, and repeat this step until you have four `Label` elements with the same `Grid.Row` value.

8. Copy the last `Label` created in *step 7* and increase the `Grid.Row` value by two and reset the value of `Grid.Column` to 0. Now, repeat *step 7* using the updated `Grid.Row` value, and stop when there are four labels with a `Grid.Row` value of 6.

9. The labels should look like the following listing:

```
<Label Grid.Row="0" Grid.Column="0" Style="{StaticResource
dotLabel}" />
<Label Grid.Row="0" Grid.Column="2" Style="{StaticResource
dotLabel}" />
<Label Grid.Row="0" Grid.Column="4" Style="{StaticResource
dotLabel}" />
<Label Grid.Row="0" Grid.Column="6" Style="{StaticResource
dotLabel}" />
<Label Grid.Row="2" Grid.Column="0" Style="{StaticResource
dotLabel}" />
<Label Grid.Row="2" Grid.Column="2" Style="{StaticResource
dotLabel}" />
<Label Grid.Row="2" Grid.Column="4" Style="{StaticResource
dotLabel}" />
<Label Grid.Row="2" Grid.Column="6" Style="{StaticResource
dotLabel}" />
<Label Grid.Row="4" Grid.Column="0" Style="{StaticResource
dotLabel}" />
<Label Grid.Row="4" Grid.Column="2" Style="{StaticResource
dotLabel}" />
```

```
<Label Grid.Row="4" Grid.Column="4" Style="{StaticResource
dotLabel}" />
<Label Grid.Row="4" Grid.Column="6" Style="{StaticResource
dotLabel}" />
<Label Grid.Row="6" Grid.Column="0" Style="{StaticResource
dotLabel}" />
<Label Grid.Row="6" Grid.Column="2" Style="{StaticResource
dotLabel}" />
<Label Grid.Row="6" Grid.Column="4" Style="{StaticResource
dotLabel}" />
<Label Grid.Row="6" Grid.Column="6" Style="{StaticResource
dotLabel}" />
```

Now that the corners are done, we can start on the game pieces: the sticks and stones. Since the sticks and stones have some similarities, we can create a common control to help display them all. However, they are visualized entirely differently. What is needed is a common interface to define the `BindableProperty` properties and use that on different layouts. .NET MAUI uses `ControlTemplate` resources to allow for the customization, or even complete replacement, of the visual elements that comprise a control. Many controls in .NET MAUI can be customized using a `ControlTemplate`, if they derive from `ContentView` or `ContentPage`. Let's get started with the sticks and stones by adding the custom control, then the `ControlTemplate` resources for sticks and stones, by following these steps:

1. Create a new class in the `Controls` folder of the `SticksAndStones.App` project named `GamePieceView`.

2. Update the class definition to match the following listing:

    ```
    namespace SticksAndStones.Controls;

    public partial class GamePieceView : ContentView
    {
    }
    ```

3. Add a `string` property and a `BindableProperty` property named `GamePiecePosition` and `GamePiecePositionProperty`, respectively, as shown in the following listing:

    ```
    public static readonly BindableProperty
    GamePiecePositionProperty = BindableProperty.
    Create(nameof(GamePiecePosition), typeof(string),
    typeof(GamePieceView), string.Empty);
    public string GamePiecePosition
    {
        get => (string)GetValue(GamePiecePositionProperty);
        set => SetValue(GamePiecePositionProperty, value);
    }
    ```

    `GamePiecePosition` is used to determine the array index in either `Sticks` or `Stones` properties on `GameViewModel`.

4. Add an int property and `BindableProperty` named `GamePieceState` and `GamePieceStateProperty`, respectively, as shown in the following listing:

```
public static readonly BindableProperty GamePieceStateProperty
= BindableProperty.Create(nameof(GamePieceState), typeof(int),
typeof(GamePieceView), 0, BindingMode.TwoWay);
public int GamePieceState
{
    get => (int)GetValue(GamePieceStateProperty);
    set => SetValue(GamePieceStateProperty, value);
}
```

GamePieceState is the owner of the piece: 1 for `PlayerOne`, 0 for no one, and -1 for `PlayerTwo`.

5. Add a string property and `BindableProperty` named `GamePieceDirection` and `GamePieceDirectionProperty`, respectively, as shown in the following listing:

```
public static readonly BindableProperty
GamePieceDirectionProperty = BindableProperty.
Create(nameof(GamePieceDirection), typeof(string),
typeof(GamePieceView), null);
public string GamePieceDirection
{
    get => (string)GetValue(GamePieceDirectionProperty);
    set => SetValue(GamePieceDirectionProperty, value);
}
```

GamePieceDirection is only needed for `Sticks` and is either `Horizontal` or `Vertical`.

6. Open the `MatchView.Xaml` file again and let's add a control template for all the sticks. Add the following snippet to the `ContentView.Resources` element:

```
<ControlTemplate x:Key="StickViewControlTemplate">
</ControlTemplate>
```

This defines a `ControlTemplate` element with a key of `StickViewControlTemplate`. The key is used to apply the `ControlTemplate` element to the control.

7. There are two elements to each stick visual: the number displayed on a label, and the stick image, which uses the `Image` control inside a border to give it the outline, colored by the player that placed the stick. The other interesting aspect is that the `Label` and `Border` controls need to be layered on top of one another. To accomplish this, a `Grid` control is used and both elements are placed in the same cell. To add the `Grid`, `Label`, `Border`, and `Image` controls, use the following listing, and add them to the `ControlTemplate` element:

```
<Grid Margin="0" Padding="0">
    <Label Text="{TemplateBinding GamePiecePosition}"
IsVisible="False" HorizontalTextAlignment="Center"
```

```
        VerticalTextAlignment="Center" TextColor="Red"
        FontAttributes="Bold" >


        </Label>
        <Border Padding="3" BackgroundColor="Transparent"
        StrokeShape="RoundRectangle 5" Stroke="Transparent">
            <Image Aspect="Fill">
            </Image>
        </Border>
    </Grid>
```

Grid has Margin and Padding values of 0 so that it doesn't occupy any screen real estate. The Label control's Text attribute is bound using TemplateBinding to the GamePiecePosition property. TemplateBinding differs slightly from Binding in that TemplateBinding uses the control this ControlTemplate is applied to as DataContext. Since this ControlTemplate will be applied to instances of GamePieceView, it will bind to the Bindable properties of those controls.

Inspecting the Image control from *step 7*, you'll find that it doesn't specify which image is displayed. For Sticks, one of two images is displayed: either hstick.jpeg for horizontal sticks or vstick.jpeg for vertical sticks, and if there is no stick at that location, then the control should not be visible. The following listing uses DataTrigger to set the values of IsVisible and Source of the Image control using TemplateBinding to the GamePieceState and GamePieceDirection properties. Add this code to the Image control of ControlTemplate:

```
<Image.Triggers>
    <DataTrigger TargetType="Image" Binding="{TemplateBinding
    Path=GamePieceState}" Value="0">
        <Setter Property="IsVisible" Value="False" />
    </DataTrigger>
    <DataTrigger TargetType="Image" Binding="{TemplateBinding
    Path=GamePieceDirection}" Value="Horizontal">
        <Setter Property="Source" Value="hstick.jpeg" />
    </DataTrigger>
    <DataTrigger TargetType="Image" Binding="{TemplateBinding
    Path=GamePieceDirection}" Value="Vertical">
        <Setter Property="Source" Value="vstick.jpeg" />
    </DataTrigger>
</Image.Triggers>
```

8. The Border control also uses DataTrigger to outline the stick in the color of the player that placed the stick. Add the following code to the Border element, after Image:

```
<Border.Triggers>
    <DataTrigger TargetType="Border" Binding="{TemplateBinding
    GamePieceState}" Value="1" >
```

```
        <Setter Property="Stroke" Value="{StaticResource
PlayerOne}" />
    </DataTrigger>
    <DataTrigger TargetType="Border" Binding="{TemplateBinding
GamePieceState}" Value="-1" >
        <Setter Property="Stroke" Value="{StaticResource
PlayerTwo}" />
    </DataTrigger>
</Border.Triggers>
```

Two triggers are needed to switch between PlayerOne (1) and PlayerTwo (-1). The Stroke attribute of the Border control is set to the color resource of the player. If neither trigger is active, then the default Stroke value of Transparent from the Border element is used. In this way, if there is no stick, GamePieceState is 0, and the border is transparent. If GamePieceState is 1, then Stroke will have the color defined by the resource named PlayerOne, and if GamePieceState is -1, then the Stroke value will be the resource named PlayerTwo.

9.   When the user is making their move during their turn, they will tap or click on the label to place their stick in that position. To call SelectStickCommand when that occurs, the Border control binds TapGestureRecognizer to the GameViewModel.SelectStickCommand property and passes GamePiecePosition along as a parameter. Add the following listing to the Border element, after the Border.Triggers element:

```
<Border.GestureRecognizers>
    <TapGestureRecognizer Command="{Binding
Source={RelativeSource AncestorType={x:Type
viewModels:GameViewModel}}, Path=SelectStickCommand}"
CommandParameter="{TemplateBinding GamePiecePosition}" />
</Border.GestureRecognizers>
```

10. Finally, take a close look at the Label element; you will see that the IsVisible attribute is set to False. If there is no stick placed at this position, then we need the label with the position displayed. That can be accomplished by using DataTrigger; the label's IsVisible property can be set to True, making the label visible if GamePieceState is 0, meaning no stick has been placed there yet. Add the following listing to the Label element:

```
<Label.Triggers>
    <DataTrigger TargetType="Label" Binding="{TemplateBinding
Path=GamePieceState}" Value="0">
        <Setter Property="IsVisible" Value="True" />
    </DataTrigger>
</Label.Triggers>
```

That completes the control template for the sticks. Next, create the control template for Stones by following these steps:

1. Right below the ControlTemplate created for the sticks, add the following code:

    ```
    <ControlTemplate x:Key="StoneViewControlTemplate">
    </ControlTemplate>
    ```

    Just like with the control template for sticks, the ControlTemplate uses a key to locate the right template.

2. The Stones template is a little less complex than the Sticks template. Here, we just have the Border control with an Image control as a child. DataTrigger is used again to select the right border color, and if the stones are not present, then the border is not visible. Use the following code sample and add it to the ControlTemplate created in *step 1*:

    ```
    <Border Margin="3" Padding="5" HorizontalOptions="Center"
    VerticalOptions="Center" StrokeShape="RoundRectangle 5"
    StrokeThickness="3">
        <Border.Triggers>
            <DataTrigger TargetType="Border"
    Binding="{TemplateBinding GamePieceState}" Value="0">
                <Setter Property="IsVisible" Value="False" />
            </DataTrigger>
            <DataTrigger TargetType="Border"
    Binding="{TemplateBinding GamePieceState}" Value="1" >
                <Setter Property="Stroke" Value="{StaticResource
    PlayerOne}" />
            </DataTrigger>
            <DataTrigger TargetType="Border"
    Binding="{TemplateBinding GamePieceState}" Value="-1" >
                <Setter Property="Stroke" Value="{StaticResource
    PlayerTwo}" />
            </DataTrigger>
        </Border.Triggers>
        <Image Source="stones.jpeg" Aspect="Fill" />
    </Border>
    ```

    You may have noticed a difference between the triggers in this listing versus the triggers in the Sticks control template. In Sticks, the IsVisible attribute was set on Image, not Border, and you may want to know why that is. The explanation is simple; if the border is not visible, it will not receive TapGuesture events. The Grid element cannot register GestureRecognizer, so the event cannot be captured there either.

The ControlTemplates that are needed for the stick and stone images are in place; now, they need to be associated with the GamePieceView control elements. A Style can set the ControlTemplate property of a GamePieceView element, but how will it determine that this element is a stick or a stone? `Style` elements have a `Class` property that can be used to further refine which styles are applied to a control. If the control has a matching class name listed in its `StyleClass` attribute, then that `Style` element is applied. Let's use sticks as an example, by following these steps:

1. Add a new `Style` element to the `ContentView.Resources` element, as shown in the following listing:

    ```
    <Style TargetType="controls:GamePieceView"
            Class="Stick">
        <Setter Property="ControlTemplate"
                Value="{StaticResource StickViewControlTemplate}" />
    </Style>
    ```

    This style is only applied to elements that are of the `GamePiece` type and have the `Stick` class listed in the `StyleClass` attribute. A matching element might look like the following:

    ```
    <controls:GamePieceView Grid.Row="0" Grid.Column="1"
    StyleClass="Stick"
                                    GamePiecePosition="01"
    GamePieceState="{Binding Game.Sticks[0]}"
    GamePieceDirection="Horizontal" />
    ```

    The highlighted sections show the parts of the control that are used to match the `Style` element. `StyleClass` can have more than one name listed; just use a comma to separate the names.

2. Add a new `Style` element. This time, it will be to apply `StoneViewControlTemplate`, as shown in the following listing:

    ```
    <Style TargetType="controls:GamePieceView"
            Class="Stone">
        <Setter Property="ControlTemplate"
                Value="{StaticResource StoneViewControlTemplate}" />
    </Style>
    ```

That is all that is required for the stick and stone elements to be added to the game board grid. To add the remaining elements, follow these steps:

1. There are seven rows of sticks: four rows of three and three rows of four. They are nearly identical, but not quite. Locate the `Grid` that defines the gameboard; it will already have the corner dots added. Right after the 16 dot elements, add the following listing for the first row of sticks:

    ```
    <controls:GamePieceView Grid.Row="0" Grid.Column="1"
    StyleClass="Stick"
                                    GamePiecePosition="01"
    GamePieceState="{Binding Game.Sticks[0]}"
    GamePieceDirection="Horizontal" />

    <controls:GamePieceView Grid.Row="0" Grid.Column="3"
    StyleClass="Stick"
    ```

```
                                 GamePiecePosition="02"
GamePieceState="{Binding Game.Sticks[1]}"
GamePieceDirection="Horizontal" />


<controls:GamePieceView Grid.Row="0" Grid.Column="5"
StyleClass="Stick"
                                 GamePiecePosition="03"
GamePieceState="{Binding Game.Sticks[2]}"
GamePieceDirection="Horizontal" />
```

Each stick in the first row is displayed horizontally. Each stick is given its own position in the `GamePiecePosition` attribute and `GamePieceState` is bound to the `Game.Sticks` object for this stick. The `Sticks` array is zero-based so the indexes for the array are one less than `GamePiecePosition`.

2. Add the code for the second row of sticks using the following listing:

```
<controls:GamePieceView Grid.Row="1" Grid.Column="0"
StyleClass="Stick"
                                 GamePiecePosition="04"
GamePieceState="{Binding Game.Sticks[3]}"
GamePieceDirection="Vertical" />


<controls:GamePieceView Grid.Row="1" Grid.Column="2"
StyleClass="Stick"
                                 GamePiecePosition="05"
GamePieceState="{Binding Game.Sticks[4]}"
GamePieceDirection="Vertical" />


<controls:GamePieceView Grid.Row="1" Grid.Column="4"
StyleClass="Stick"
                                 GamePiecePosition="06"
GamePieceState="{Binding Game.Sticks[5]}"
GamePieceDirection="Vertical" />


<controls:GamePieceView Grid.Row="1" Grid.Column="6"
StyleClass="Stick"
                                 GamePiecePosition="07"
GamePieceState="{Binding Game.Sticks[6]}"
GamePieceDirection="Vertical" />
```

These elements are all `Vertical` instead of `Horizontal`; otherwise, they follow the same pattern as the previous step. Keep on going to add the remaining rows.

3. Use the following listing to add the third row of sticks:

```
<controls:GamePieceView Grid.Row="2" Grid.Column="1"
StyleClass="Stick"
                                 GamePiecePosition="08"
GamePieceState="{Binding Game.Sticks[7]}"
GamePieceDirection="Horizontal" />
```

```
<controls:GamePieceView Grid.Row="2" Grid.Column="3"
StyleClass="Stick"

                        GamePiecePosition="09"
GamePieceState="{Binding Game.Sticks[8]}"
GamePieceDirection="Horizontal" />


<controls:GamePieceView Grid.Row="2" Grid.Column="5"
StyleClass="Stick"

                        GamePiecePosition="10"
GamePieceState="{Binding Game.Sticks[9]}"
GamePieceDirection="Horizontal" />
```

4. Use the following listing to add the fourth row of sticks:

```
<controls:GamePieceView Grid.Row="3" Grid.Column="0"
StyleClass="Stick"

                        GamePiecePosition="11"
GamePieceState="{Binding Game.Sticks[10]}"
GamePieceDirection="Vertical" />


<controls:GamePieceView Grid.Row="3" Grid.Column="2"
StyleClass="Stick"

                        GamePiecePosition="12"
GamePieceState="{Binding Game.Sticks[11]}"
GamePieceDirection="Vertical" />


<controls:GamePieceView Grid.Row="3" Grid.Column="4"
StyleClass="Stick"

                        GamePiecePosition="13"
GamePieceState="{Binding Game.Sticks[12]}"
GamePieceDirection="Vertical" />


<controls:GamePieceView Grid.Row="3" Grid.Column="6"
StyleClass="Stick"

                        GamePiecePosition="14"
GamePieceState="{Binding Game.Sticks[13]}"
GamePieceDirection="Vertical" />
```

5. Use the following listing to add the fifth row of sticks:

```
<controls:GamePieceView Grid.Row="4" Grid.Column="1"
StyleClass="Stick"

                        GamePiecePosition="15"
GamePieceState="{Binding Game.Sticks[14]}"
GamePieceDirection="Horizontal" />


<controls:GamePieceView Grid.Row="4" Grid.Column="3"
StyleClass="Stick"
```

```
                                GamePiecePosition="16"
GamePieceState="{Binding Game.Sticks[15]}"
GamePieceDirection="Horizontal" />


<controls:GamePieceView Grid.Row="4" Grid.Column="5"
StyleClass="Stick"
                                GamePiecePosition="17"
GamePieceState="{Binding Game.Sticks[16]}"
GamePieceDirection="Horizontal" />
```

6. Use the following listing to add the sixth row of sticks:

```
<controls:GamePieceView Grid.Row="5" Grid.Column="0"
StyleClass="Stick"
                                GamePiecePosition="18"
GamePieceState="{Binding Game.Sticks[17]}"
GamePieceDirection="Vertical" />


<controls:GamePieceView Grid.Row="5" Grid.Column="2"
StyleClass="Stick"
                                GamePiecePosition="19"
GamePieceState="{Binding Game.Sticks[18]}"
GamePieceDirection="Vertical" />


<controls:GamePieceView Grid.Row="5" Grid.Column="4"
StyleClass="Stick"
                                GamePiecePosition="20"
GamePieceState="{Binding Game.Sticks[19]}"
GamePieceDirection="Vertical" />


<controls:GamePieceView Grid.Row="5" Grid.Column="6"
StyleClass="Stick"
                                GamePiecePosition="21"
GamePieceState="{Binding Game.Sticks[20]}"
amePieceDirection="Vertical" />
```

7. Use the following listing to add the seventh row of sticks:

```
<controls:GamePieceView Grid.Row="6" Grid.Column="1"
StyleClass="Stick"
                                GamePiecePosition="22"
GamePieceState="{Binding Game.Sticks[21]}"
GamePieceDirection="Horizontal" />


<controls:GamePieceView Grid.Row="6" Grid.Column="3"
StyleClass="Stick"
                                GamePiecePosition="23"
GamePieceState="{Binding Game.Sticks[22]}"
GamePieceDirection="Horizontal" />
```

```
<controls:GamePieceView Grid.Row="6" Grid.Column="5"
StyleClass="Stick"
                          GamePiecePosition="24"
GamePieceState="{Binding Game.Sticks[23]}"
GamePieceDirection="Horizontal" />
```

8. The sticks are all added, so now we need to add the stones. Use the following listing to add the nine `Stone` elements to the game board `Grid` control following the sticks:

```
<controls:GamePieceView Grid.Row="1" Grid.Column="1"
StyleClass="Stone" GamePieceState="{Binding Game.Stones[0]}" />
<controls:GamePieceView Grid.Row="1" Grid.Column="3"
StyleClass="Stone" GamePieceState="{Binding Game.Stones[1]}" />
<controls:GamePieceView Grid.Row="1" Grid.Column="5"
StyleClass="Stone" GamePieceState="{Binding Game.Stones[2]}" />
<controls:GamePieceView Grid.Row="3" Grid.Column="1"
StyleClass="Stone" GamePieceState="{Binding Game.Stones[3]}" />
<controls:GamePieceView Grid.Row="3" Grid.Column="3"
StyleClass="Stone" GamePieceState="{Binding Game.Stones[4]}" />
<controls:GamePieceView Grid.Row="3" Grid.Column="5"
StyleClass="Stone" GamePieceState="{Binding Game.Stones[5]}" />
<controls:GamePieceView Grid.Row="5" Grid.Column="1"
StyleClass="Stone" GamePieceState="{Binding Game.Stones[6]}" />
<controls:GamePieceView Grid.Row="5" Grid.Column="3"
StyleClass="Stone" GamePieceState="{Binding Game.Stones[7]}" />
<controls:GamePieceView Grid.Row="5" Grid.Column="5"
StyleClass="Stone" GamePieceState="{Binding Game.Stones[8]}" />
```

This is a wrap on the game app. You can now test out the project in the next section.

Testing the completed project

This project has spanned two chapters, with *Chapter 9, Setting Up a Backend for a Game Using Azure Services* , *and this chapter, Building a Real-Time Game*. Since this is a two-player turn-based game, getting all the components configured correctly can be a challenge. Follow these steps to test out your game locally on Windows:

1. The first step is to get the service running in the background. In Visual Studio, right-click the `SticksAndStones.Functions` project and select **Debug | Start Without Debugging** or press *Ctrl + F5*.

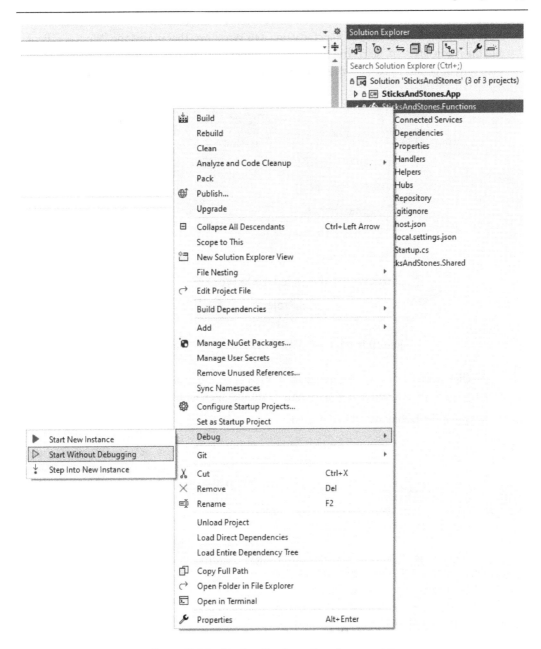

Figure 10.15 – Starting the Azure Functions service

That should launch a terminal window with the Azure Functions service running.

2. Now, two clients are needed to play the game. On Windows, that means the Windows client and the Android client. Start with the Windows client first, and use the same method that was used for Functions. Make sure the Windows target is selected in the **Debug** options:

Figure 10.16 – Selecting Windows as the Debug target

3. Right-click the `SticksAndStones.App` project and then select **Debug | Start Without Debugging** or press *Ctrl + F5*. A new window should open with the login page displayed.

4. Now, switch the **Debug** target to Android:

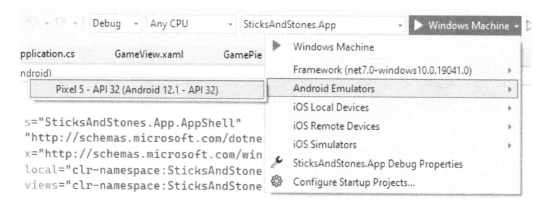

Figure 10.17 – Selecting Android as the Debug target

5. Now, either use *F5* to debug the app in an Android emulator, or *Ctrl + F5* to just run the app.

6.   Log in to each app using a different email and gamer tag.

Figure 10.18 – Log in to the game

7.   Challenge the other player to a match!

Figure 10.19 – Challenge issued

8. Try to best yourself in a game of *Sticks and Stones*!

Figure 10.20 – The match has started

Android: cleartext http traffic to 10.0.2.2 not permitted

If you are attempting to test the game using the Android client, you will probably hit this error when you try to send a move to the server. Fortunately, the resolution is easy. Open the `MainApplication.cs` file in the `Platforms/Android` folder and modify the `Application` attribute on the `MainApplication` class to match the following:

```
[Application(UsesClearextTraffic = true)]
```

If you encounter any errors or something just doesn't work the way you expect it to, go back through all the steps and make sure you didn't miss anything. Otherwise, congratulations on making it through this project.

Summary

That's that! Excellent work! There is so much in the chapter that it is hard to keep this summary short. In this chapter, we created a game app that connects to our backend. We created a service that managed the calls to the backend service and handled errors, all asynchronously. We have learned how to respond to messages from SignalR, and how to send and receive messages within the app using the `IMessenger` interface. We created custom controls and used them across multiple pages. We learned how to style an app with XAML styling, how to use control templates, and how to select them using styles. We explored routes and how they work in a multi-page .NET MAUI app. We examined triggers and how we can use them to update the interface without using C# code and converters.

Now, reward yourself and challenge a friend to a match in your new game.

In the next chapter, we will dive into Blazor and .NET MAUI together.

11

Building a Calculator Using .NET MAUI Blazor

In this chapter, we will explore a .NET **Blazor** app embedded within a .NET MAUI app. We will write part of the calculator app using Blazor and host that in .NET MAUI using `BlazorWebView`. We will also communicate between Blazor and .NET MAUI.

In this chapter, we are going to cover the following topics:

- What is Blazor?
- Exploring the differences between .NET MAUI projects and .NET MAUI Blazor projects
- Using HTML and CSS to define a UI
- Using XAML controls with HTML in a WebView
- Writing C# code that will integrate with XAML controls and HTML controls
- Working with the main .NET MAUI window to resize it to its contents

Technical requirements

You will need to have Visual Studio for Mac or PC installed, as well as the .NET MAUI components. See *Chapter 1, Introduction to .NET MAUI*, for more details on how to set up your environment. The source code for this chapter is available in this book's GitHub repository: `https://github.com/PacktPublishing/MAUI-Projects-3rd-Edition`.

Project overview

In this chapter, you will discover .NET Blazor and how to use it with .NET MAUI to develop a UI for an app. We will explore the different options available for hosting a Blazor app within a .NET MAUI app. Communication between the two apps is a crucial part of the interoperability and this chapter's project will show you how to send data from .NET MAUI to Blazor and vice versa.

What is Blazor?

.NET Blazor is a web framework built on top of .NET. Blazor applications run in the browser by using **WebAssembly** (**WASM**), or on a server using SignalR. Blazor is one part of the whole ASP.NET ecosystem, and it leverages Razor pages for developing the UI. Blazor uses HTML and CSS to render a rich UI. Blazor uses a component-based UI, where each component is a Razor markup page. Within a Razor page, you can mix and match HTML, CSS, and C# code. There are three deployment models for a Blazor app:

- **Blazor Server**: In a Blazor Server deployment, the application code runs on the server in an ASP.NET Core app, and it communicates with the UI running in a browser via SignalR.

- **Blazor WebAssembly**: For Blazor WebAssembly, the entire app runs in the browser via WASM. It is an open web standard that makes it possible to run .NET code safely in the browser. WASM provides interoperability with JavaScript.

- **Blazor Hybrid**: Blazor Hybrid is a blend of native .NET and web technologies. Blazor Hybrid apps can be hosted in .NET MAUI, **WPF**, and **Windows Forms** applications. Since the hosts are all .NET, the Blazor runtime runs natively within the same .NET process and renders the Razor page web UI into a WebView control.

Now that we have some general knowledge of Blazor, let's look at the app we will build in this chapter!

Creating the calculator app

In this chapter, we are going to build a calculator app. The UI for the calculator is built using Razor pages in Blazor, but the actual mechanics of the calculator reside in the .NET MAUI app.

Setting up the project

This project, like all the rest, is a **File** | **New** | **Project…**-style project. This means that we will not be importing any code at all. So, this first section is all about creating the project and setting up the basic project structure.

Creating the new project

The first step is to create a new .NET MAUI project. Follow these steps:

1. Open Visual Studio 2022 and select **Create a new project**:

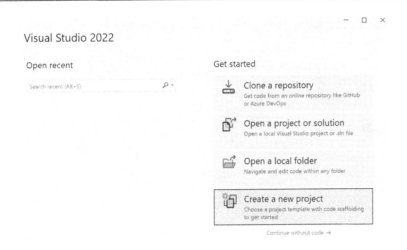

Figure 11.1 – Visual Studio 2022

This will open the **Create a new project** wizard.

2. In the search field, type in `blazor` and select the **.NET MAUI Blazor App** item from the list:

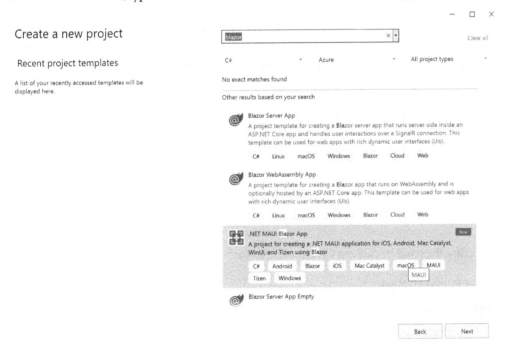

Figure 11.2 – Create a new project

3. Click **Next**.

4. Enter `Calculator` as the name of the app, as shown in the following screenshot:

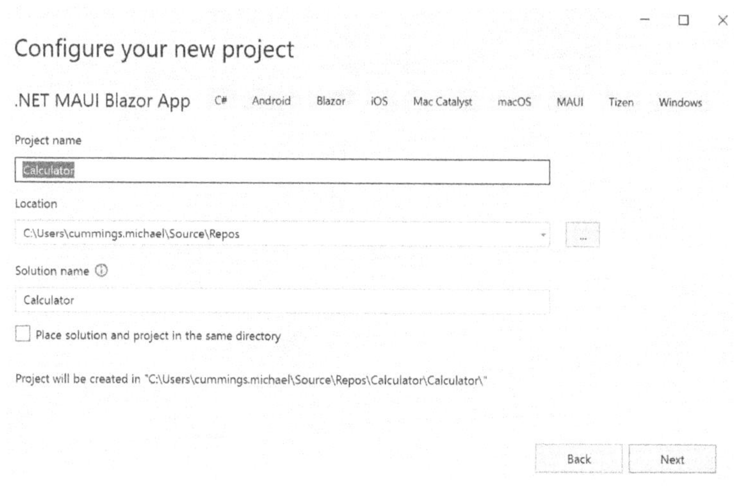

Figure 11.3 – Configure your new project

5. Click **Next**.

The last step will prompt you for the version of .NET Core to support. At the time of writing, .NET 6 is available as **Long-Term Support** (**LTS**), and .NET 7 is available as **Standard Term Support**. For this book, we will assume that you will be using .NET 7:

Figure 11.4 – Additional information

6. Finalize the setup by clicking **Create** and wait for Visual Studio to create the project.

That's it for project creation.

Let's continue by reviewing the structure of the app.

Exploring a .NET MAUI Blazor Hybrid project

If you run the project, you will see an app, as shown in *Figure 11.5*. It does not resemble the .NET MAUI app template and has a distinctive web feel to it. Poke around the app a little bit to see how all the visual elements work together. Then, close the application, return to Visual Studio, and continue exploring the project:

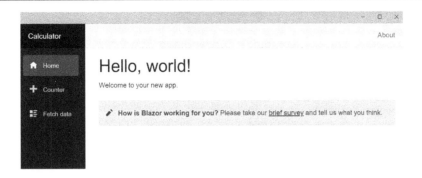

Figure 11.5 – Running the .NET MAUI Blazor template project

The structure of the .NET MAUI Blazor app is a mashup of a .NET MAUI app and a Blazor app. If you look at the **Solution Explorer** window, as shown in *Figure 11.6*, you will see the `Platforms` and `Resources` folders and the `App.xaml`, `MainPage.xaml`, and `MauiProgram.cs` files that are normally present in the .NET MAUI template. The `wwwroot`, `Data`, `Pages`, and `Shared` folders all support the Blazor app. Additionally, you'll find `_Imports.razor` and `Main.razor` at the root of the project:

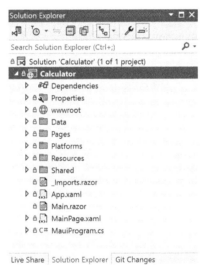

Figure 11.6 – Solution Explorer view of a .NET MAUI Blazor project

If you need a refresher on how a .NET MAUI app is structured and works, refer to *Chapter 1*. Ignoring how a Blazor app works for the moment, let's take a look at how a Blazor app is hosted by .NET MAUI.

Since all .NET MAUI programs start with the `MauiProgram.cs` file, that seems like a good place to start. Open the `MauiProgram.cs` file and inspect its contents. The following snippet highlights the differences for a .NET MAUI Blazor app:

```
builder.Services.AddMauiBlazorWebView();
```

```
#if DEBUG
    builder.Services.AddBlazorWebViewDeveloperTools();
    builder.Logging.AddDebug();
#endif
```

The first highlighted line enables the hosting services for the Blazor app, specifically the WebView controller. The second highlighted line enables the developer tools (*F12*) within the WebView control, but only in a debug configuration.

App.xaml and App.xaml.cs are essentially the same as in a .NET MAUI template project, but MainPage.xaml is not. Open the MainPage.xaml file to inspect its contents, as shown here:

```
<?xml version="1.0" encoding="utf-8" ?>
<ContentPage
    xmlns="http://schemas.microsoft.com/dotnet/2021/maui"
    xmlns:x="http://schemas.microsoft.com/winfx/2009/xaml"
    xmlns:local="clr-namespace:Calculator"
    x:Class="Calculator.MainPage"
    BackgroundColor="{DynamicResource PageBackgroundColor}">
    <BlazorWebView x:Name="blazorWebView" HostPage="wwwroot/index.html">
        <BlazorWebView.RootComponents>
            <RootComponent Selector="#app" ComponentType="{x:Type local:Main}" />
        </BlazorWebView.RootComponents>
    </BlazorWebView>
</ContentPage>
```

MainPage has a single control, BlazorWebView. This is a wrapper around the native controls to host web pages in apps. The HostPage attribute points to the starting page – in this case, wwwroot/index.html. The BlazorWebView.RootComponents element identifies the starting points for the Blazor application and the location where they are hosted on the page. In this case, RootComponent Main is rooted in the element with an ID of app.

To see where the app element is located, open the index.html file in the wwwroot folder and inspect its contents, as shown here:

```
<!DOCTYPE html>
<html lang="en">
<head>
  <meta charset="utf-8" />
  <meta name="viewport" content="width=device-width, initial-scale=1.0, maximum-scale=1.0, user-scalable=no, viewport-fit=cover" />
  <title>Calculator</title>
  <base href="/" />
```

```
    <link rel="stylesheet" href="css/bootstrap/bootstrap.min.css" />
    <link href="css/app.css" rel="stylesheet" />
    <link href="Calculator.styles.css" rel="stylesheet" />
</head>

<body>
    <div class="status-bar-safe-area"></div>
    <div id="app">Loading...</div>
    <div id="blazor-error-ui">
      An unhandled error has occurred.
      <a href="" class="reload">Reload</a>
      <a class="dismiss">✕</a>
    </div>
    <script src="_framework/blazor.webview.js" autostart="false"></
script>
</body>
</html>
```

The key parts of the file are highlighted in the preceding code. First, the `div` element with `id` set to app is where the `Main` component is rooted, or loaded. In the header of the page, the stylesheets are identified. The first stylesheet, `app.css`, is located in the `wwwroot/css` folder of the project. The second stylesheet, `Calculator.Styles.css`, is created during the build process from isolated CSS files. **Isolated CSS** is a feature of ASP.NET that allows you to create your styles close to the elements that use them. They are isolated from other components, so they won't inadvertently bleed into other parts of your application. Finally, the `_framework/blazor.webview.js` file is imported and is responsible for doing all the heavy lifting of getting your Blazor components rendered at the right location on the page.

The final piece for us to review before we move on to creating the rest of the application is the Blazor components. `Main.razor` is a routing file that directs the Blazor runtime to the starting component, `MainLayout.razor`, as shown in the following code:

```
<Router AppAssembly="@typeof(Main).Assembly">
  <Found Context="routeData">
    <RouteView RouteData="@routeData" DefaultLayout="@
typeof(MainLayout)" />
    <FocusOnNavigate RouteData="@routeData" Selector="h1" />
  </Found>
  <NotFound>
    <LayoutView Layout="@typeof(MainLayout)">
      <p role="alert">Sorry, there's nothing at this address.</p>
    </LayoutView>
  </NotFound>
</Router>
```

The `MainLayout.razor` file defines the basic layout of the page, with a navigation bar on the left, and the main body consuming the remainder of the page, as shown in the following code:

```
@inherits LayoutComponentBase
<div class="page">
  <div class="sidebar">
    <NavMenu />
  </div>
  <main>
    <div class="top-row px-4">
      <a href="https://docs.microsoft.com/aspnet/" target="_
blank">About</a>
    </div>
    <article class="content px-4">
      @Body
    </article>
  </main>
</div>
```

The `@Body` content is provided by the page that satisfies the route. For the first page, that would be `/`. If you look at the files in the `Pages` folder, the `Index.razor` file has a `@page` directive with an argument of `/`. So, by default, that is the page that would be displayed. The `@page` directive is a Razor construct that allows the route to be satisfied without using a controller. The `NavLink` entries in the `Shared/NavMenu.razor` file reference the route using the `href` attribute. That value is looked up in the list of `@page` directives to find a match. If no match is found, then the content in the `<NotFound>` element in the `Main.razor` file is rendered.

Open the `Pages/Counter.razor` page to see how a Razor page works:

```
@page "/counter"
<h1>Counter</h1>
<p role="status">Current count: @currentCount</p>
<button class="btn btn-primary" @onclick="IncrementCount">Click me</
button>

@code {
    private int currentCount = 0;
    private void IncrementCount()
    {
        currentCount++;
    }
}
```

After the `@page` directive, the page is made up of HTML mixed with some Razor directives. Then, there's the `@code` directive, which contains the C# code for the page. The HTML `button` element has its click event mapped to the C# `IncrementCount` method using the `@onclick` directive.

> **Learn more**
>
> To learn more about Razor pages, take a look at the official documentation at `https://learn.microsoft.com/en-us/aspnet/core/blazor/?view=aspnetcore-7.0`.

Now, let's start creating the project.

Preparing the project

.NET MAUI and Blazor integrate seamlessly – so seamlessly that it is sometimes hard to differentiate what is executing where. That makes it very easy to render data in both XAML and HTML.

Let's get started by preparing the project for the calculator. We will start by removing much of the template code we won't be using. Follow these steps to prepare the template:

1. In Visual Studio, using **Solution Explorer**, delete the `Data` folder.

2. In the `Pages` folder, delete `Index.razor`, `Counter.Razor`, and `FetchData.razor`.

3. In the `Shared` folder, delete `NavMenu.razor`, `NavMenu.razor.css`, and `SurveyPrompt.razor`.

4. Right-click on the `Pages` folder, then select **Add | Razor Component…**, as shown in the following screenshot:

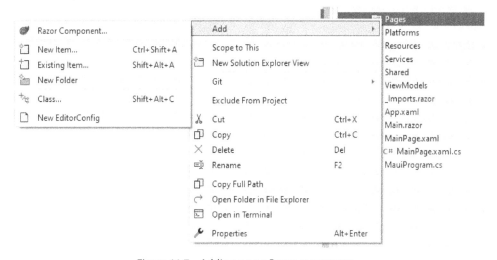

Figure 11.7 – Adding a new Razor component

5.   In the **Add New Item** dialog, name the component `Keypad.razor` and click **Add**:

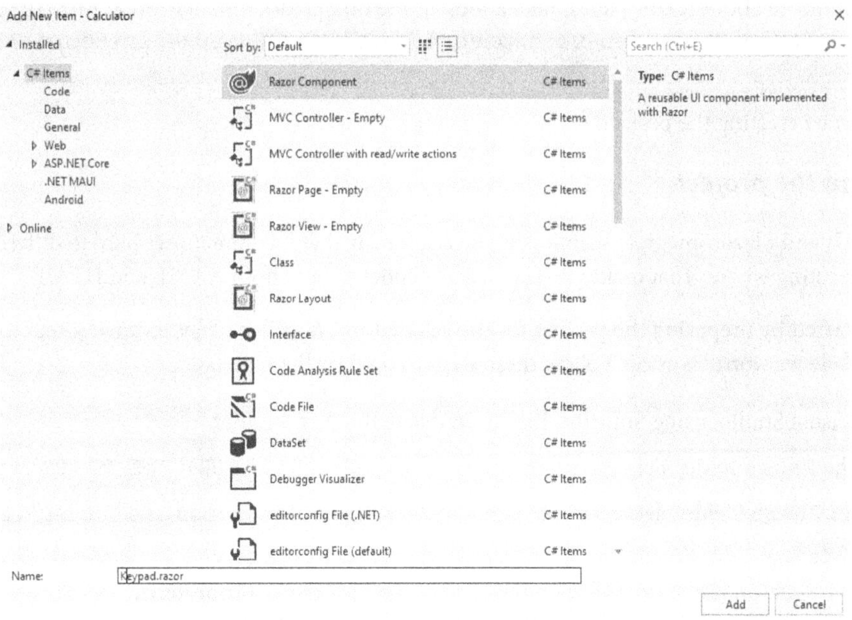

Figure 11.8 – Razor component

6.   In the new `Keypad.razor` file, add the following highlighted lines:

```
@page "/"
<h3>Keypad</h3>
<div>Keypad goes here</div>

@code {
}
```

7.   Open the `MainLayout.razor` file in the `Shared` folder and remove the following highlighted sections:

```
@inherits LayoutComponentBase

<div class="page">
    <div class="sidebar">
      <NavMenu />
    </div>
    <main>
      <div class="top-row px-4">
        <a href="https://docs.microsoft.com/aspnet/" target="_
blank">About</a>
      </div>
```

```
            <article class="content px-4">
                @Body
            </article>
        </main>
    </div>
```

8. Open `MauiProgram.cs` and remove the highlighted lines of code:

```
using Calculator.Data;
using Microsoft.Extensions.Logging;
namespace Calculator
{
    public static class MauiProgram
    {
        public static MauiApp CreateMauiApp()
        {
            var builder = MauiApp.CreateBuilder();
            builder
                .UseMauiApp<App>()
                .ConfigureFonts(fonts =>
                {
                    fonts.AddFont("OpenSans-Regular.ttf",
"OpenSansRegular");
                });
            builder.Services.AddMauiBlazorWebView();

#if DEBUG
            builder.Services.AddBlazorWebViewDeveloperTools();
            builder.Logging.AddDebug();
#endif

            builder.Services.AddSingleton<WeatherForecastService>();

            return builder.Build();
        }
    }
}
```

9. Now, run the project to see the effects of the changes, as shown in *Figure 11.9*:

Keypad

Keypad goes here

Figure 11.9 – Running the blank calculator app on Windows

Now that the project is ready to be an app, let's start with the Keypad view.

Creating the Keypad view

The Keypad view is the basic functionality of the calculator. It has buttons for each digit, 0 through 9, the decimal separator, ., clear, *C*, clear everything, *CE*, and, finally, addition, +, subtraction, -, multiplication, *x*, division, /, = to get the result, and < to erase the last character. The view has three basic components to it – HTML and CSS for styling and the C# code. Each of the following sections will guide you through adding each of these components for the view, starting with the HTML.

Adding the HTML

Razor files and .NET MAUI XAML files do not have a designer for you to visualize your changes. You have to run your app to see the changes you have made, which usually involves building, deploying, and then navigating to the view in your app. .NET has a time-saving feature called Hot Reload. It works by applying the changes you make to your Razor, XAML, CSS, and C# files to the running application without you having to stop and restart the app. Try using **Hot Reload** while you go through the next set of steps. To apply the changes, use the **Hot Reload** button in the **Debug** toolbar. It's easy to find – it's the flames icon:

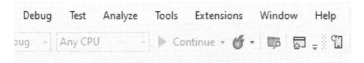

Figure 11.10 – The Hot Reload toolbar button

If at any point you receive a "Rude Edit" dialog like the following, it just means that the change cannot be applied by Hot Reload, so you need to stop, rebuild, and restart debugging:

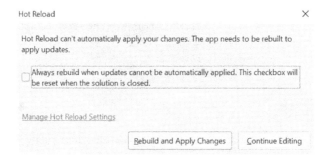

Figure 11.11 – Hot Reload dialog

Hot Reload is constantly getting better, so there will be fewer and fewer of these cases in the future.

To add the HTML that will provide you with the UI to interact with the keypad, follow these steps:

1. Start the app by using the debugger or pressing *F5*.

2. If you want to see your app in Visual Studio, you can open the **XAML Live Preview** pane. To open the **XAML Live Preview** pane, use the Visual Studio menu and select **Debug | Windows | XAML Live Preview**. I usually pin it open so that it is always available.

3. Open the Keypad.razor file in the Pages folder.

4. Remove the highlighted content shown in the following code block:

```
@page "/"
<h3>Keypad</h3>
<div>Keypad goes here</div>
@code {

}
```

5. Add the following HTML after @page but before the @code directive:

```
<div class="keypad">
  <div class="keypad-body">
    <div class="keypad-screen">
      <div class="keypad-typed"></div>
    </div>
    <div class="keypad-row">
      <div class="keypad-button wide command">C</div>
        <div class="keypad-button command">CE</div>
        <div class="keypad-button operator">/</div>
    </div>
    <div class="keypad-row">
      <div class="keypad-button">7</div>
      <div class="keypad-button">8</div>
      <div class="keypad-button">9</div>
      <div class="keypad-button operator">X</div>
    </div>
    <div class="keypad-row">
      <div class="keypad-button">4</div>
      <div class="keypad-button">5</div>
      <div class="keypad-button">6</div>
      <div class="keypad-button operator">-</div>
    </div>
    <div class="keypad-row">
      <div class="keypad-button">1</div>
      <div class="keypad-button">2</div>
      <div class="keypad-button">3</div>
      <div class="keypad-button operator">+</div>
```

```
      </div>
      <div class="keypad-row">
        <div class="keypad-button">.</div>
        <div class="keypad-button">0</div>
        <div class="keypad-button">&lt;</div>
        <div class="keypad-button operator">=</div>
      </div>
    </div>
  </div>
```

Don't forget to press the **Hot Reload** button to see your changes! The keypad is implemented using div and not table since it is easier to style div elements than table elements. The keypad is arranged in rows of four buttons, except for the top two rows. The top row is the display where the expression or result is shown. The second row only has three buttons. The class names have already been added to the HTML elements, but since they don't exist yet, they don't alter the appearance.

The keypad doesn't quite look like a keypad yet, but it will once we add some style to it.

Adding style to the HTML

CSS has been around for a very long time and is the best way to make your HTML look its best. Follow these steps to add the styles we used in the previous section:

1. Make sure your app is still running and use **Solution Explorer** to add a new CSS stylesheet file named `Keypad.razor.css` in the `Pages` folder. Click the **Show All Templates** button if Visual Studio doesn't display them by default:

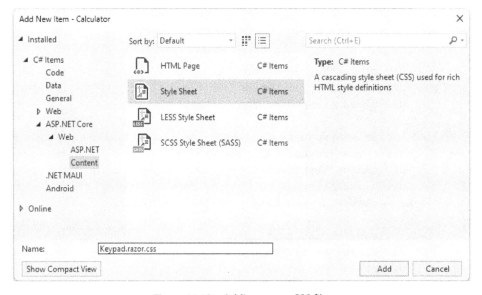

Figure 11.12 – Adding a new CSS file

If you look in **Solution Explorer** under the Pages folder, you will notice that your new file is now under the Keypad.razor file:

Figure 11.13 – A Razor page with an isolated CSS file

Visual Studio automatically recognized that you wanted to add an **isolated CSS** file and made the association. Isolated CSS files are a feature of ASP.NET that allows you to write CSS that only affects the Razor file it's associated with. None of the styles you add to this file can be used outside of the Keypad.razor file.

2. Add the keypad style to the Keypad.razor.css file:

    ```
    .keypad {
        width: 300px;
        margin: auto;
        margin-top: -1.1em;
    }
    ```

    This style will set the width of the element to 300px, with automatic margins, except for the top, which is at -1.1em. -1.1em moves the top edge of the keypad right up against the top of the web view control.

3. Now, add the keypad-body style using the following code:

    ```
    .keypad-body {
        border: solid 1px #3A4655;
    }
    ```

    This style just gives the entire element a border one pixel wide in a dark gray color.

4. We will save the keypad-screen and keypad-typed styles for last, so add the keypad-row style shown in the following code:

    ```
    .keypad-row {
        width: 100%;
        background: #3C4857;
    }
    ```

    This style sets the element width to 100% of its parent, which would be keypad-body, and sets the background to a pleasing dark gray color.

5. The next style to add is keypad-button. Use the following code to add the style:

    ```
    .keypad-button {
    ```

```
    width: 25%;
    background: #425062;
    color: #fff;
    padding: 20px;
    display: inline-block;
    font-size: 25px;
    text-align: center;
    vertical-align: middle;
    margin-right: -4px;
    border-right: solid 2px #3C4857;
    border-bottom: solid 2px #3C4857;
    transition: all 0.2s ease-in-out;
}
```

This style is the base for all the buttons on the keypad, hence it has the most attributes. Elements that have this style applied have a border on the right and bottom that is 2 pixels wide and uses the same color as the background of the row. The `background` attribute of the button is a bit darker than the border color, which provides a bit of depth. The text is aligned in the center both vertically and horizontally and uses a 25-pixel font size. The width is set to 25% since there are four buttons in a row, mostly. The `transition` attribute imposes a 200-millisecond transition using `ease-in-out`, which accelerates from the start to the middle and then decelerates from the middle to the end. `transition` is applied to all attributes, so any time an attribute of this style changes, it will slowly change from the starting value to the end value.

6. If the button is an action button, such as an operator, then an additional style is applied, called `operator`. The definition of this style is slightly different from the rest of the styles that have been created so far. Instead of just being named `operator`, this style is named `keypad-button.operator`. In CSS, `.` is a selector; it is used to locate which attributes to apply. In this case, we want all elements that have both the `keypad-button` class and the `operator` class applied. To add the `keypad-button.operator` class, use the following code:

```
.keypad-button.operator {
    color: #AEB3BA;
    background: #404D5E;
}
```

These buttons will appear with a slightly darker background with slightly less white text.

7. The clear (*C*) and clear everything (*CE*) buttons also have their own class, as follows:

```
.keypad-button.command {
    color: #D95D4E;
    background: #404D5E;
}
```

These buttons will appear with slightly darker backgrounds with red text.

8. Now, for the desktop, we can add hover highlighting by using the :hover pseudo selector. Add the hover styles using the following code:

```
.keypad-button:hover {
    background: #E0B612;
}

.keypad-button.command:hover,
.keypad-button.operator:hover {
    background: #E0B612;
    color: #fff;
}
```

The background is altered to an orange color. The change isn't immediate due to the transition attribute on the keypad-button style, so it will transition from dark gray to orange in two-tenths of a second.

9. The last button-related style is wide, or keypad-button.wide. This style makes the button twice as wide as a normal button. To add this style, use the following code:

```
.keypad-button.wide {
    width: 50%;
}
```

10. The final two styles, keypad-screen and keypad-typed, are for displaying the expression and the result. Use the following code to add the remaining two styles:

```
.keypad-screen {
    background: #3A4655;
    width: 100%;
    height: 75px;
    padding: 20px;
}

.keypad-typed {
    font-size: 45px;
    text-align: right;
    color: #fff;
}
```

Now, the keypad looks like a real calculator keypad; refer to *Figure 11.14* for an example. Were you able to keep adding styles and seeing the changes in your app without having to restart? Remember to click the **Hot Reload** button or set the **Hot Reload on File Save** option under the **Hot Reload** button menu; Visual Studio will attempt to apply the changes as soon as you save the file. Next, we will add the code to enable the buttons to work:

Figure 11.14 – HTML keypad with style

Wiring up the controls

In most web pages, you would use **JavaScript** to write the code to interact with the page. In Blazor, you use C#. To add the C# functions to the Keypad.razor file, follow these steps:

1. Within the @code directive block, add the following:

    ```
    string inputDisplay = string.Empty;
    bool clearInputBeforeAppend = false;
    ```

    This declares a field named inputDisplay as a string element and initializes it to the empty string. It also declares a bool element and initializes it to false. clearInputBeforeAppend is a flag to keep inputDisplay clean. After a result is displayed, when the user taps a button, inputDisplay should be cleared before the character is added to the screen.

2. Update the element with the keypad-typed class, as in the following code:

    ```
    <div class="keypad-typed">@inputDisplay</div>
    ```

    This will render the contents of the inputDisplay variable into the div element. Note the use of @ to reference the C# field.

3. To help with validating the input, add the following:

    ```
    readonly char[] symbols = { '/', 'X', '+', '-', '.' };
    ```

4. When any of the numbers (0 through 9) or the operator buttons are tapped, `inputDisplay` will be updated by appending a character to the display. Add the `AppendInput` method using the following code:

```
void AppendInput(string inputValue)
{
    double numValue;
    if (clearInputBeforeAppend)
    {
        inputDisplay = string.Empty;
    }
    if (string.IsNullOrEmpty(inputDisplay) && inputValue.
IndexOfAny(symbols) != -1)
    {
        return;
    }
    if (!double.TryParse(inputValue, out numValue) && !string.
IsNullOrEmpty(inputDisplay) && $"{inputDisplay[^1]}".
IndexOfAny(symbols) != -1)
    {
        return;
    }
    if (inputDisplay.Trim() == "0" && inputValue == "0")
    {
        return;
    }
    clearInputBeforeAppend = false;
    inputDisplay += inputValue;
}
```

Let's review the code. The first check is to see whether `inputDisplay` needs to be cleared; if so, it's cleared. Then, a check for operators is made. The next check is more complex as it doesn't allow multiple operators in a row. This check uses the `Range` syntax of `^1` to indicate the last character. String interpolation is used to convert the last character back into a string, so `IndexOfAny` can be used to find that character in the array of symbols. Multiple leading 0s are checked and rejected if found. If all the checks pass, the input is appended to `inputDisplay`, and the `clearInputBeforeAppend` flag is reset to `false`.

5. When the user presses the < button, it will remove the last character from the input. Add the `Undo` method using the following code:

```
void Undo()
{
    if (!clearInputBeforeAppend && inputDisplay.Length > 0)
    {
        inputDisplay = inputDisplay[0..^1];
```

```
        return;
    }
}
```

This method uses the `Range` syntax again, and the capability of strings being able to be indexed as arrays. It uses the array syntax to get the elements from index 0 to the next to last index and returns it.

6. When the user presses the **C** button, it will clear the input display. Add the `ClearInput` method using the following code:

```
void ClearInput()
{
    inputDisplay = string.Empty;
}
```

7. When the user presses the **CE** button, it will clear everything. For now, that is just the input display. Use the following code to add the `ClearAll` method:

```
void ClearAll()
{
    ClearInput();
}
```

8. Finally, the = button will take the expression that was entered and evaluate the result. Use the following code to add the `EvaluateExpression` method:

```
void EvaluateExpression()
{
    var expression = inputDisplay;
    clearInputBeforeAppend = true;
}
```

This method doesn't evaluate the expression that's entered yet; that will happen in the *Creating the Compute service* section.

The next step in the `Keypad.razor` file is to hook the methods that were just defined so that they are called when the user taps or clicks on that element. Like in standard HTML, events are hooked up via attributes on the element that reference code, either inline or methods. The attributes in Razor pages use a page directive with the event name. For example, the DOM event for handling taps or clicks is `click`, so the Razor event name would be `@onclick`. That attribute is then assigned to a delegate, which could be any method. The full attribute might look like `@onclick="DoSomething"`, where `DoSomething` is a C# method defined in the page's `@code` directive.

The `AppendInput` method accepts a single string argument, so the delegate cannot be just `AppendInput` – it has to be wrapped in an expression so that the parameter can be passed along.

Expressions in Razor pages are contained in the @(...) directive. All calls to AppendInput from an event directive would look like @onclick="@(AppendInput("0"))".

Use the highlighted lines in the following code to update the HTML in the Keypad.razor file:

```
<div class="keypad">
    <div class="keypad-body">
        <div class="keypad-screen">
            <div class="keypad-typed">@inputDisplay</div>
        </div>
        <div class="keypad-row">
            <div class="keypad-button wide command" @
onclick="ClearInput">C</div>
            <div class="keypad-button command" @
onclick="ClearAll">CE</div>
            <div class="keypad-button operator" @onclick="@
(()=>AppendInput("/"))">/</div>
        </div>
        <div class="keypad-row">
            <div class="keypad-button" @onclick="@
(()=>AppendInput("7"))">7</div>
            <div class="keypad-button" @onclick="@
(()=>AppendInput("8"))">8</div>
            <div class="keypad-button" @onclick="@
(()=>AppendInput("9"))">9</div>
            <div class="keypad-button operator" @onclick="@
(()=>AppendInput("X"))">X</div>
        </div>
        <div class="keypad-row">
            <div class="keypad-button" @onclick="@
(()=>AppendInput("4"))">4</div>
            <div class="keypad-button" @onclick="@
(()=>AppendInput("5"))">5</div>
            <div class="keypad-button" @onclick="@
(()=>AppendInput("6"))">6</div>
            <div class="keypad-button operator" @onclick="@
(()=>AppendInput("-"))">-</div>
        </div>
        <div class="keypad-row">
            <div class="keypad-button" @onclick="@
(()=>AppendInput("1"))">1</div>
            <div class="keypad-button" @onclick="@
(()=>AppendInput("2"))">2</div>
            <div class="keypad-button" @onclick="@
(()=>AppendInput("3"))">3</div>
            <div class="keypad-button operator" @onclick="@
(()=>AppendInput("+"))">+</div>
```

```
        </div>
        <div class="keypad-row">
            <div class="keypad-button" @onclick="@
((()=>AppendInput(".")) ">.</div>
            <div class="keypad-button" @onclick="@
((()=>AppendInput("0")) ">0</div>
            <div class="keypad-button" @onclick="Undo">&lt;</div>
            <div class="keypad-button operator" @
onclick="EvaluateExpression">=</div>
        </div>
    </div>
</div>
```

> **More about Razor event handling**
>
> To learn more about Razor event handling, visit https://learn.microsoft.com/en-us/
> aspnet/core/blazor/components/event-handling?view=aspnetcore-7.0.

The keypad is nearly complete. At this point, you should be able to enter a full expression to evaluate and then clear the display. In the next section, you will create a service to evaluate the expression.

Creating the Compute service

The Compute service evaluates the expression and returns the result. To illustrate how .NET MAUI and the Blazor application can interact with one another, this service will be injected from the .NET MAUI dependency injection container into the Blazor page. To implement the Compute service, follow these steps:

1. Create a new folder in the root of the project named Services.

2. In the Services folder, add a new C# class file named Compute.

3. Modify the Compute.cs file so that it matches the following code:

```
namespace Calculator.Services;

internal class Compute
{
    public string Evaluate(string expression)
    {
        System.Data.DataTable dataTable = new System.Data.
DataTable();
        var finalResult = dataTable.Compute(expression, string.
Empty);
        return finalResult.ToString();
    }
}
```

This code might be shorter than you expected. Instead of writing a lot of code to parse the expression and build an expression tree to evaluate, there is already a built-in way of evaluating simple expressions: `DataTable`. The `DataTable.Compute` method can evaluate all the expressions that can be built from the calculator.

4. Open the `MauiProgram.cs` file and add the following highlighted line of code to register the class with dependency injection:

```
#if DEBUG
    builder.Services.AddBlazorWebViewDeveloperTools();
    builder.Logging.AddDebug();
#endif
    builder.Services.AddSingleton<Compute>();
    return builder.Build();
```

5. To allow the `Keypad.razor` page to use the `Compute` type, it needs to be declared with Razor. Open the `_Imports.razor` file and add the following highlighted line of code to the end of the file:

```
@using Calculator.Services
```

Without this line, you could still use the type, but you would have to fully qualify it as `Calculator.Services.Compute`. This is the Razor equivalent of `global using` directives in C# files.

6. Now, open the `Keypad.razor` file and add the following line of code after the `@page` directive:

```
@inject Compute compute
```

The `@inject` directive will use the .NET MAUI dependency injection container to resolve the type provided as the first parameter and assign it to the variable defined in the second parameter.

7. In the `EvaluateExpression` method, the `compute` variable can be used to evaluate the expression contained in `inputDisplay`, as shown in the following highlighted code:

```
void EvaluateExpression()
{
    var expression = inputDisplay;
    var result = compute.Evaluate(inputDisplay.Replace('X',
'*'));
    inputDisplay = result;
    clearInputBeforeAppend = true;
}
```

Here, the `Evaluate` method is called with `inputDisplay` as the parameter. `inputDisplay` is first modified to replace all the X values in the string with * as that is what the `DataTable` expects for multiplication. The result is then assigned to `inputDisplay`.

At this point, the calculator app can accept input of a combination of numbers and operators and evaluate the result, displaying it to the user. The user can also clear out the display. In the next section, we will explore more interoperability by giving the calculator memory.

Adding memory functions

Most calculators can store previous calculations. In this section, you will add a list of previous calculations to the calculator app and recall previous calculations to the `inputDisplay` parameter of the keypad.

The code will use .NET MAUI controls to render right next to `WebViewControl`. A new class, `Calculations`, will be used to manage the listing.

To add the memory functions to the calculator app, follow these steps:

1. Create a new folder named `ViewModels`.

2. Add a new class named `Calculations` in the `ViewModels` folder and modify the file so that it matches the following code:

```
using System.Collections.ObjectModel;
namespace Calculator.ViewModels;

public class Calculations : ObservableCollection<Calculation>
{
}

public class Calculation : Tuple<string, string>
{
    public Calculation(string expression, string result) :
base(expression, result) { }

    public string Expression => this.Item1;
    public string Result => this.Item2;
}
```

This code adds two classes – `Calculations`, which is just shorthand for `Observable Collection<Calculation>`, and `Calculation`, which is `Tuple<string, string>` and defines two properties to reference `Item1` as `Expression` and `Item2` as `Result`.

3. Add a reference to the `CommunityToolkit.Mvvm` NuGet package.

4. Add a new class named `MainPageViewModel` to the `ViewModels` folder and modify the file, as shown in the following code:

```
using CommunityToolkit.Mvvm.Input;
using CommunityToolkit.Mvvm.Messaging;
namespace Calculator.ViewModels;
```

```
public partial class MainPageViewModel
{
    IMessenger messenger;
    public MainPageViewModel(Calculations results, IMessenger
messenger)
    {
        Results = results;
        this.messenger = messenger;
    }

    public Calculations Results { get; init; }
    [RelayCommand]
    public void Recall(Calculation sender)
    {
        messenger.Send(sender);
    }
}
```

The MainViewModel class uses two features of CommunityToolkit: RelayCommand and IMessenger. As in the other chapters, RelayCommand is used to bind methods to XAML actions. IMessenger is an interface for sending messages between parts of your application. It is helpful when you don't want to take a hard dependency between two classes, especially if it creates a circular reference. CommunityToolkit provides a default implementation of IMessenger called WeakReferenceMessenger.

5. Open MauiProgram.cs and add the following using declarations to the top of the file:

```
using Calculator.ViewModels;
using CommunityToolkit.Mvvm.Messaging;
```

6. In the CreateMauiApp method make the following highlighted changes:

```
builder.Services.AddSingleton<Compute>();
builder.Services.AddSingleton<Calculations>();
builder.Services.AddSingleton<MainPage>();
builder.Services.AddSingleton<MainPageViewModel>();

builder.Services.
AddSingleton<IMessenger>(WeakReferenceMessenger.Default);

return builder.Build();
```

This adds MainPage, MainPageViewModel, and the default instance of WeakReferenceMessenger to the dependency injection container. The next few steps will enable MainPage to be initialized via dependency injection.

7. Open `MainView.xaml.cs` and make the following highlighted changes:

```
using Calculator.ViewModels;
namespace Calculator;

public partial class MainPage : ContentPage
{
    public MainPage(MainPageViewModel vm)
    {
        InitializeComponent();
        BindingContext = vm;
    }
}
```

As in other chapters, the constructor for the view is updated to accept the view model as a parameter, and it is assigned as `BindingContext`.

8. Open `App.xaml.cs`, modify the constructor, and add the `OnHandlerChanging` event handler, as shown in the following code:

```
public App()
{
    InitializeComponent();
}
protected override void
OnHandlerChanging(HandlerChangingEventArgs args)
{
    base.OnHandlerChanging(args);
    MainPage = args.NewHandler.MauiContext.Services.
GetService<MainPage>();
}
```

Since a .NET MAUI Blazor app does not use `Shell` by default, the views cannot be initialized via dependency injection the same way as is done with `Shell`. Instead, the instance of `MainPage` is obtained after `Handler` has been set. The `OnHandlerChanging` event is used to get the reference to the new `Handler`, which then provides `MauiContext` to the dependency injection container.

9. Open `_Imports.razor` and add the following lines to the end of the file:

```
@using Calculator.ViewModels
@using CommunityToolkit.Mvvm.Messaging
```

10. Open `Keypad.razor` and add the following highlighted lines:

```
@inject Compute compute
@inject Calculations calculations
@inject IMessenger messenger
```

```
<div class="keypad">
```

This will inject the `Calculations` instance as `calculations` and `WeakReference` `Messenger` as `messenger` from the .NET MAUI dependency injection container.

11. Modify the `ClearAll` and `EvaluateExpression` methods and add an `OnAfter RenderAsync` method, as shown in the following code:

```
void ClearAll()
{
    ClearInput();
    calculations.Clear();
}

void EvaluateExpression()
{
    var expression = inputDisplay;
    var result = compute.Evaluate(inputDisplay.Replace('X',
'*'));
    calculations.Add(new(expression, result));
    inputDisplay = result;
    clearInputBeforeAppend = true;
}

protected override Task OnAfterRenderAsync(bool firstRender)
{
    if (firstRender)
    {
        messenger.Register<Calculation>(this, (sender, er) =>
        {
            inputDisplay = er.Expression;
            clearInputBeforeAppend = true;
            StateHasChanged();
        });
    }
    return base.OnAfterRenderAsync(firstRender);
}
```

`ClearAll` will just clear the collection, and `EvaluateExpression` will add the new `Calulation` to the collection. `OnAfterRenderAsync` is used to `register` this class to receive messages for any `Calculation` objects. When a message is received, `inputDisplay` is set to the `Expression` value of `Calculation`, and `StateHasChanged` is called to force the UI to refresh with the updated value.

12. Open the `Shared/MainLayout.razor.css` file and add the following line to the page class:

```
background-color: black;
```

This is simply for aesthetics and makes the areas surrounding the calculator black.

13. Open the `MainPage.xaml` file and modify it so that it matches the following code:

```
<ContentPage
    xmlns="http://schemas.microsoft.com/dotnet/2021/maui"

    xmlns:x="http://schemas.microsoft.com/winfx/2009/xaml"
    xmlns:local="clr-namespace:Calculator"
    xmlns:viewModels="clr-namespace:Calculator.ViewModels"
    x:Class="Calculator.MainPage"
    x:DataType="viewModels:MainPageViewModel">
    <Grid BackgroundColor="Black">
      <Grid.RowDefinitions>
        <RowDefinition Height="1*" />
        <RowDefinition Height="1*" />
      </Grid.RowDefinitions>

        <ScrollView Grid.Row="0" BackgroundColor="Bisque"
WidthRequest="400" VerticalScrollBarVisibility="Always">
        <CollectionView ItemsSource="{Binding Results}"
ItemsUpdatingScrollMode="KeepLastItemInView">
          <CollectionView.ItemTemplate>
            <DataTemplate
x:DataType="viewModels:ExpressionResult">
              <SwipeView>
                <SwipeView.LeftItems>
                  <SwipeItems Mode="Execute">
                    <SwipeItem
Text="Recall"                            BackgroundColor=
"LightPink"
                                          Command="
{Binding Source=
{RelativeSource AncestorType={x:Type
viewModels:MainPageViewModel}},
Path=RecallCommand}"
CommandParameter="{Binding}"/>
                  </SwipeItems>
                </SwipeView.LeftItems>
                <VerticalStackLayout>
                  <HorizontalStackLayout Padding="10"
HorizontalOptions="EndAndExpand">
                    <Label Text="{Binding Expression}"
FontSize="Large" TextColor="Black" HorizontalTextAlignment="End"
```

```
                HorizontalOptions="EndAndExpand"/>
                              <Label Text="=" TextColor="Blue"
        FontSize="Large" HorizontalTextAlignment="End"
        HorizontalOptions="EndAndExpand"/>
                              <Label Text="{Binding Result}"
        FontSize="Large" TextColor="Black" HorizontalTextAlignment="End"
        HorizontalOptions="EndAndExpand"/>
                        </HorizontalStackLayout>
                        <Line Stroke="LightSlateGray" X2="400" />
                        <Line Stroke="Black" X2="400" />
                      </VerticalStackLayout>
                    </SwipeView>
                  </DataTemplate>
                </CollectionView.ItemTemplate>
              </CollectionView>
          </ScrollView>

          <BlazorWebView Grid.Row="1" x:Name="blazorWebView"
        HostPage="wwwroot/index.html" HeightRequest="540">
              <BlazorWebView.RootComponents>
                <RootComponent Selector="#app" ComponentType="{x:Type
        local:Main}" />
              </BlazorWebView.RootComponents>
          </BlazorWebView>
        </Grid>
      </ContentPage>
```

A `Grid` element with two rows defined is used to contain the new display for previous calculations
and the original `BlazorWebView` control. The calculations are rendered using `ScrollView`,
which contains `CollectionView`. The `CollectionView.ItemTemplate` property
contains `DataTemplate` for each `Calculation`. The `SwipeView` control allows the
user to swipe in either the up, down, left, or right direction to expose additional commands.
Each direction has an element to define those actions. For `Calculation` items, when the
user swipes right, it exposes a `Recall` item, which is bound to the `Recall` command of
`MainPageViewModel`. The display for `Calculation` uses a combination of horizontal
and vertical `StackLayout` controls to stack `Expression` on top of `Result` with =, all
aligned to the left.

That completes the main functionality of our calculator app. The next section deals with some aesthetics
of the main window when running on a desktop such as Windows or macOS.

Resizing the main window

The calculator app is defined to be a fixed size. Most of the projects in this book have allowed the
controls to grow or shrink with changes to the window size. For our calculator app, the main window

should be fixed so that it appears in the best way possible. To fix the window size when the app launches, open the App.xaml.cs file and add the following method to the App class:

```
protected override Window CreateWindow(IActivationState
activationState)
{
    var window = base.CreateWindow(activationState);
    if (OperatingSystem.IsWindows() || OperatingSystem.
IsMacCatalyst())
    {
        window.Created += Window_Created;
    }
    return window;
}

private async void Window_Created(object sender, EventArgs e)
{
    const int defaultWidth = 450;
    const int defaultHeight = 800;

    var window = (Window)sender;
    window.Width = defaultWidth;
    window.Height = defaultHeight;
    window.X = -defaultWidth;
    window.Y = -defaultHeight;

    await window.Dispatcher.DispatchAsync(() => { });

    var displayInfo = DeviceDisplay.Current.MainDisplayInfo;
    window.X = (displayInfo.Width / displayInfo.Density - window.
Width) / 2;
    window.Y = (displayInfo.Height / displayInfo.Density - window.
Height) / 2;

    window.Created -= Window_Created;
}
```

The CreateWindow method is overridden so that the Window.Created event can have a custom handler attached to it if the app is running on Windows or macOS. The work of resizing the window is done in the Window_Created method. It uses the defaultHeight and defaultWidth constants to set the new windows' height, width, and location on the screen. The method then waits for all threads to have finished before altering the X and Y properties of the window again, but this time accounting for the screen pixel density. Finally, it detaches the event handler

Figure 11.15 – The completed calculator project

Summary

Excellent work! In this chapter, you completed a project that used the .NET MAUI Blazor template. You created a UI using HTML, updated it with C# code, and then implemented a service that was managed by .NET MAUI and injected it into a Razor page. Then, you used `CollectionView` to display a list of the previous calculations. Within `CollectionView` `ItemTemplate`, the `SwipeView` control was used to recall a previous calculation to the keypad for additional editing and reevaluation.

To extend this project even further, consider the following:

- Add an additional swipe action to delete a calculation from the collection
- Add an additional keypad layout for scientific calculations

In the next – and final – chapter, you will discover the world of AI as you build an object recognition app.

12
Hot Dog or Not Hot Dog Using Machine Learning

In this chapter, we will learn how to use machine learning to create a model that we can use for image classification. We will export the model as an **Onnx** model that we can use on all platforms – that is, Android, iOS, macOS, and Windows. To train and export models, we will use Azure Cognitive Services and the Custom Vision service.

Once we have exported the models, we will learn how to use them in a .NET MAUI app.

The following topics will be covered in this chapter:

- Training a model with Azure Cognitive Services and the Custom Vision service
- Using Onnx models for image classification using ML.NET
- Using custom routes in .NET MAUI for navigation

Technical requirements

To be able to complete this project, you need to have Visual Studio for Mac or PC installed, as well as the .NET MAUI components. See *Chapter 1, Introduction to .NET MAUI,* for more details on how to set up your environment. You also need an Azure account. If you have a Visual Studio subscription, there are a specific amount of Azure credits included each month. To activate your Azure benefits, go to `https://my.visualstudio.com`.

You can also create a free account, where you can use selected services for free over 12 months. You will get $200 worth of credit to explore any Azure service for 30 days, and you can also use the free services at any time. Read more at `https://azure.microsoft.com/en- us/free/`.

If you do not have and do not want to sign up for a free Azure account, the trained model is available in the source code for this chapter. You can download and use the pre-trained model instead.

The source code for this chapter is available at the GitHub repository for the book at `https://github.com/PacktPublishing/MAUI-Projects-3rd-Edition`.

Machine learning

The term **machine learning** was coined in 1959 by Arthur Samuel, an American pioneer in **artificial intelligence** (**AI**). Tom M. Mitchell, an American computer scientist, provided the following more formal definition of machine learning later:

> *"A computer program is said to learn from experience E with respect to some class of tasks T and performance measure P if its performance at tasks in T, as measured by P, improves with experience E."*

In simpler terms, this quote describes a computer program that can learn without being explicitly programmed. In machine learning, algorithms are used to build a mathematical model of sample data or training data. The models are used for computer programs to make predictions and decisions without being explicitly programmed for the task in question.

In this section, we will learn about a few different machine learning services and APIs that are available when developing a .NET MAUI application. Some APIs are only available for specific platforms, such as Core ML, while others are cross-platform.

Azure Cognitive Services – Custom Vision

Custom Vision is a tool or service that can be used to train models for image classification and to detect objects in images. With Custom Vision, we can upload our own images and tag them so that they can be trained for image classification. If we train a model for object detection, we can also tag specific areas of an image. Because models are already pre-trained for basic image recognition, we don't need a large amount of data to get a great result. The recommendation is to have at least 30 images per tag.

When we have trained a model, we can use it with an API, which is part of the Custom Vision service. We can also export models for **Core ML (iOS)**, **TensorFlow (Android)**, the **Open Neural Network Exchange** (**ONNX**), and a **Dockerfile** (**Azure IoT Edge**, **Azure Functions**, and **Azure ML**). These models can be used to carry out classification or object detection without being connected to the Custom Vision service.

You will need an Azure subscription to use it – go to `https://azure.com/free` to create a free subscription, which should be enough to complete this project.

Core ML

Core ML is a framework that was introduced in iOS 11. Core ML makes it possible to integrate machine learning models into iOS apps. On top of Core ML, we have high-level APIs, as follows:

- Vision APIs for image analysis

- Natural language APIs for natural language processing

- Speech to convert audio to text

- Sound analysis to identify sounds in audio

- GameplayKit to evaluate learned decision trees and strategies

> **More information**
>
> More information about Core ML can be found in the official documentation from Apple at `https://developer.apple.com/ documentation/coreml`.

TensorFlow

TensorFlow is an open source machine learning framework. However, TensorFlow can be used for more than simply running models on mobile devices – it can also be used to train models. To run it on mobile devices, we have TensorFlow Lite. The models that are exported from Azure Cognitive Services are for TensorFlow Lite. There are also C# bindings for TensorFlow Lite that are available as a NuGet package.

> **More information**
>
> More information about TensorFlow can be found in the official documentation at `https://www.tensorflow.org/`.

ML.Net

ML.NET is an open source and cross-platform machine learning framework with support for iOS, macOS, Android, and Windows, all from a familiar environment – C#. ML.NET provides **AutoML**, a set of productivity tools that make building, training, and deploying custom models easy. ML.NET can be used in the following scenarios and more:

- Sentiment analysis and product recommendation

- Object detection and image classification

- Price prediction, sales spike detection, and forecasting
- Fraud detection
- Customer segmentation

Now that we have a broad overview of the technologies at play, we will focus on using ML.NET, since it is a cross-platform framework and built for C#. Let's look at the project we are going to build next.

The project overview

If you have seen the TV series *Silicon Valley*, you have probably heard of the *Not Hotdog* application. In this chapter, we will learn how to build that app. The first part of this chapter will involve collecting the data that we will use to create a machine learning model that can detect whether a photo contains a hot dog.

In the second part of the chapter, we will build an app using .NET MAUI and ML.NET, where the user can either take a new photo or pick a photo in the photo library, analyzing it to see whether it contains a hot dog. The estimated time for completing this project is 120 minutes.

Getting started

We can use either Visual Studio 2022 on a PC or Visual Studio for Mac to do this project. To build an iOS app using Visual Studio for PC, you must have a Mac connected. If you don't have access to a Mac at all, you can choose to just do the Android and Windows parts of this project.

Similarly, if you only have a Mac, you can choose to just do the iOS and macOS or Android parts of this project.

Building the Hot Dog or Not Hot Dog application using machine learning

Let's get started! We will first train a model for image classification that we can use later in the chapter to decide whether a photo contains a hot dog.

> **Note**
>
> If you do not want to go through the effort of training a model, you can download a pre-trained model from the following URL: `https://github.com/PacktPublishing/MAUI-Projects-3rd-Edition/tree/main/Chapter12/HotdogOrNot/Resources/Raw`.

Training a model

To train a model for image classification, we need to collect photos of hot dogs and photos that aren't of hot dogs. Because most items in the world are not hot dogs, we need more photos that don't contain hot dogs. It's better if the photos of hot dogs cover a lot of different hot dog scenarios – with bread, ketchup, or mustard. This is so that the model will be able to recognize hot dogs in different situations. When we collect photos that aren't of hot dogs, we also need to have a large variety of photos that are both of items that are like hot dogs and that are completely different from hot dogs.

The model that is in the solution on GitHub was trained with 240 photos, 60 of which were of hot dogs, and 180 of which were not.

Once we have collected all the photos, we will be ready to start training the model by following these steps:

1. Go to `https://customvision.ai`.

2. Log in and create a new project.

3. Give the project a name – in our case, `HotDogOrNot`.

4. Select a resource or create a new one by clicking **Create new**. Fill in the dialog box, and select **CustomVision.Training** in the **Kind** dropdown.

   The project type should be **Classification**, and the classification type should be **Multiclass (Single tag per image)**.

5. Select **General (compact)** as the domain. We use a compact domain if we want to export models and run them on a mobile device.

6. Click **Create project** to continue, as shown in the following screenshot:

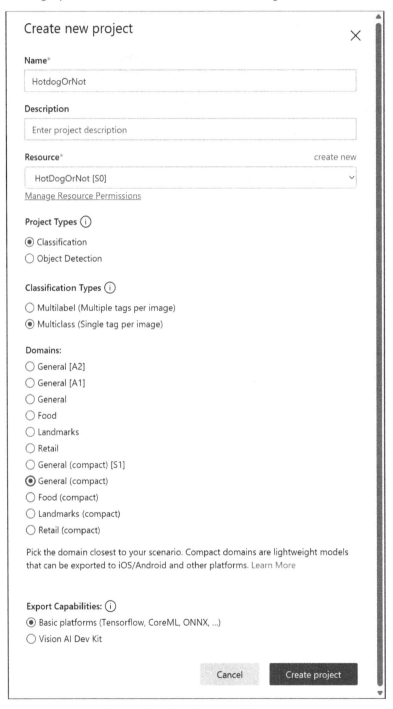

Figure 12.1 – Creating a new AI project

Once we have created a project, we can start to upload images and tag them.

Tagging images

The easiest way to get images is to go to Google and search for them. We will start by adding photos of hot dogs by following these steps:

1. Click **Add images**.
2. Select the photos of hot dogs that should be uploaded.
3. Tag the photos with hotdog, as shown in the following screenshot:

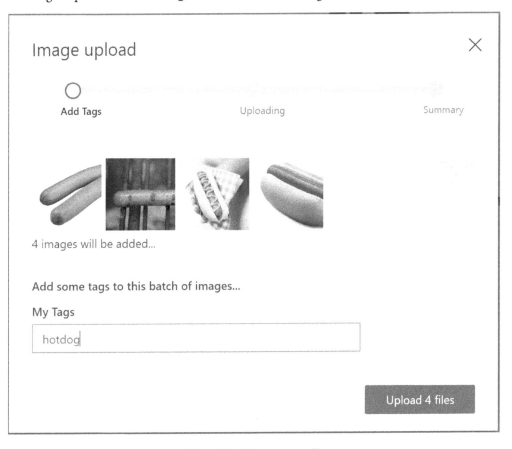

Figure 12.2 – Uploading images of hot dogs

Once we have uploaded all the photos of hot dogs, it's time to upload photos that aren't of hot dogs by following the following steps. For best results, we should also include photos of objects that look similar to hot dogs but are not:

1. Click the **Add images** button above the gallery of uploaded images.

2. Select the photos that aren't of hot dogs.

3. Tag the photos with `Negative`.

    A Negative tag is used for photos that don't contain any objects that we have created other tags for. In this case, none of the photos we will upload contain hot dogs, as can be seen in the following screenshot:

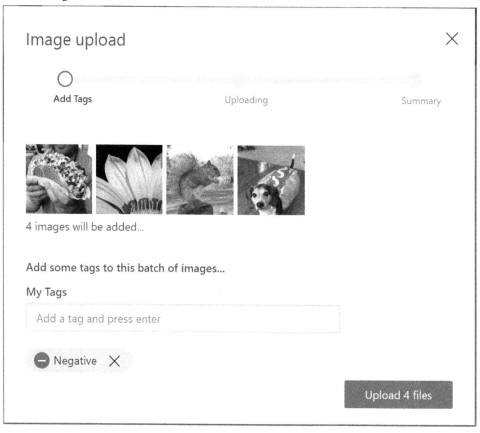

Figure 12.3 – Uploading images that aren't hot dogs

Once we have uploaded the photos, it's time to train a model.

Training a model

Not all the photos that we are uploading will be used for training; some will be used for verification to give us a score about how good the model is. If we upload photos in chunks and train the model after each chunk, we will be able to see our scores improving. To train a model, click the green **Train** button at the top of the page, as illustrated in the following screenshot:

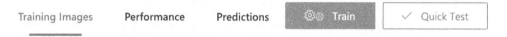

Figure 12.4 – Training the model

The following screenshot shows the result of a training iteration where the precision of the model is **91.7%**:

Iteration 2

Finished training on **2020-01-30 08:11:11** using **General (compact)** domain
Iteration id: **e435fe7f-33ca-4f4d-89d0-a262aeac618b**
Classification type: **Multiclass (Single tag per image)**

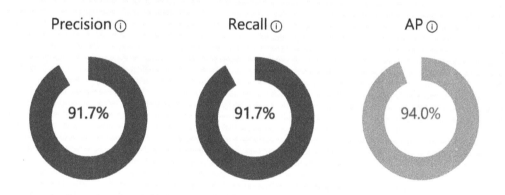

Precision ⓘ	Recall ⓘ	AP ⓘ
91.7%	91.7%	94.0%

Performance Per Tag

Tag	Precision ∧	Recall	A.P.	Image count ⚠
Negative	92.1%	97.2%	97.4%	180
hotdog	90.0%	75.0%	80.4%	60

Figure 12.5 – Model verification results

Once we have trained a model, we will export it so that it can be used on a device.

Exporting a model

We can use the APIs if we want to, but to make fast classifications and to be able to do this offline, we will add the models to the app packages. Click the **Export** button and then on **ONNX** to download the model, as shown in the following screenshot:

Figure 12.6 – Exporting the model

Once we have downloaded the ONNX model, it's time to build the app.

Building the app

Our app will use the trained models to classify photos, according to whether they are photos of hot dogs. We will use the same ONNX model for all platforms in the .NET MAUI app.

Creating the new project

Let's begin, as follows.

The first step is to create a new .NET MAUI projec:

1.   Open Visual Studio 2022, and select **Create a new project**:

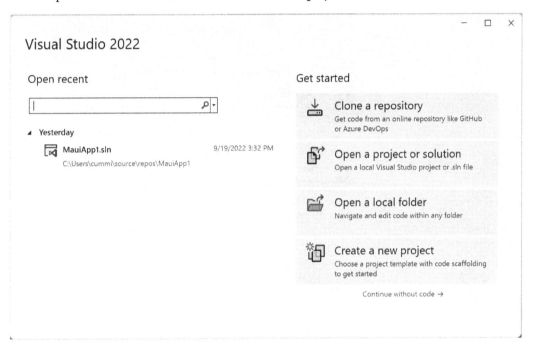

Figure 12.7 – Visual Studio 2022

This will open the **Create a new project** wizard.

2. In the search field, type in `maui`, and select the **.NET MAUI App** item from the list:

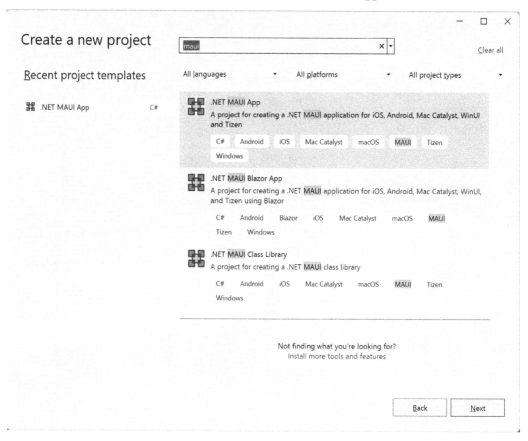

Figure 12.8 – Create a new project

3. Click **Next**.

4. Complete the next step of the wizard by naming your project. We will call our application `HotdogOrNot` in this case. Move on to the next dialog box by clicking **Next**, as illustrated in the following screenshot:

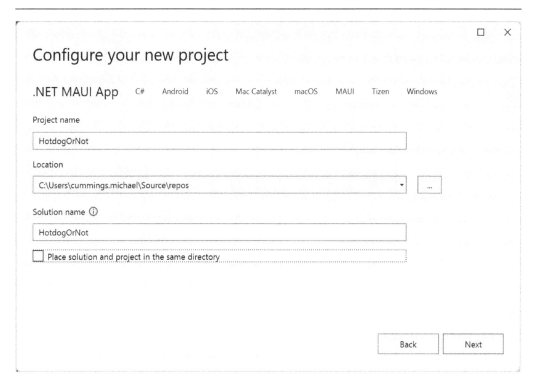

Figure 12.9 – Configure your new project

5.   The last step will prompt you for the version of .NET Core to support. At the time of writing, .NET 6 is available as **Long-Term Support** (**LTS**), and .NET 7 is available as **Standard Term Support**. For the purposes of this book, we will assume that you are using .NET 7.

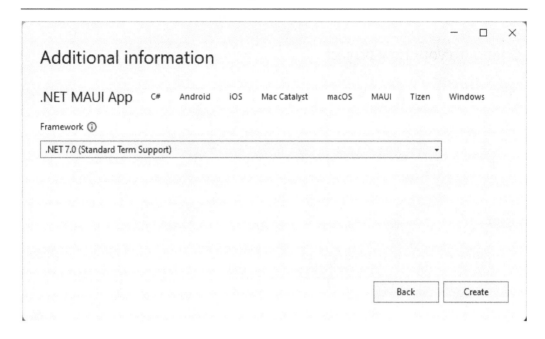

Figure 12.10 – Additional information

6.   Finalize the setup by clicking **Create**, and wait for Visual Studio to create the project.

If you run the app now, you should see something like the following:

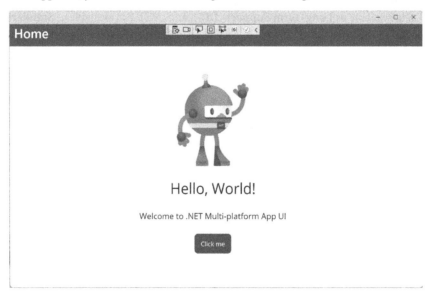

Figure 12.11 – The HotdogOrNot applicaton

Just like that, the app is created. Next, let's start creating the image classifier.

Classifying images with machine learning

The first thing we will do is add the ONNX ML model to the project, by following these steps:

1. Extract the `.zip` file that we got from the Custom Vision service.

2. Find the `.onnx` file, and rename it `hotdog-or-not.onnx`.

3. Add it to the `Resources/Raw` folder in the project.

Once we add the file to the project, we are ready to create the implementation of the image classifier. The code that we will use for image classification will be shared between the .NET MAUI-supported platforms. We can create an interface for the classifier by following these steps:

1. Create a new folder named `ImageClassifier`.

2. Create a new class called `ClassifierOutput` in the `ImageClassifier` folder.

3. Modify the `ClassifierOutput` class to look like the following:

```
namespace HotdogOrNot.ImageClassifier;

internal sealed class ClassifierOutput
{
    ClassifierOutput() { }
}
```

4. Create a new interface called `IClassifier` in the `ImageClassifier` folder.

5. Add a method called `Classify` that returns `ClassifierOutput` and takes `byte[]` as an argument.

6. Your interface should look like the following code block:

```
namespace HotdogOrNot.ImageClassifier;

public interface IClassifier
{
    ClassifierOutput Classify(byte[] bytes);
}
```

Now that we have the interface for the classifier, we can move on to the implementation.

Using ML.NET for image classification

We are now ready to create the implementation of the `IClassifier` interface. Before we jump right into the implementation, let's take a look at the high-level steps that will need to happen so that we understand the flow a little better.

Our trained model, `hotdog-or-not.onnx`, has specific input and output parameters, and we will need to convert the image that we want to classify into the input format before submitting it to the ML.NET framework. Additionally, we need to ensure that the image is in the correct shape before submitting it. The shape of the image is defined by the size, width, height, and color format. If the image does not match the input format, then it needs to be resized and converted before submission, or you will run the risk of the image being classified incorrectly. For image classification models that are generated by the Custom Vision service, such as the *hotdog-or-not* model, the inputs and outputs look like the following:

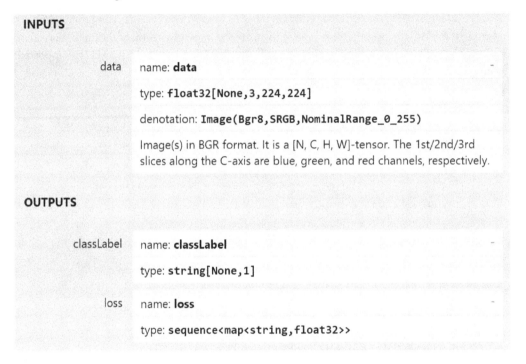

Figure 12.12 – Model inputs and outputs from Netron

The input to the model is formatted into a multidimensional array named `data`. There are four dimensions that make up the array:

- **The image**: The format allows you to submit multiple images at once; however, for this app, we will only submit one image at a time

- **The color**: Each index in this dimension represents a single-color channel – 0 is blue, 1 is green, and 2 is red

- **The height**: Each index is a position along the *y*, or vertical, axis of an image, in the range between 0 and 223

- **The width**: Each index is a position along the *x*, or horizontal, axis of an image, in the range between 0 and 223

The value is the color value for that specific image, color, and *x* and *y* positions. For example, data[0,2,64,64] would be the value of the green channel in the first image at a position of 64 pixels from the left and 64 pixels from the bottom of the image.

To reduce the number of incorrect classifications, we need to scale all the images for submission to 224 x 224 pixels and order the color channels properly.

We can do that by following these steps:

1. Create a new class called MLNetClassifier in the ImageClassifier folder of the project.
2. Add the IClassifier interface.
3. Implement the Classify method from the interface, as shown in the following code block:

```
namespace HotdogOrNot.ImageClassifier;

internal class MLNetClassifier : Iclassifier
{
    public MLNetClassifier(byte[] model)
    {
        // Initialize Model here
    }

    public ClassifierOutput Classify(byte[] imageBytes)
    {
        // Code will be added here
    }
}
```

So far, we have not referenced any classes from ML.NET. To use the ML.NET APIs, we will need to add a reference to the NuGet package, by following these steps:

1. In the project, install the Microsoft.ML.OnnxRuntime NuGet package.
2. Accept any license dialog boxes.

This will install the relevant NuGet packages.

Now that we are referencing the ML.NET package, we can compile the ONNX ML model by following these steps:

1. At the top of the MLNetClassifier file, add the using Microsoft.ML.Onnx Runtime; declaration.

2. In the MLNetClassifier class, add the following fields:

```
readonly InferenceSession session;
readonly bool isBgr;
readonly bool isRange255;
readonly string inputName;
readonly int inputSize;
```

3. In the MLNetClassifier constructor, add the following lines of code to initialize the OnnxRuntime session, replacing the // Initialize Model here comment:

```
Session = new InferenceSession(model);
isBgr = session.ModelMetadata.CustomMetadataMap["Image.
BitmapPixelFormat"] == "Bgr8";
isRange255 = session.ModelMetadata.
CustomMetadataMap["Image.NominalPixelRange"] ==
"NominalRange_0_255";
inputName = session.InputMetadata.Keys.First();
inputSize = session.InputMetadata[inputName].
Dimensions[2];
```

Let's discuss the preceding code before moving on. The constructor for the MLNetClassifier class accepts byte[] as a parameter. This represents the ML model file. byte[] is then passed into a new instance of InferenceSession, which is the main entry point into the ML.NET API. Once the model is loaded into the session, we can then inspect the model for certain properties, such as the image format (isBGR), the color value range (isRange255), the input name, and the input size. We cache these values in the class fields for use during classification. Your MLNetClassifier class should now look like the following:

```
using Microsoft.ML.OnnxRuntime;

namespace HotdogOrNot.ImageClassifier;

internal class MLNetClassifier : Iclassifier
{
    readonly InferenceSession session;
    readonly bool isBgr;
    readonly bool isRange255;
    readonly string inputName;
    readonly int inputSize;

    public MLNetClassifier(byte[] model)
    {
        session = new InferenceSession(model);
        isBgr = session.ModelMetadata.CustomMetadataMap["Image.
BitmapPixelFormat"] == "Bgr8";
```

```
        isRange255 = session.ModelMetadata.
CustomMetadataMap["Image.NominalPixelRange"] ==
"NominalRange_0_255";
        inputName = session.InputMetadata.Keys.First();
        inputSize = session.InputMetadata[inputName].
Dimensions[2];
    }

    public ClassifierOutput Classify(byte[] imageBytes)
    {
        // Code will be added here
    }
}
```

We can now move on to implementing the `Classify` method of the `MLNetClassifier` class.

The first step in running a classification is to get the input into the correct format. For image classification, that means resizing the image to the right dimensions and organizing the color values into the expected format. The image data is then loaded into `Tensor`, which is how we pass data into an ML.NET model. The following steps will create a method named `LoadInputTensor` to do just that:

1. Add a new method named `LoadInputTensor` after the `Classify` method in the `MLNetClassifier` class. This method will accept four parameters, `byte[]`, `int`, and two Booleans, and return a tuple of `Tensor<float>` and `byte[]`. Your method should look like the following:

   ```
   static (Tensor<float>, byte[] resizedImage)
   LoadInputTensor(byte[] imageBytes, int imageSize, bool isBgr,
   bool isRange255)
       {
       }
   ```

2. Inside `LoadInputTensor`, we will create the `return` objects and add the following highlighted lines of code:

   ```
       {
           var input = new DenseTensor<float>(new[] { 1, 3,
       imageSize, imageSize });
           byte[] pixelBytes;

           // Add code here

           return (input, pixelBytes);
       }
   ```

The next step is to resize the image; we will use the **ImageSharp** NuGet library to make this very easy.

3. Add the ImageSharp NuGet package to the project.

4. Add the following lines of code to resize the image, replacing the \\ Add code here comment:

```
using (var image = Image.Load<Rgb24>(imageBytes))
{
    image.Mutate(x => x.Resize(imageSize, imageSize));
    pixelBytes = new byte[image.Width * image.Height * Unsafe.
SizeOf<Rgba32>()];

    image.ProcessPixelRows(source =>
    {
        // Add Code here
    });
}
```

This code uses the ImageSharp library to load the image from byte[]. The image is then resized to the size required by the model. We use the imageSize field, whose value captures the model requirement from the constructor. Finally, we set up a call to the ProcessPixelRows method that will allow us to manipulate the individual pixels in the image.

5. Due to conflicts in naming between .NET MAUI and ImageSharp, we must add a declaration that tells the compiler which class we really want to use at the top of the file:

```
using Image = SixLabors.ImageSharp.Image;
```

6. The next section of code will also need the following highlighted declarations:

```
using Microsoft.ML.OnnxRuntime;
using SixLabors.ImageSharp.Formats.Png;
using Microsoft.ML.OnnxRuntime.Tensors;
using Image = SixLabors.ImageSharp.Image;
```

7. To get the input image into the correct color format required by the model, we use the ProcessPixelRows method from the ImageSharp library. This method provides a writable buffer for us to manipulate. Use the following highlighted code, in place of the // Add Code here comment, to iterate over the resized image data, putting the color values into the right order and clamping the values between 0 and 255, if required:

```
image.ProcessPixelRows(source =>
{
    for (int y = 0; y < image.Height; y++)
    {
        Span<Rgb24> pixelSpan = source.GetRowSpan(y);
```

```
        for (int x = 0; x < image.Width; x++)
        {
            if (isBgr)
            {
                input[0, 0, y, x] = pixelSpan[x].B;
                input[0, 1, y, x] = pixelSpan[x].G;
                input[0, 2, y, x] = pixelSpan[x].R;
            }
            else
            {
                input[0, 0, y, x] = pixelSpan[x].R;
                input[0, 1, y, x] = pixelSpan[x].G;
                input[0, 2, y, x] = pixelSpan[x].B;
            }

            if (!isRange255)
            {
                input[0, 0, y, x] = input[0, 0, y, x] / 255;
                input[0, 1, y, x] = input[0, 1, y, x] / 255;
                input[0, 2, y, x] = input[0, 2, y, x] / 255;
            }
        }
    }
});
```

What this code does is simple – using the provided source variable, it iterates over each row in the image, and each pixel in the row. If the model expects the colors to be in the blue, green, and red order, isBGR is true, and then the extracted color values are placed in the input tensor in that order; otherwise, they are added to the input tensor in the red, green, and blue order. The tricky part here is accessing the correct element for each pixel. The tensor is organized into four dimensions, as explained previously. The first element will always be zero for this model, since we are only processing one image at a time. The second dimension is the color channel, so you will see that change for the red, green, and blue color values.

Finally, if the model expects color values to be in the range of 0 to 255, isRange255, then each color channel is clamped to that range.

8. The last thing that we will do is copy the contents of the resized image to the pixelBytes array so that we can display the image to the user. Add the following highlighted code to do this; note that the previous code has been omitted for brevity:

```
});

var outStream = new MemoryStream();
image.Save(outStream, new PngEncoder());
```

```
        pixelBytes = outStream.ToArray();
    }
    return (input, pixelBytes);
```

Now that we have written the code to process the image and populate the input tensor, we can complete the `Classify` method by following these steps:

1. Replace the `// Code will be added here` comment with a call to the `LoadInputTensor` method:

   ```
   public ClassifierOutput Classify(byte[] imageBytes)
   {
       (Tensor<float> tensor, byte[] resizedImage) =
   LoadInputTensor(imageBytes, inputSize, isBgr, isRange255);
   }
   ```

2. Next, we can run the session, passing in the newly created input tensor and capturing the result:

   ```
   public ClassifierOutput Classify(byte[] imageBytes)
   {
       (Tensor<float> tensor, byte[] resizedImage) =
   LoadInputTensor(imageBytes, inputSize, isBgr, isRange255);

       var resultsCollection = session.Run(new List<NamedOnnxValue>
       {
           NamedOnnxValue.CreateFromTensor<float>(inputName,
   tensor)
       });
   }
   ```

3. We grab the label from the output result, which will be used to determine whether this image contains a hotdog or not:

   ```
   public ClassifierOutput Classify(byte[] imageBytes)
   {
       (Tensor<float> tensor, byte[] resizedImage) =
   LoadInputTensor(imageBytes, inputSize, isBgr, isRange255);

       var resultsCollection = session.Run(new List<NamedOnnxValue>
       {
           NamedOnnxValue.CreateFromTensor<float>(inputName,
   tensor)
       });

       var topLabel = resultsCollection
   ```

```
        ?.FirstOrDefault(i => i.Name == "classLabel")
        ?.AsTensor<string>()
        ?.First();
}
```

4. Then, we can get the confidence level of the result, which tells us how sure the model is of the classification. This will be used when we display the result:

```
public ClassifierOutput Classify(byte[] imageBytes)
{
    (Tensor<float> tensor, byte[] resizedImage) =
LoadInputTensor(imageBytes, inputSize, isBgr, isRange255);

    var resultsCollection = session.Run(new List<NamedOnnxValue>
    {
        NamedOnnxValue.CreateFromTensor<float>(inputName,
tensor)
    });

    var topLabel = resultsCollection
        ?.FirstOrDefault(i => i.Name == "classLabel")
        ?.AsTensor<string>()
        ?.First();

    var labelScores = resultsCollection
        ?.FirstOrDefault(i => i.Name == "loss")
        ?.AsEnumerable<NamedOnnxValue>()
        ?.First()
        ?.AsDictionary<string, float>();
}
```

5. Finally, we can return the result of the classification using the `ClassifierOutput` class:

```
public ClassifierOutput Classify(byte[] imageBytes)
{
    (Tensor<float> tensor, byte[] resizedImage) =
LoadInputTensor(imageBytes, inputSize, isBgr, isRange255);

    var resultsCollection = session.Run(new List<NamedOnnxValue>
    {
        NamedOnnxValue.CreateFromTensor<float>(inputName,
tensor)
    });
```

```
      var topLabel = resultsCollection
          ?.FirstOrDefault(i => i.Name == "classLabel")
          ?.AsTensor<string>()
          ?.First();

      var labelScores = resultsCollection
          ?.FirstOrDefault(i => i.Name == "loss")
          ?.AsEnumerable<NamedOnnxValue>()
          ?.First()
          ?.AsDictionary<string, float>();

    return ClassifierOutput.Create(topLabel, labelScores,
resizedImage);
    }
```

6. The last step is to finish the MLNetClassifier implementation by implementing the ClassifierOutput class. Update your ClassifierOutput class by adding the highlighted code:

```
internal sealed class ClassifierOutput
{
    public string TopResultLabel { get; private set; }
    public float TopResultScore { get; private set; }
    public IDictionary<string, float> LabelScores { get; private
set; }

    public byte[] Image { get; private set; }

    ClassifierOutput() { }

    public static ClassifierOutput Create(string topLabel,
IDictionary<string, float> labelScores, byte[] image)
    {
        var topLabelValue = topLabel ?? throw new
ArgumentException(nameof(topLabel));
        var labelScoresValue = labelScores ?? throw new
ArgumentException(nameof(labelScores));

        return new ClassifierOutput
        {
            TopResultLabel = topLabelValue,
            TopResultScore = labelScoresValue.First(i => i.Key
== topLabelValue).Value,
            LabelScores = labelScoresValue,
            Image = image,
```

```
        };
    }
}
```

The `ClassifierOutput` class is used to encapsulate the four values that will be used in the UI and expose them as public properties. The `Create` static method is used to create an instance of the class. The `Create` method validates the arguments provided and sets the public properties appropriately for use by the UI.

We have now written the code to recognize hot dogs in an image.

Now, we can build the user interface for the application and call `MLNetClasssifier` to classify an image.

Requesting app permissions

Before we dive right into building the rest of the app functionality, we need to address permissions. This app will have two buttons that the user will use, one to take a photo and another to select a photo from the device. This is similar to the functionality that we saw in *Chapter 6, Building a Photo Gallery App Using CollectionView and CarouselView*, where we needed to request permission from the user before accessing the camera or device storage. However, we will implement the permissions differently than we did in that chapter. Since gaining access to the camera and accessing photos on the user's device requires separate permissions, we will request them from each button handler.

Follow these steps to add a class to help us with the permission checks:

1. Create a new class named `AppPermissions` in the project.

2. Modify the class definition to add a `partial` modifier, and remove the default constructor:

    ```
    namespace HotdogOrNot;

    internal partial class AppPermissions
    {
    }
    ```

3. Add the following method to the `AppPermissions` class:

    ```
    public static async Task<PermissionStatus>
    CheckRequiredPermission<TPermission>() where TPermission
    : Permissions.BasePermission, new() => await Permissions.
    CheckStatusAsync<TPermission>();
    ```

The `CheckRequiredPermission` method is used to ensure that our app has the right permissions before we attempt any operations that might fail if we don't. Its implementation is to call the .NET MAUI `CheckSyncStatus` with the provided permission type in `TPermission`. It returns `PermissionStatus`, which is enum. We are mostly interested in the `Denied` and `Granted` values.

4. Add the `CheckAndRequestRequiredPermission` method to the `AppPermissions` class:

```
public static async Task<PermissionStatus>
CheckAndRequestRequiredPermission() <TPermission>() where
TPermission : Permissions.BasePermission, new()
{
    PermissionStatus status = await Permissions.
CheckStatusAsync< TPermission >();

    if (status == PermissionStatus.Granted)
        return status;

    if (status == PermissionStatus.Denied && DeviceInfo.Platform
== DevicePlatform.iOS)
    {
        // Prompt the user to turn on in settings
        // On iOS once a permission has been denied it may not
be requested again from the application
        await App.Current.MainPage.DisplayAlert("Required App
Permissions", "Please enable all permissions in Settings for
this App, it is useless without them.", "Ok");
    }

    if (Permissions.ShouldShowRationale< TPermission >())
    {
        // Prompt the user with additional information as to why
the permission is needed
        await App.Current.MainPage.DisplayAlert("Required App
Permissions", "This app uses photos, without these permissions
it is useless.", "Ok");
    }

    status = await MainThread.
InvokeOnMainThreadAsync(Permissions.RequestAsync<TPermission>);
    return status;
}
```

The `CheckAndRequestRequiredPermission` method handles the intricacies of requesting access from the user. The first step is to simply check and see whether the permission has already been granted and, if it has, return the status. Next, if we are on iOS and the permission has been denied, it cannot be requested again, so you must instruct the user on how to grant permission to the app by using the settings panel. Android includes in the request behavior the ability to nag the user if they have denied access. This behavior is exposed through .NET MAUI with the `ShouldShowRationale` method. It will return `false` for any platform that does not support this behavior, and on Android, it will return `true` the first time after the user denies access and `false` if the user denies it a second time. Finally, we request access to the permission from the user. Again, .NET MAUI hides all the platform implementation details from us, making checking and requesting access to certain resources very straightforward.

> **Look familiar?**
>
> If the preceding code looks familiar, then you are right. It is based on the implementation that is described in the .NET MAUI documentation. You can find it at `https://learn.microsoft.com/en-us/dotnet/maui/platform-integration/appmodel/permissions`.

Now that we have the shared `AppPermissions` in place, we can start with the platform configuration. Before we can use the media picker, however, we need to do some configuration for each platform. We will start with Android.

In Android API version 33, three new permissions were added to enable read access to media files – `ReadMediaImages`, `ReadMediaVideos`, and `ReadMediaAudio`. Prior to API version 33, all that was required was the `ReadExternalStorage` permission. To access the camera, we will need both `Camera` and `WriteExternalStorage` permissions. To properly request the correct permission for the API version of the device, open `MauiApplication.cs` in the `Platform/Android` folder and modify it to look like the following:

```
using Android.App;
using Android.Runtime;

// Needed for Picking photo/video
[assembly: UsesPermission(Android.Manifest.Permission.
ReadExternalStorage, MaxSdkVersion = 32)]
[assembly: UsesPermission(Android.Manifest.Permission.
ReadMediaImages)]

// Needed for Taking photo/video
[assembly: UsesPermission(Android.Manifest.Permission.Camera)]
[assembly: UsesPermission(Android.Manifest.Permission.
WriteExternalStorage)]
```

Finally, for Android, we need to declare usage of the `IMAGE_CAPTURE` intent as follows in the `AndroidManifest.xml` file:

```
<queries>
  <intent>
    <action android:name="android.media.action.IMAGE_CAPTURE" />
  </intent>
</queries>
```

For iOS and Mac Catalyst, the only thing we need to do is add the following four usage descriptions to the `info.plist` file in the `platform/ios` and `platform/maccatalyst` folders:

```
<key>NSCameraUsageDescription</key>
<string>This app needs access to the camera to take photos.</string>
<key>NSPhotoLibraryUsageDescription</key>
<string>This app needs access to photos.</string>
<key>NSMicrophoneUsageDescription</key>
<string>This app needs access to microphone.</string>
<key>NSPhotoLibraryAddUsageDescription</key>
<string>This app needs access to the photo gallery.</string>
```

For Windows, we need to add the following highlighted code to the `Capabilities` section of the `package.appxmanifest` file in the `platforms/windows` folder:

```
<Capabilities>
  <rescap:Capability Name="runFullTrust" />
  <DeviceCapability Name="webcam"/>
</Capabilities>
```

Now that we have declared the permissions we need for each platform, we can implement the remaining functionality to take a photo or pick an existing image.

Building the first view

The first view in this app will be a simple view with two buttons. One button will be to start the camera so that users can take a photo of something to determine whether it is a hot dog. The other button will be to pick a photo from the photo library of the device. We will continue to use the MVVM pattern in this chapter, so we will split the view into two classes, `MainView` for the UI visible to the user and `MainViewModel` for the actual implementation.

Building the ViewModel class

We will start by creating the `MainViewModel` class, which will handle what will happen when a user taps one of the buttons. Let's set this up by going through the following steps:

1. Create a new folder called `ViewModels`.

2. Add a NuGet reference to `CommunityToolkit.Mvvm`; we use `CommunityToolkit.Mvvm` to implement the `INotifyPropertyChanged` interface and commands, as we did in other chapters.

3. Create a new partial class called `MainViewModel` in the `ViewModels` folder, using `ObservableObject` from the `CommunityToolkit.Mvvm.ComponentModel` namespace as a base class.

4. Create a private field of the `IClassifier` type and call it `classifier`, as shown in the following code block:

```
using CommunityToolkit.Mvvm.ComponentModel;
using HotdogOrNot.ImageClassifier;

namespace HotdogOrNot.ViewModels;

public partial class MainViewModel : ObservableObject
{
    private IClassifier classifier;

    public MainViewModel()
    {
    }
}
```

Initializing the ONNX model requires the use of asynchronous methods, so we need to handle them carefully, since we will be calling them from the constructor and the button handlers. The following steps will create the model initializer:

1. Create an `InitTask` property that is of the `Task` type.

2. Use a property initializer to set it to a new `Task`, using `Task.Run`.

3. Initialize the model from the raw resources of the .NET MAUI app. The method should look like the following code:

```
Task InitTask() => Task.Run(async () =>
{
    using var modelStream = await FileSystem.
OpenAppPackageFileAsync("hotdog-or-not.onnx");
    using var modelMemoryStream = new MemoryStream();

    modelStream.CopyTo(modelMemoryStream);
    var model = modelMemoryStream.ToArray();

    _classifier = new MLNetClassifier(model);
});
```

The `InitTask` property holds a reference to `Task` that does the following:

- Loads the `hotdog-or-not.onnx` file into `Stream`

- Copies the bytes from the original stream to an array of bytes so that the original stream can be closed and any native resources, such as file handles, can be released.

- Creates and returns a new instance of the `MLNetClassifier` class using the loaded model.

4. To ensure that `InitTask` will only run successfully once, add the following highlighted code:

```
public partial class MainViewModel : ObservableObject
{
    IClassifier _classifier;
    Task initTask;

    public MainViewModel()
    {
        _ = InitAsync();
    }

    public Task InitAsync()
    {
        if (initTask == null || initTask.IsFaulted)
            initTask = InitTask();

        return initTask;
    }

    // Code omitted for brevity
}
```

In `InitAsync`, the initialization task is captured by a field only if the field is `null` or its value has faulted. This ensures that we only run the initialization successfully once. The value of the field is then returned to the caller, which, in this case, is the constructor. Unwinding this, the constructor calls `InitAsync` and throws away the return value. `InitAsync`, meanwhile, captures the value returned by the `InitTask` property, which is `Task` that has already been queued for execution. Since `InitAsync` and `InitTask` and their closure are all asynchronous, they complete sometime after the constructor completes.

Now that we have initialized the `hotdog-or-not` ONNX model, we can now implement the two buttons, one that takes a photo and another that allows the user to pick a photo from their device storage. Let's start by implementing a couple of helper methods to use in both use cases.

The first helper method is used to convert `FileResult` to `byte[]`. To implement `ConvertPhotoToBytes`, follow these steps:

1. Open the `MainViewModel.cs` file.

2. Add a new method named `ConvertPhotoToBytes`, which takes `FileResult` as a parameter and returns `byte []`. Since the method is `async`, you'll need to return `Task` and use the `async` modifier.

3. In the method, check whether `FileResult` is `null` and that it returns an empty array.

4. Next, open a stream from `FileResult` using the `OpenStreamAsync` method.

5. Create a new variable of the MemoryStream type and initialize it using the default constructor.

6. Use the Copy method to copy stream to MemoryStream.

7. Finally, return MemoryStream as byte[]; your method should look like the following:

```
private async Task<byte[]> ConvertPhotoToBytes(FileResult photo)
{
    if (photo == null) return Array.Empty<byte>();

    using var stream = await photo.OpenReadAsync();
    using MemoryStream memoryStream = new();
    stream.CopyTo(memoryStream);

    return memoryStream.ToArray();
}
```

The other helper method we will need is to use our classification model to get the results of a photo and return the results. We will need a new type to return the results. Follow these steps to implement the new class:

1. Create a new folder named Models in the project.

2. In the Models folder, create a new class, Result, in a file named Result.cs.

3. Add a public property, IsHotdog, as bool.

4. Add a public property, Confidence, as float.

5. Add a public property, PhotoBytes, as byte[]; the class should now look like the following:

```
namespace HotdogOrNot.Models;

public class Result
{
    public bool IsHotdog { get; set; }
    public float Confidence { get; set; }
    public byte[] PhotoBytes { get; set; }
}
```

The IsHotdog property is used to capture whether the label returned from the model is "hotdog." Confidence is a score of how sure the model is that this is a hotdog or not. Finally, since we transform the image prior to processing, we store the transformed image in the PhotoBytes property.

Now, we can implement the method that will run and process the classification result, by following these steps:

1. Open the `MainViewModel.cs` file.

2. In the `MainViewModel` class, add a new field, `isClassifying`, with a `bool` type.

3. Add the `ObservableAttribute` attribute to the field; it should look like the following:

    ```
    [ObservableProperty]
    private bool isClassifying;
    ```

4. Add a new method to the `MainViewModel` class, named `RunClassificationAsync`. The method will accept a `byte[]` parameter and return `Result`, wrapped in `Task`, since it is `async`:

    ```
    async Task<Result> RunClassificationAsync(byte[]
    imageToClassify)
    {

    }
    ```

5. In the method, the first thing we do is set the `IsClassifying` property to `true`; this will be used to disable the buttons later in the chapter.

6. Add a `try..catch..finally` statement.

7. Inside the `try` statement, ensure the model is initialized by calling `InitAsync`.

8. Then, call `Classify` on the `classifier` field passing `byte[]`, representing the image as a parameter and storing the result.

9. The last statement in the `try` statement block is to return a new `Result`, setting `IsHotdog` to `true` only if the classification result's `TopResultLabel` property is "hotdog," `Confidence` is set to the classification result's `TopResultScore` property, and `PhotoBytes` is set to the classification result's `Image` property. The `try` portion should look like the following:

    ```
    try
    {
        await InitAsync().ConfigureAwait(false);
        var result = _classifier.Classify(imageToClassify);
        return new Result()
        {
            IsHotdog = result.TopResultLabel == "hotdog",
            Confidence = result.TopResultScore,
            PhotoBytes = result.Image
        };
    }
    catch
    ```

10. Now, in the `catch` statement block, return a new `Result`, setting the `IsHotdog` property to `false`, `Confidence` to `0.0f`, and the `PhotoBytes` property to the bytes passed into the method. The `catch` block should look like the following:

```
catch
{
    return new Result
    {
        IsHotdog = false,
        Confidence = 0.0f,
        PhotoBytes = imageToClassify
    };
}
finally
```

11. Lastly, for the `finally` block, we want to set the `IsClassiying` property back to `false`; however, we will need to do this on the main UI thread using the `MainThread.BeginInvokeOnMainThread` method from .NET MAUI, as shown in the following code:

```
finally
{
    MainThread.BeginInvokeOnMainThread(() => IsClassifying =
false);
}
```

Now that we have written the helper methods, we can create two methods, one to handle capturing an image from the camera and another to pick a photo from user storage. We will start with the camera capture method.

Let's set this up by following these steps:

1. Open the `MainViewModel.cs` file.

2. Create a public async void method called `TakePhoto`.

3. Add the `RelayCommand` attribute to make the method bindable.

4. Add an `if` statement to check whether the `MediaPicker.Default.IsCaptureSupported` parameter is `true`.

5. In the `true` statement block of `if`, get the status of the `Camera` permission using the `CheckAndRequestPermission` method.

6. If the status is `Granted`, then use `CheckAndRequestMethod` again to check the `WriteExternalStorage` permission.

7. If the status is `Granted`, use `MediaPicker` to capture a photo using the `CapturePhotoAsync` method.

8. Call a method named `ConvertPhotoToBytes`, passing in the file returned from `MediaPicker`.

9. Pass the photo bytes to the `RunClassificationAsync` method.

10. Finally, we will dynamically navigate to the `Result` view, which we will create in the next section, passing the result from `RunClassificationAsync` as a parameter. We do this by using `Shell.Current.GotoAsync` and ensuring that the app uses the main thread to do so, as shown in the following code block:

```
[RelayCommand()]
public async void TakePhoto()
{
    if (MediaPicker.Default.IsCaptureSupported)
    {
        var status = await AppPermissions.
CheckAndRequestRequiredPermissionAsync<Permissions.Camera>();
        if (status == PermissionStatus.Granted) {
            status = await AppPermissions.
CheckAndRequestRequiredPermissionAsync<Permissions.
StorageWrite>();
        }
        if (status == PermissionStatus.Granted)
        {
            FileResult photo = await MediaPicker.Default.
CapturePhotoAsync(new MediaPickerOptions() { Title = "Hotdog or
Not?" });
            var imageToClassify = await
ConvertPhotoToBytes(photo);
            var result = await
RunClassificationAsync(imageToClassify);
            await MainThread.InvokeOnMainThreadAsync(async () =>
await
                Shell.Current.GoToAsync("Result", new
Dictionary<string, object>() { { "result", result } })
            );
        }
    }
}
```

`Shell.Current.GotoAsync` takes two parameters – the first is the route that `Shell` is to navigate to, and the second is a dictionary of key-value pairs to send to the destination view. Later in this chapter, we will see how to configure a route to a view without using XAML and, when we create the `Result` view, how to access the parameters passed to it.

We will now create the `PickPhoto` method to allow a user to use an image from their device. Use the following steps to create the method:

1. Create a public async void method called `PickPhoto`.

2. Add the `RelayCommand` attribute to make the method bindable.

3. Grant the status of the `Photos` permission using the `CheckAndRequestPermission` method.

4. If the status is `Granted`, use `MediaPicker` to capture a photo using the `PickPhotoAsync` method.

5. Call a method named `ConvertPhotoToBytes`, passing in the file returned from `MediaPicker`.

6. Pass the photo bytes to the `RunClassificationAsync` method.

7. Finally, we will dynamically navigate to the `Result` view, which we will create in the next section, passing the result from `RunClassificationAsync` as a parameter. We will do this by using `Shell.Current.GotoAsync` and ensuring that the app uses the main thread to do so, as shown in the following code block:

```
[RelayCommand()]
public async void PickPhoto()
{
    var status = await AppPermissions.
CheckAndRequestRequiredPermissionAsync<Permissions.Photos>();
    if (status == PermissionStatus.Granted)
    {
        FileResult photo = await MediaPicker.Default.
PickPhotoAsync();
        var imageToClassify = await ConvertPhotoToBytes(photo);
        var result = await
RunClassificationAsync(imageToClassify);
        await MainThread.InvokeOnMainThreadAsync(async () =>
await
            Shell.Current.GoToAsync("Result", new
Dictionary<string, object>() { { "result", result } })
        );
    }
}
```

When a user clicks on a button, the classification could take a noticeable amount of time. To prevent the user from clicking the button again because they think it's not working, we will disable the buttons until the operation completes. The `IsClassifying` property is already set; we just need to use that value to restrict `RelayCommands`, by following these steps:

1. Add a new method that returns a Boolean named `CanExecuteClassification`, and return the inverse of the `IsClassifying` property, as shown in the following code:

```
private bool CanExecuteClassification() => !IsClassifying;
```

2. Update the `RelayCommand` attribute for the `TakePhoto` method, as highlighted here:

```
[RelayCommand(CanExecute = nameof(CanExecuteClassification))]
public async void TakePhoto()
```

3. Update the `RelayCommand` attribute for the `PickPhoto` method, as highlighted here:

```
[RelayCommand(CanExecute = nameof(CanExecuteClassification))]
public async void PickPhoto()
```

Now that ViewModel for the main page is complete, we can build View for the main page.

Building the view

Now, once we have created the `MainViewModel` class, it is time to create the code for the **graphical user interface (GUI)**. Go through the following steps to create the GUI for the `MainView` view:

1. Create a new folder called `Views`.

2. Add a new **.NET MAUI ContentPage (XAML)** called `MainView`.

3. Set the `Title` property of `ContentPage` as `Hotdog or Not hotdog`.

4. Add `HorizontalStackLayout` to the page, and set its `VerticalOptions` property to `Center` and its `HorizontalOptions` property to `CenterAndExpand`.

5. Add `Button` to the `HorizontalStackLayout`, with the text `Take Photo`. For the `Command` property, add a binding to the `TakePhoto` property in the `MainViewModel` class.

6. Add `Button` to `HorizontalStackLayout`, with the text `Pick Photo`. For the `Command` property, add a binding to the `PickPhoto` property in the `MainViewModel` class, as shown in the following code block:

```
<ContentPage xmlns="http://schemas.microsoft.com/dotnet/2021/
maui"
             xmlns:x="http://schemas.microsoft.com/winfx/2009/
xaml"
             xmlns:viewModels="clr-namespace:HotdogOrNot.
ViewModels"
             x:Class="HotdogOrNot.Views.MainView"
```

```
                    x:DataType="viewModels:MainViewModel"
                    Title="Hotdog or Not hotdog">
         <HorizontalStackLayout VerticalOptions="Center"
    HorizontalOptions="CenterAndExpand">
              <Button Text="Take Photo" Command="{Binding
    TakePhotoCommand}" WidthRequest="150" HeightRequest="150"
    Margin="20" FontSize="Large"/>
              <Button Text="Pick Photo" Command="{Binding
    PickPhotoCommand}" WidthRequest="150" HeightRequest="150"
    Margin="20" FontSize="Large"/>
         </HorizontalStackLayout>
    </ContentPage>
```

In the code-behind `MainView.xaml.cs` file, we will set the binding context of the view by following these steps:

1. Add `MainViewModel` as a parameter of the constructor.

2. After the `InitialComponent` method call, set the `BindingContext` property of the view to the `MainViewModel` parameter.

3. Use the `SetBackButtonTitle` static method on the `NavigationPage` class so that an arrow to navigate back to this view will be shown in the navigation bar on the result view, as shown in the following code block:

    ```
    public MainView(MainViewModel viewModel)
    {
        InitializeComponent();
        BindingContext = viewModel; NavigationPage.
    SetBackButtonTitle(this, string.Empty);
    }
    ```

Building the result view

The last thing we need to do in this project is to create the result view. This view will show the input photo and the classification of a hot dog or not.

Building the ResultViewModel class

Before we create the view, we will create a `ResultViewModel` class that will handle all the logic for the view, by following these steps:

1. Create a `partial` class called `ResultViewModel` in `ViewModels`.

2. Add `ObservableObject` as a base class to the `ResultViewModel` class.

3. Create a `private` field of the `string` type, called `title`. Add the `ObservableProperty` attribute to the field to make it a bindable property.

4. Create a `private` field of the `string` type, called `description`. Add the `ObservableProperty` attribute to the field to make it a bindable property.

5. Create a `private` field of the `string` type, called `Title`. Add the `ObservableProperty` attribute to the field to make it a bindable property, as shown in the following code block:

```
using CommunityToolkit.Mvvm.ComponentModel;
using HotdogOrNot.Models;

namespace HotdogOrNot.ViewModels;

public partial class ResultViewModel : ObservableObject
{
    [ObservableProperty]
    private string title;

    [ObservableProperty]
    private string description;

    [ObservableProperty]
    byte[] photoBytes;

    public ResultViewModel()
    {
    }
}
```

The next thing we will do in `ResultViewModel` is to create an `Initialize` method that will have the result as a parameter. Let's set this up by following these steps:

1. Add a `private` method named `Initialize` to the `ResultViewModel` class that accepts a parameter of the `Result` type, named `result`, and returns `void`.

2. In the `Initialize` method, set the `PhotoBytes` property to the value of the `PhotoBytes` property of the `result` parameter.

3. Add an `if` statement that checks whether the `IsHotDog` property of the `result` parameter is `true` and whether `Confidence` is higher than `90%`. If this is the case, set `Title` to `"Hot dog"` and `Description` to `"This is for sure a hotdog"`.

4. Add an `else if` statement to check whether the `IsHotdog` property of the `result` parameter is `true`. If this is the case, set `Title` to `"Maybe"` and `Description` to `"This is maybe a hotdog"`.

5. Add an `else` statement that sets `Title` to `"Not a hot dog"` and `Description` to `"This is not a hot dog"`, as shown in the following code block:

```
public void Initialize(Result result)
{
    PhotoBytes = result.PhotoBytes;

    if (result.IsHotdog && result.Confidence > 0.9)
    {
        Title = "Hot dog";
        Description = "This is for sure a hot dog";
    }
    else if (result.IsHotdog)
    {
        Title = "Maybe";
        Description = "This is maybe a hot dog";
    }
    else
    {
        Title = "Not a hot dog";
        Description = "This is not a hot dog";
    }
}
```

The final thing we need to do is call the `Initialize` method with the result. If you recall from the previous section on building the main view, we navigated to the `Result` view and passed the `Result` object as a parameter. To access the parameter and call the `Initialize` method properly, follow these steps:

1. Add the `IQueryAttributable` interface to the list of inherited interfaces:

    ```
    public partial class ResultViewModel : ObservableObject,
    IQueryAttributable
    ```

2. Add a new `void` method, `ApplyQueryAttributes`, that accepts a parameter named query of the `IDictionary<string, object>` type:

    ```
    public void ApplyQueryAttributes(IDictionary<string, object>
    query)
    {
    }
    ```

3. Now, in the method, call the `Initialize` method, passing the `"result"` object from the query dictionary and casting it to a `Result` type, as shown in the following code:

```
public void ApplyQueryAttributes(IDictionary<string, object>
query)
{
    Initialize(query["result"] as Result);
}
```

`ViewModel` is now complete, and we are ready to create `View`.

Building the view

Because we want to show the input photo in the result view, we need to convert it from `byte []` to `Microsft.Maui.Controls.ImageSource`. We will do this in a value converter that we can use together with the binding in the **XAML**, by following these steps:

1. Create a new folder called `Converters`.
2. Create a new class called `BytesToImageConverter` in the `Converters` folder.
3. Add and implement the `IValueConverter` interface, as shown in the following code block:

```
using System.Globalization;

namespace HotdogOrNot.Converters;

public class BytesToImageConverter : IvalueConverter
{
    public object Convert(object value, Type targetType, object
parameter, CultureInfo culture)
    {
        throw new NotImplementedException();
    }

    public object ConvertBack(object value, Type targetType,
object parameter, CultureInfo culture)
    {
        throw new NotImplementedException();
    }
}
```

The `Convert` method will be used when `ViewModel` updates a view. The `ConvertBack` method will be used in two-way bindings when `View` updates `ViewModel`. In this case, we only need to write code for the `Convert` method, by following these steps:

1. First, check whether the `value` parameter is `null`. If so, we should return `null`.
2. If the value is not `null`, cast it as `byte []`.

3. Create a `MemoryStream` object from the `byte` array.

4. Return the result of the `ImageSource.FromStream` method to which we will pass the stream, as shown in the following code block:

```
public object Convert(object value, Type targetType, object
parameter, CultureInfo culture)
{
    if(value == null)
    {
        return null;
    }
    var bytes = (byte[])value;
    var stream = new MemoryStream(bytes);

    return ImageSource.FromStream(() => stream);
}
```

The view will contain the photo, which will take up two-thirds of the screen. Under the photo, we will add a description of the result. Let's set this up by going through the following steps:

1. In the `Views` folder, create a new file using the .NET MAUI ContentPage (XAML) file template, and name it `ResultView`.

2. Import the namespace for the converter.

3. Add `BytesToImageConverter` to `Resources` for the page and give it "`ToImage`" key.

4. Bind the `Title` property of `ContentPage` as the `Title` property of `ViewModel`.

5. Add `Grid` to the page with two rows. The `Height` value for the first `RowDefinition` should be `2*`. The height of the second row should be `*`. These are relative values that mean that the first row will take up two-thirds of `Grid`, while the second row will take up one-third of `Grid`.

6. Add `Image` to `Grid`, and bind the `Source` property to the `PhotoBytes` property in `ViewModel`. Use the converter to convert the bytes to an `ImageSource` object and set the `Source` property.

7. Add `Label`, and bind the `Text` property to the `Description` property of `ViewModel`, as shown in the following code block:

```
<ContentPage xmlns="http://xamarin.com/schemas/2014/
forms" xmlns:x="http://schemas.microsoft.com/winfx/2009/
xaml"    xmlns:converters="clr-
    namespace:HotdogOrNot.Converters"
    x:Class="HotdogOrNot.Views.ResultView" Title="{Binding
Title}">
    <ContentPage.Resources>
```

```xml
      <converters:BytesToImageConverter x:Key="ToImage" />
    </ContentPage.Resources>
    <Grid>
      <Grid.RowDefinitions>
        <RowDefinition Height="2*" />
        <RowDefinition Height="*" />
      </Grid.RowDefinitions>

      <Image Source="{Binding PhotoBytes, Converter=
{StaticResource ToImage}}" Aspect="AspectFill" />
      <Label Grid.Row="1" HorizontalOptions="Center"
FontAttributes="Bold" Margin="10" Text="{Binding Description}"
/>
    </Grid>
</ContentPage>
```

We also need to set `BindingContext` of the view. We will do this in the same way as we did in `MainView` – in the code-behind file (`ResultView.xaml.cs`), as shown in the following code snippet:

```csharp
public ResultView (ResultViewModel viewModel)
{
    InitializeComponent ();
    BindingContext = viewModel;
}
```

We are now ready to write the initialization code for the app.

Initializing the app

We will set up **inversion of control** (**IoC**) and configure the main page for `Shell`.

Open `App.xaml.cs`, and set `MainPage` to `MainView` by following these steps:

1. Delete the `MainPage.xaml` and `MainPage.xaml.cs` files from the root of the project, since we won't be needing those.

2. Open the `AppShell.xaml` file in the root of the project, and modify it to look like the following code:

```xml
<?xml version="1.0" encoding="UTF-8" ?>
<Shell
    x:Class="HotdogOrNot.AppShell"
    xmlns="http://schemas.microsoft.com/dotnet/2021/maui"
    xmlns:x="http://schemas.microsoft.com/winfx/2009/xaml"
    xmlns:views="clr-namespace:HotdogOrNot.Views"
```

```
        Shell.FlyoutBehavior="Disabled">

        <ShellContent
            Title="Home"
            ContentTemplate="{DataTemplate views:MainView}"
            Route="MainView" />

</Shell>
```

Now, configure the `View` and `ViewModel` classes in the IoC container by following these steps:

1. Open the `MauiProgram.cs` file.

2. In the `CreateMauiApp` method before the `return` statement, add the following highlighted lines of code:

```
#if DEBUG
    builder.Logging.AddDebug();
#endif
    builder.Services.AddTransient<Views.MainView>();
    builder.Services.AddTransient<Views.ResultView>();
builder.Services.AddTransient<ViewModels.MainViewModel>();
        builder.Services.AddTransient<ViewModels.
ResultViewModel>();

        return builder.Build();
```

The very last thing we need to do is add the route to `ResultView` to enable navigation from `MainView`. We will do this by adding the following highlighted code to the constructor of `AppShell` in `AppShell.xaml.cs`:

```
public AppShell()
{
    Routing.RegisterRoute("Result", typeof(HotdogOrNot.Views.
ResultView));
    InitializeComponent();
}
```

Now, we are ready to run the app. If we use the simulator/emulator, we can just drag and drop photos to it if we need photos to test with. When the app has started, we can now pick a photo and run it against the model. The following screenshot shows how the app will look if we upload a photo of a hot dog:

Figure 12.13 – HotdogOrNot running in an Android emulator

> **Note**
> The prediction result for Android may not be as accurate as the web portal at https://
> github.com/Azure-Samples/cognitive- services-android-customvision-
> sample/issues/12. If you desire better, more consistent results, you can use the REST APIs.

Summary

In this chapter, we built an app that can recognize whether a photo contains a hot dog or not. We accomplished this by training a machine learning model for image classification, using Azure Cognitive Services and the Custom Vision service.

We exported models for ML.NET, and we learned how to use it in an MAUI app that targets iOS, Mac Catalyst, Windows, and Android. In the app, a user can take a photo or pick one from their photo library. This photo will be sent to the model to be classified, and we will get a result that tells us whether the photo is of a hot dog.

Now, we can continue to build other apps and use what we have learned in this chapter regarding machine learning, both on-device and in the cloud using Azure Cognitive Services. Even if we are building other apps, the concept will be the same.

Now, we have completed all the chapters in this book. We have learned the following:

- What .NET MAUI is and how we can get started building apps

- How to use the basic layouts and controls of .NET MAUI

- How to work with navigation

- How to make the user experience better with animations

- How to use sensors such as the **Global Positioning System** (**GPS**) in the background

- How to build apps for multiple form factors

- How to build real-time apps powered by Azure

- How to make apps smarter with machine learning

The next step is to start to build your own apps. To stay up to date and learn more about .NET MAUI, our recommendation is to read the official Microsoft dev blogs and watch live streams on Twitch and YouTube videos from the .NET MAUI team.

Thank you for reading the book!

Index

Symbols

.NET Blazor 516
  deployment models 516
  project, overview 515
.NET for Android 4
.NET for iOS/tvOS/Mac Catalyst 4
.NET MAUI
  architecture 10, 11
  defining, with XAML 11
  files, examining 42, 43, 44
  new project, creating 39-42
  packages, updating 48-51
  page in C#, creating 13
  page in XAML, creating 12
  project, setting up 38
  usage, considerations 16
  versus traditional .NET mobile 15, 16
  Xamarin.Forms app, converting to 96
.NET MAUI app
  creating 101, 102, 103
.NET MAUI Blazor Hybrid project
  exploring 518-522
.NET MAUI ContentPage (XAML) 582
.NET MAUI development
  project overview 38
.NET MAUI, files
  Android-specific platform code 46
  iOS and Mac Catalyst platform files 47

  shared code 44, 45, 46
  Tizen 47
  Windows 48
.NET MAUI framework
  exploring 9
.NET MAUI Markup Community Toolkit 14
.NET MAUI project
  creating 213
.NET MAUI Shell 444
.NET mobile 4, 5
  code sharing 5
  existing knowledge, using 6
  installing, for Visual Studio
    2022 or later 23, 24
.NET mobile platforms 6
  for Androids 8, 9
  for iOS/tvOS/Mac Catalyst 6
  for macOS 7
.NET mobile productivity tooling 34
  Hot Reload 35
  Windows Subsystem, for Android 35, 36
  Xamarin Hot Restart 34
.NET Upgrade Assistant
  BuildChat.Android project,
    upgrading 130, 131
  BuildChat project, upgrading 125-130
  installing 119-121

preparing, to run 121-125

running 119, 125

upgrades to .NET MAUI,
    completing 131-136

.NET workloads 48

A

AddItemsAsync() method

implementing 58, 59

Advanced Micro Devices (AMD) 27

Advanced RISC Machine (ARM) 6

ahead-of-time (AOT) 6

Android

Windows Subsystem 35

Android Emulator

configuring 19- 33

Android Runtime (ART) 8

Android SDK versions 115

Android-specific platform code 46

application, contents

application-wide background
    color, setting 91, 92

filter button 92, 93

laying out 91

ListView items, laying out 92

ListView, touching up 93

MainView, laying out 92

application-wide background color

setting 91, 92

artificial intelligence (AI) 548

AsyncLazy<T>

reference link 420

Async suffix 258

Automatic Mac Provisioning 24

Azure Cognitive Services

Custom Vision 548

Azure Functions

deploying 403, 404

game play 391

game player, connecting 359

game return 387

lobby page, refreshing 372

players challenge 373

project, creating 353-358

scheduled job, creating to clear
    storage photos 402

service, creating 351

using 351

Azure serverless services 348

Functions 349

SignalR service 348

Azure SignalR AckableChatRoom 380

B

background location tracking

background job, creating 283

background job, scheduling 284-286

background job, subscribing 286-289

location updates, enabling 278

location updates, subscribing 278-282

with Android 283

base view model

creating 237

Bing Image Creator 454

blank .NET MAUI template

MauiMigration app, migrating
    to MyMauiApp 103

migrating, by creating.NET
    MAUI app 101, 102, 103

migrating, by creating Xamarin.
    Forms app 97, 99, 100

migrating into 96

Blazor Hybrid 516
Blazor Server 516
Blazor WebAssembly 516

C

CalculatePanState() method 202
calculator app
.NET MAUI Blazor Hybrid project,
  exploring 518-522
Compute service, creating 536-538
creating 516
Keypad view, creating 526
main window, resizing 543, 544
memory functions, adding 538-543
new project, creating 516-518
project, preparing 523-526
project, setting up 516
calculator app, Keypad view
controls, wiring up 532- 536
HTML, adding 526-528
style, adding to HTML 528-531
C# Hot Reload 35
Common Language Runtime (CLR) 4, 51
CommunityToolkit.Mvvm library 60
ObservableObject 61-63
ObservableProperty 61-63
reference, adding to 146
Compute service
creating 536-538
Connect view
ActivityButton control, creating 447-454
adding 447
creating 458-464
images, creating with Bing Image
  Creator 454- 458

core ML 549
custom control map
creating 294-297
Custom Vision 548

D

data bindings
adding 74, 75
command, used for making item
  as complete 87, 88
command, used for navigating to item 85-87
filter toggle function, creating
  with command 88-91
new items, adding to list 77-79
data visualization
adding 294
custom control map, creating 294-297
preparing 290-293
dependency injection (DI) 27
Services, registering 71, 73
used, for registering PhotoImporter 235, 236
ViewModels, registering 71-73
View, registering 71-73
wiring up 70, 71
dependency injection (DI),
  news app structure
used, for registering Views and
  ViewModels 152, 153
wiring up 152
developer mode
configuring 33, 34
development machine
Mac, setting up 17
platform 17
setting up 17
Windows machine, setting up 23
DoToo 42

F

filter toggle function
creating, with command 88-91
**first view, Hot Dog or Not Hot
Dog application**
building 574
MainViewModel class, building 574-582
view, building 582, 583
FlexLayout
using 321, 322
Fluent API 44, 45
Font Awesome
.NET MAUI, configuring 154
configuring 153
downloading 154
icons, defining in resource dictionary 155
function apps 349
Functions 349

G

gallery view
carousels, creating for MainView 247
creating 238
favorite photo, displaying 251, 252
favorites, saving 243-247
GalleryView, creating 239-242
GalleryViewModel, creating 238, 239
photos, displaying 249, 250
photos incrementally, loading 242, 243
view model, creating for MainView 247-249
game play, Azure Functions
ProcessTurn function 392-401
ProcessTurn function, creating 391, 392
game player
Connect function, creating 364- 371
models, creating 359- 364

game return, Azure Functions
GetGame function 388, 389
messages, creating 387
GetAllAsync method
adding, to LocationRepository 289
GetAllPlayers function 372
GetItemsAsync() method
implementing 58, 59
gold bar indicator 123
graphical user interface (GUI) 582
Gravatar service
reference link 462

H

handlers 10
**Hardware Accelerated Execution
Manager (HAXM) 19**
hardware acceleration
configuring 27-33
head-to-head game application
overview 407
heat map
creating 289
data visualization, adding 294
data visualization, preparing 290-293
GetAllAsync method, adding to
LocationRepository 289
Hot Dog or Not Hot Dog application
building 550
first view, building 574
image classification, with ML.NET 561-571
images, classifying with
machine learning 561
images, tagging 553, 554
initializing 588, 589
model, training 551-553
permissions, requesting 571-574

project, creating 557-561

result view, building 583

Hot Reload 34, 35, 526

HTML

adding 526-528

style, adding 528-531

hue 292

I

image classification

Hot Dog or Not Hot Dog application,
building 550, 556

images, tagging 553, 554

model, exporting 556

model, training 551-555

project overview 550

ImageSharp NuGet library 566

indows UI (WinUI) library 9

**integrated development
environment (IDE) 17**

IntelliSense 74

inversion of control (IoC) 588

iOS and Mac Catalyst platform files 47

IQueryAttributable 487

Isolated CSS 521

ItemViewModel class

creating 65

ItemView view

creating 69, 70

J

Java Native Interface (JNI) bridge

Android Callable Wrapper (ACW) 8

Manage Callable Wrapper (MCW) 8

JavaScript 532

JavaScript Object Notation (JSON) 161

L

Language Integrated Query (LINQ) 45, 58

lightness 292

Linq extension 363

ListView

binding, in MainViewModel 79-82

ListView item

laying out 92

lobby page

GetAllPlayers function 372, 373

messages, creating 372

refreshing 372

LocationRepository

GetAllAsync method, adding 289

location tracking service

creating 261, 262

logging providers

reference link 419

logic, MeTracker app

classes, adding to DI container 277, 278

location permission, requesting 270- 272

map view, creating 263-267

platform-specific location permissions,
declaring 267-269

user location map, centering 272, 273

ViewModel class, creating 273-277

long polling 348

Long-Term Support (LTS) 41, 215, 412, 559

M

Mac (Macintosh) 17, 212

Android Emulator, configuring 19-22

setting up 17

Visual Studio, installing 18

Visual Studio, pairing with 24-27

Xcode, installing 17

machine learning (ML) 548
   Azure Cognitive Services 548
   Core ML 549
   ML.Net 549
   TensorFlow 549
MainView
   navigating, to ItemView to
      add new item 75-77
   view, creating 66-69
MainView item
   laying out 92
MainViewModel
   creating 63, 64
   ListView, binding 79-82
matchmaking app, Swiper control
   code, adding to calculate state 202, 203
   count labels, adding 208, 209
   decision zones, creating 199
   events, adding 205
   events, raising 206, 207
   events, subscribing 209
   events, wiring up 210
   exit logic, adding 203-205
   grid, extending 200
   images, adding 207
   initial photos, adding 207, 208
   methods, adding to update GUI and
      respond to events 209, 210
   screen size, determining 201
   StackLayout, adding for denying photos 200
   StackLayout, adding for liking photos 200
   testing 199
   two events, declaring 206
   wiring up 207
Match view
   creating 488-490
   Font Awesome, configuring 490, 491

   Font Awesome, downloading 490, 491
   game board, creating 496-508
   layout, defining 491-493
   scoreboard, creating 493-496
MAUI.Markup 14
**MauiMigration app, migrating
   to MyMauiApp 103**
   app startup, modifying 108-110
   files, copying to project 104, 105
   namespaces, updating 105-107
memory functions
   adding 538-543
MeTracker app
   background location tracking, on iOS 278
   background location tracking,
      on Mac Catalyst 278
   building 254-256
   heat map, creating 289
   location tracking service, creating 261, 262
   logic, setting up 263
   map app resume, refreshing 297
   project overview 254
   repository, creating to save
      user locations 257
Microsoft Azure 345
**Microsoft.Extensions.
   DependencyInjection 51, 235**
ML.Net 549
Model-View-ViewModel (MVVM)
   executing 73, 74
   using 59
   ViewModels, creating 59, 60
   views, creating 59, 60
Mono 4
MVVMCross 60
MVVMLight library 60

N

native API access 4
native applications
  defining 4
native performance 4
native UI 4
navigation service, news app
  INavigate interface, adding to
    ViewModel class 175, 176
  INavigation interface, creating 174
  INavigation interface, implementing
    with .NET MAUI Shell 174, 175
  Navigator class, registering with
    dependency injection 175
news app
  building 142
  creating 142-145
  navigation, handling 173
  navigation service, creating 173
  navigation, with routes 176
  news service, creating 158
  overview 142
  running 158
  setting up 142
  structure, creating 145
news app navigation, with routes
  ArticleView, extending 176, 177
  ArticleView, extending with WebView 178
  ItemSelected command, finalizing 176
  route, registering 178
news app structure
  basic structure, defining 156, 157
  Font Awesome, downloading 153
  shell, defining 156
  ViewModel base class 146
  ViewModel base class, creating 145

news service
  API key, obtaining 158
  creating 158
  models, creating 159-161
NewsService class
  ContentTemplate attribute,
    adding to 171, 172
  ContentView of ArticleItem,
    extending 169, 170
  dependency injection, adding to 171
  HeadlinesView, extending 166-169
  HeadlinesViewModel class, extending 165
  wiring up 165

O

object-relational mapper (ORM) 258
OpenWeatherMap 305
output binding 366

P

PanCompleted
  updating 205
persist data
  SQLite, connecting to 54
phone view
  creating 327-329
  grouped CollectionView, using 329, 330
  pull-to-refresh functionality, adding 331
photo gallery app
  .NET MAUI project, creating 213-215
  building 212
  photos, importing 216, 217
PhotoImporter
  registering, with dependency
    injection 235, 236

photos
    app-initializing code, writing 234
    app permissions, requesting 217-219
    base view model, creating 237
    dependency injection, wiring 234
    importing, from Android photo
        gallery 224-229
    importing, from iOS photo gallery 220-223
    importing, from Mac Catalyst
        photo gallery 224
    importing, from Windows
        photo gallery 229-234
    shell, creating 236, 237
Plain Old CLR Object (POCO) 51
players challenge, Azure Functions
    AcknowledgeChallenge function,
        creating 385, 386
    IssueChallenge function, creating 378-385
    messages and models, creating 374-377
Portable Class Library (PCL) 54
Postman 306
Prism 60

Q

quicktype tool 306

R

repository
    creating 51-54
    creating, to save user locations 257
    interface 52- 54
    interface, creating 257-261
    location data model, creating 257

result view, Hot Dog or Not
        Hot Dog application
    building 583
    view, building 586-588
    ViewModel class, building 583-586

S

saturation 292
SemaphoreSlim
    reference link 430
separation of concerns 59
serverless backend
    Azure Functions, using 351
    building 349
    SignalR service, creating 349, 350
Server-Sent Events (SSEs) 348
service class
    creating 161
    NewsScope enum, creating 162
    NewsService class, creating 162-165
shared code 44-46
shell
    creating 236, 237
SignalR service 348
    creating 349, 351
    examples 349
SQLite
    connecting, to persist data 54
SQLite database
    connection, creating to 56-58
SQLite.NET 38
SQLite NuGet package
    adding 54, 55
Standard Term Support , 41, 412, 559

Sticks & Stones game app
  building 409, 410, 412, 413, 414
  connect page 409
  Connect page, creating 439, 440
  game page 409
  game services, creating 415
  lobby page 409
  Lobby page, creating 464
  Match page, creating 479
  overview 347, 408, 409
  project, testing 508-512
  services 409
Sticks & Stones game app, Connect page
  Connect view, adding 447
  ConnectViewModel, adding 440-447
Sticks & Stones game app, game services
  GameService class, creating 427-438
  ServiceConnection class, creating 418-427
  settings service, creating 415-418
Sticks & Stones game app, Lobby page
  Lobby view, adding 473-479
  LobbyViewModel, adding 468-472
  PlayerViewModel, adding 465-468
Sticks & Stones game app, Match page
  Match view, adding 487
  ViewModel classes, creating 479

T

tablet and desktop computer view
  background image, setting 326, 327
  creating 319, 320
  FlexLayout 322
  FlexLayout, using 321, 322
  toolbar item, adding to weather
      data 323, 324
Target Framework Monikers (TFMs) 112
TensorFlow 549

TinyMvvm 60
Tizen 47
TodoItem class
  updating 55
TodoItem model
  creating 51
TodoItemViewModel
  creating 64, 65
to-do list item 51, 52
toolbar item
  adding, to refresh weather data 323, 324
  data loading indicator, adding 325, 326
traditional .NET mobile
  versus .NET MAUI 15, 16

U

UI, with XAML
  defining 12
  Label control, defining 11
  page, creating 12, 13
**Universal Windows Platform
    (UWP) apps 48**
UpdateItemsAsync() method
  implementing 58, 59

V

ValueConverter object
  creating, for item's status 82, 83
  using 83- 85
ViewModel
  base class 60, 61
ViewModel base class, news app structure
  AboutView, creating 151
  ArticleItem, creating 149
  ArticleView, creating 150, 151
  CommunityToolkit.Mvvm 146

HeadlinesView, creating 147-149

HeadlinesViewModel class,
    creating 147, 148

INotifyPropertyChanged,
    implementing 146, 147

ViewModel class

creating 315, 479

MatchPlayerViewModel, adding 479-481

MatchViewModel, adding 481-487

weather data, grouping 316-319

weather data, obtaining 315

VisualStateManager states

change behavior, creating 336-340

handling 334, 335

Visual Studio

installing 18

pairing, with Mac 24-27

Visual Studio Installer 23

W

weather app

building 302-305

form factor view, navigating to 332-334

models, creating 305, 306

phone view, creating 327-329

platform configuration for
    location services 313

services and ViewModels, adding to
    dependency injection 332

states, handling with
    VisualStateManager 335

tablet and desktop computer
    view, creating 319, 320

ViewModel class, creating 315

VisualStateManager states, handling 334

weather service, creating 310-312

weather app, models

app-specific models, adding 308

ForecastItem model, adding 309

Forecast model, adding 309, 310

Weather API model, adding
    manually 306, 308

weather app, platform for location services

Android platform, configuring 313

iOS platform, configuring 313

WinUI platform, configuring 314

WebAssembly (WASM) 516

WebSockets 348

Windows 48

Windows Forms (WPF) 516

Windows machine

.NET mobile, installing for Visual
    Studio 2022 or later 23, 24

Android emulator, configuring 27-33

developer mode, configuring 33, 34

hardware acceleration, configuring 27-33

setting up 23

Visual Studio, pairing with Mac 24-27

Windows Subsystem

for Android 35

Windows Subsystem for Android (WSA) 34

Window UI Library (WinUI) 48, 302

WinUI app 48

X

Xamarin.Forms app

converting, to .NET MAUI 96

creating 97-100

**Xamarin.Forms app to .NET MAUI,
    manual migration 110**

breaking API changes, addressing 118

code, updating 113-115

converted app, running 119

custom renderers and effects 118

functionality, verifying 119

incompatible NuGet packages,
    updating 116-118

projects, converting from .NET Framework
    into .NET SDK style 110-113

Xamarin Hot Restart 34

XAML 586

XAML Hot Reload 35

Xcode

installing 17

X is Not Unix (XNU) 6

www.packtpub.com

Subscribe to our online digital library for full access to over 7,000 books and videos, as well as industry leading tools to help you plan your personal development and advance your career. For more information, please visit our website.

Why subscribe?

- Spend less time learning and more time coding with practical eBooks and Videos from over 4,000 industry professionals

- Improve your learning with Skill Plans built especially for you

- Get a free eBook or video every month

- Fully searchable for easy access to vital information

- Copy and paste, print, and bookmark content

Did you know that Packt offers eBook versions of every book published, with PDF and ePub files available? You can upgrade to the eBook version at packtpub.com and as a print book customer, you are entitled to a discount on the eBook copy. Get in touch with us at customercare@packtpub.com for more details.

At www.packtpub.com, you can also read a collection of free technical articles, sign up for a range of free newsletters, and receive exclusive discounts and offers on Packt books and eBooks.

Other Books You May Enjoy

If you enjoyed this book, you may be interested in these other books by Packt:

.NET MAUI Cross-Platform Application Development

Roger Ye

ISBN: 978-1-80056-922-5

- Discover the latest features of .NET 6 that can be used in mobile and desktop app development
- Find out how to build cross-platform apps with .NET MAUI and Blazor
- Implement device-specific features using .NET MAUI Essentials
- Integrate third-party libraries and add your own device-specific features
- Discover .NET class unit test using xUnit.net and Razor components unit test using bUnit
- Deploy apps in different app stores on mobile as well as desktop

Unity 2022 Mobile Game Development

John P. Doran

ISBN: 978-1-80461-372-6

- Design responsive UIs for your mobile games
- Detect collisions, receive user input, and create player movements
- Create interesting gameplay elements using mobile device input
- Add custom icons and presentation options
- Keep players engaged by using Unity s mobile notification package
- Integrate social media into your projects
- Add augmented reality features to your game for real-world appeal
- Make your games juicy with post-processing and particle effects

Packt is searching for authors like you

If you're interested in becoming an author for Packt, please visit authors.packtpub.com and apply today. We have worked with thousands of developers and tech professionals, just like you, to help them share their insight with the global tech community. You can make a general application, apply for a specific hot topic that we are recruiting an author for, or submit your own idea.

Hi!

We are Michael Cummings, Daniel Hindrikes, and Johan Karlsson, authors of *.NET MAUI Projects*. We really hope you enjoyed reading this book and found it useful for increasing your productivity and efficiency using .NET MAUI.

It would really help us (and other potential readers!) if you could leave a review on Amazon sharing your thoughts on this book.

Go to the link below or scan the QR code to leave your review:

https://packt.link/r/1837634912

Your review will help us to understand what's worked well in this book, and what could be improved upon for future editions, so it really is appreciated.

Best Wishes,

Michael Cummings

Daniel Hindrikes

Johan Karlsson

Download a free PDF copy of this book

Thanks for purchasing this book!

Do you like to read on the go but are unable to carry your print books everywhere?

Is your eBook purchase not compatible with the device of your choice?

Don't worry, now with every Packt book you get a DRM-free PDF version of that book at no cost.

Read anywhere, any place, on any device. Search, copy, and paste code from your favorite technical books directly into your application.

The perks don't stop there, you can get exclusive access to discounts, newsletters, and great free content in your inbox daily

Follow these simple steps to get the benefits:

1. Scan the QR code or visit the link below

https://packt.link/free-ebook/9781837634910

2. Submit your proof of purchase
3. That's it! We'll send your free PDF and other benefits to your email directly

www.ingramcontent.com/pod-product-compliance
Lightning Source LLC
Chambersburg PA
CBHW060634060326
40690CB00020B/4399